Water–Energy–Food Security Nexus in Large Asian River Basins

Special Issue Editors

Marko Keskinen
Shokhrukh Jalilov
Olli Varis

Guest Editors

Marko Keskinen
Aalto University
Finland

Shokhrukh Jalilov
Aalto University
Finland

Olli Varis
Aalto University
Finland

Editorial Office
MDPI AG
St. Alban-Anlage 66
Basel, Switzerland

This edition is a reprint of the Special Issue published online in the open access journal *Water* (ISSN 2073-4441) from 2015–2016 (available at: http://www.mdpi.com/journal/water/special_issues/Water-Energy-Food).

For citation purposes, cite each article independently as indicated on the article page online and as indicated below:

Author 1; Author 2; Author 3 etc. Article title. *Journal Name*. **Year**. Article number/page range.

ISBN 978-3-03842-344-7 (Pbk)
ISBN 978-3-03842-345-4 (PDF)

Table of Contents

About the Guest Editors

Marko Keskinen is a University Lecturer at Aalto University, Finland. He has been working as a researcher and consultant as well as a civil servant at the Finnish Ministry for Foreign Affairs, and has led several multidisciplinary and multicultural projects. Dr Keskinen has a long experience in sustainability, water resources management, transboundary cooperation, integrated approaches and science–policy–stakeholder interaction.

Shokhrukh Jalilov is a Postdoctoral Fellow at the Institute for the Advanced Study of Sustainability of the United Nations University. His current research focuses on the development of modern methodological approaches for the economic evaluation of clean water environment in urban contexts. His past research examined the interdependencies among water, food, energy and climatic resources, including a policy analysis and designing of hydro-economic models to optimally allocate water resources in transboundary river basins. He has also consulted on a number of projects for international and bilateral donor organizations. Originally from Tashkent, Uzbekistan, he did his Master's and PhD Degrees in USA, followed by one and half years spent as a postdoctoral researcher at Aalto University, Finland. At Aalto, Dr Jalilov continued his research on the diverse connections between water, energy and food security in large Asian river basins, with a specific focus on Central Asia (Amu Darya) and Southeast Asia (Mekong).

Olli Varis is the Matti Pursula Professor for Water and Development and Vice Dean at Aalto University, Finland. Dr Varis is an internationally distinguished and recognised researcher in water and environmental issues, with a specialisation in global water resources as well as the interaction between development and sustainable use of natural resources. He is especially renowned for his research related to Asia's growth economies. Dr Varis has led numerous international research projects and has a broad publication track record of high quality. He has also acted as an expert in numerous significant positions both in Finland and internationally.

Preface to "Water–Energy–Food Security Nexus in Large Asian River Basins"

Is our world secure?

This question is now, at the end of the tumultuous year 2016, extremely relevant. The political, economic and climate-related developments around the world have underlined the interconnectedness of our world, but also fundamentally contested the idea of one global village. At the same time, increasing migration is challenging the idea of free movement of people. As a result, nationalistic views are on the rise, and many countries strongly highlight their sovereignty and also the security of their critical resources.

Water, energy and food are the most critical resources of the whole of humankind. The three are also closely linked, with water acting as an enabler for both food and energy production and at the same time feeling adverse effects of activities in those sectors. All three also regularly cross national borders, with energy and food flowing from one country to another with the help of trade. Water, energy and food are present in today's world; thus all are, in essence, transboundary.

While all this is well known, the wide-ranging societal and economic impacts of increasing water and resource scarcity have only been properly recognized in recent years. This is exemplified by the World Economic Forum's Global Risks Report 2015, which for the first time placed "water crises" as the most important global risk in terms of impact. The increased awareness about scarcities of resources and their interconnectedness has also translated into heightened attention given to security-related aspects of water, energy and food. Hence, instead of just using and managing the different resources, attention is also increasingly given to securing both availability and access to them—and the related inequities and politics.

The water–energy–food security nexus aims to "enhance water, energy and food security by increasing efficiency, reducing trade-offs, building synergies and improving governance across sectors". That was the description in the influential Background Document for *Bonn2011 Conference on the Water, Energy and Food Security Nexus*.

Since the Bonn2011 Conference, the number of nexus-related activities, projects and publications has boomed, with a variety of actors using the nexus concept to guide, frame and reflect upon their activities. At the same time, there is still no commonly agreed definition for the nexus. Several articles in this Special Issue engage in this discussion, providing their views on how to define the nexus and what kind of elements and methods it could entail.

Yet, any discussion about the characteristics of a nexus as an approach remain theoretical unless linked to actual planning and decision-making processes on the ground. That was also the aim of our Special Issue: to see what kind of practical benefits and challenges the emerging nexus approach could accrue when actually examined, analyzed and applied in differing management contexts.

The geographical focus of our Special Issue is Asia and its large river basins. The majority of these basins are transboundary, shared by more than one nation state. While recognizing that the nexus approach should not be applied only within water-bound boundaries, i.e., catchments, we decided to set the focus on large Asian river basins as they present particularly challenging management settings due to their scale and variety of actors. In addition, there is already a rich literature on their water, energy and food-related characteristics and management processes, making comparison between the nexus and other management approaches easier. Finally, despite all these activities, the situation in many transboundary basins remains contentious, with different sectoral actors and nation states not able to agree on the most suitable ways forward. We were thus interested in seeing what kind of potential benefits this nexus could bring to such contexts.

Editing the Special Issue has been an important learning experience for us, and provided new insights (see our Editorial). One important point of learning relates to the concept of security and its inclusion or exclusion in the nexus discussion. When initially drafting the call for the Special Issue, we decided, after long debate, to leave the term "security" out from the title: the idea was that the articles would, in this way, focus more on the three sectors and their interrelations, rather than broader security-related aspects. However, after going through the articles and related cases included in the Special Issue, it has become apparent that the nexus is, obviously, very much about security—whether about 'separate' securities related to water, energy and food, or more comprehensive forms of security. Further, in many cases, it is exactly these security-related aspects that have led to the biggest challenges, whether between sectors or riparian states. The inclusion of security into the nexus is therefore crucial for almost any possible solutions. For this reason, we have also added the term to the title of this Special Issue book.

This Special Issue includes 11 articles and an editorial from a total of 35 different authors. Geographically, the majority of the articles look at Central Asia and the Mekong, both of which are regions seeing rapid development and include several important transboundary river basins. Methodologically, the articles address both theoretical and practical aspects related to the nexus. We would like to thank all the authors, reviewers and MDPI editors for their contribution and engagement during the process. Special thanks to the Academy of Finland (#269901 NexusAsia) and Strategic Research Council (SRC) of Finland (#303623 Winland), whose financial support enabled the research and editorial work related to this Special Issue.

During this editorial process, some of our reviewers challenged the very idea of the nexus, questioning whether it actually brings anything new to the discussion about water, energy and food security. This is a valid point, and it can well be argued that the interconnections between these three sectors have already been known for centuries and that over time the nexus may indeed wade from fashion. Yet, we already see that the impressive set of articles in this Special Issue exemplifies that the nexus helps both researchers and practitioners to think about relationships between water, energy and food security in new—sometimes perhaps forgotten—ways. With its focus on three critical sectors, their interlinkages and related security aspects, we believe that the nexus has potential as a complementary approach to support water resources management, transboundary cooperation and, ultimately, sustainable development.

Marko Keskinen, Shokhrukh Jalilov and Olli Varis
Guest Editors

water

MDPI

Editorial

Water-Energy-Food Nexus in Large Asian River Basins

Marko Keskinen * and Olli Varis

Water & Development Research Group, Aalto University, P.O. Box 15200, Aalto 00076, Finland; olli.varis@aalto.fi
* Correspondence: marko.keskinen@aalto.fi; Tel.: +358-50-563-8030

Academic Editor: Arjen Y. Hoekstra
Received: 13 July 2016; Accepted: 29 August 2016; Published: 12 October 2016

Abstract: The water-energy-food nexus ("nexus") is promoted as an approach to look at the linkages between water, energy and food. The articles of *Water*'s Special Issue "Water-Energy-Food Nexus in Large Asian River Basins" look at the applicability of the nexus approach in different regions and rivers basins in Asia. The articles provide practical examples of the various roles and importance of water-energy-food linkages, but also discuss the theoretical aspects related to the nexus. While it is evident that any application of the nexus must be case-specific, some general lessons can be learnt as well. Firstly, there are a variety of interpretations for the nexus. These include three complementary perspectives that see nexus as an analytical approach, governance framework and emerging discourse. Secondly, nexus is—despite its name—a predominantly water-sector driven and water-centered concept. While this brings some benefits by, e.g., setting systemic boundaries, it is also the nexus' biggest challenge: If the nexus is not able to ensure buy-in from food and energy sector actors, its added value will stay limited. Ultimately, however, what really matters is not the approach itself but the processes it helps to establish and outcomes it helps to create. Through its focus on water-energy-food linkages—rather than on those themes separately—the nexus is well positioned to help us to take a more systemic view on water, energy and food and, hence, to advance sustainable development.

Keywords: water-energy-food nexus; transboundary water-energy-food nexus; river basin; transboundary river; water security; energy security; food security; Central Asia; South Asia; Southeast Asia; Mekong

1. Introduction: The Emergence of Water-Energy-Food Nexus

Water-energy-food (security) nexus aims to "enhance water, energy and food security by increasing efficiency, reducing trade-offs, building synergies and improving governance across sectors" [1] (p. 4). That was the description in the Background Document for *Bonn2011 Conference on The Water, Energy and Food Security Nexus*, held in November 2011 [1]. Since then, the number of nexus-related activities, projects and publications have boomed, with a variety of actors using nexus concept to guide, frame and reflect upon their activities. For example, international Future Earth research platform included the management of water-energy-food synergies and trade-offs as one of the eight global sustainability challenges [2], the Food and Agriculture Organization of the United Nations (FAO) has looked at the applicability of the nexus in different contexts [3,4], and the European Commission added food-energy-water-climate linkages among the topics for its Horizon 2020 research and innovation program [5].

The rapid emergence of the nexus can be seen as rather surprising, given that the nexus as a concept is still not clearly defined and—despite efforts, in particular by Germany [6,7]—it was included neither in the Outcome Document of Rio+20 nor in the Sustainable Development Goals.

The nexus as an approach thus lacks an officially recognized status that for example Integrated Water Resources Management (IWRM) has received.

There must therefore be something else that has attracted attention to the nexus. We argue that the emergence of the nexus is related to three interlinked drivers: crucial importance that water, energy and food have for humanity and related need to ensure their security; enhanced awareness of the societal and economic risks included in the mounting resource scarcities; and the failure of existing water-related management approaches—most notably IWRM—to engage other sectors to address water-related challenges.

The rationale behind the first driver is clear: water, energy and food are the key prerequisites for the existence of humankind. The three are also closely linked, with water acting as an enabler for both food and energy production and being at the same time a major target for adverse environmental effects of activities in those sectors [8,9]. While all this is well known, the wide-ranging societal and economic impacts of increasing water and resource scarcity (e.g., [10,11]) have been properly recognized only during the past years, and can thus be considered as the second driver for the nexus [1]. This is exemplified by the World Economic Forum's Global Risks Report 2015 [12], which for the first time placed "water crises" as the most important global risk in terms of impact—ahead of infectious diseases, weapons of mass destruction, interstate conflicts, and even failure of climate change adaptation (in 2016 water crisis ranked third, after climate change and weapons of mass destruction [13]). The World Economic Forum has also broadened its view on water-related risks, changing the term "water supply crisis" to "water crisis", and revising the risk category of water crisis from environmental to societal risk [13]. The increased awareness about resources scarcities has also translated into heightened attention to security-related aspects of water, energy and food: Instead of just using and managing the different resources and related sectors, the attention is increasingly also on securing both availability and access to them.

The third driver is linked to the planning and management processes of water and related resources. In terms of water, IWRM is arguably the most commonly accepted global framework due to its recognition in several UN conferences and agreements—most recently within the context of Sustainable Development Goals [14]. While IWRM has been successful in promoting integrated management of water resources, it has not always been that successful in engaging water-related sectors—most importantly food and energy—into its implementation. Despite the critical role that those sectors have on water use and degradation of aquatic ecosystems, IWRM has remained water sector orientated and often rather technical management process. We see two main reasons for this: first of all IWRM is by its definition all-encompassing, linking water to all relevant policy sectors. Secondly, the IWRM processes have not been very strong in pushing the debate "out of the water box", but other sectors are instead often considered as kind of externalities that are to be managed under IWRM framework [15]. The nexus changes this by its very name, recognizing that it is first and foremost about water, energy and food (referring to entire sectors, not just their water-using parts) and focusing on their specific linkages, synergies and trade-offs.

Link to Practice: Nexus in Large Asian River Basins

Any discussion about the characteristics of a nexus approach remain theoretical unless linked to actual planning and decision-making processes on the ground. That was also the aim of our Special Issue: to see what kind of practical benefits and challenges the emerging nexus approach could accrue when actually examined, analyzed and applied in differing management contexts.

The geographical focus of our Special Issue is very wide, covering Asia and its large river basins. Majority of these basins are transboundary, i.e., shared by more than one nation state. While recognizing that the nexus approach should not be applied only within water-bound boundaries, i.e., catchments, we decided to set the focus on large Asian river basins as they present particularly challenging management settings due to their scale and variety of actors. In addition, there is already a rich literature on their water, energy and food-related characteristics and management processes:

This makes it easier to compare the nexus with other management approaches and thus to discuss its potential value added. Finally, despite all these activities, the situation in many transboundary basins remains contentious and even 'deadlocked', with different sectoral actors and nation states failing to agree on the most suitable ways forward. The geographical focus also corresponded with the Academy of Finland—funded NexusAsia project (http://www.wdrg.fi/nexusasia), which built on our Water and Development Research Group's earlier research related to the water resources management in large Asian river basins (e.g., [16–20]).

Of particular focus in this Special Issue were three regions and their related rivers: Central Asia, South Asia and the Mekong Region. Over 3.3 billion people live in these three regions, i.e., well over 40% of the world's population, with over 1.1 billion living within the seven main transboundary river basins of Amu Darya, Syr Darya, Ganges-Brahmaputra-Meghna, Indus, Irrawaddy, Salween and Mekong [9]. The regions are, however, very different, with South Asia clearly the most highly populated [20]. Also the river basins have differing characteristics in terms of their hydrology, connection to food and energy production, and future plans for development. At the same time, all regions have plans for the development of water infrastructure, including large hydropower dams, with direct implications to water, energy and food security.

2. Contributions from the Articles

Water's Special Issue, "Water-Energy-Food Nexus in Large Asian River Basins" (http://www.mdpi.com/journal/water/special_issues/Water-Energy-Food), include 11 articles from altogether 35 different authors. Geographically, a majority of the articles look at Central Asia [21–25] and the Mekong [26–29], with one of the articles looking at Central Asia, South Asia and the Mekong [9] and two others including case study areas outside these three regions [25,30]. Methodologically, the articles address both theoretical and practical aspects related to the nexus. While all articles include elements from both of these, six of them can be seen to be more practical, i.e., context-driven [21–24,26,28], while five articles focus more on the theoretical discussion about the nexus as an approach [9,25,27,29,30]. Next, we will provide a concise summary for all the articles, organized according to their geographical (Central Asia and Mekong) and methodological (theory) focus.

In their article on Central Asia's Syr Darya River Basin, Soliev et al. [21] emphasise the importance of historical perspective and more rigorous institutional analysis on water-energy-food nexus and benefit sharing in transboundary river basins. Through their insightful analysis of water-related development from 1917 until present day in the Ferghana Valley, the authors recognize five different periods for the cooperation between Kyrgyzstan and Uzbekistan. Each of these periods has differing benefit sharing mechanisms and related types of benefits. The analysis powerfully shows the evolution and implications of changing institutional settings for shaping water-related management decisions as well as transboundary cooperation.

The article by Guillaume et al. [22] provides two significant views to this Special Issue. First, it gives an in-depth historical assessment of the trajectories of water scarcity and consumption in Central Asia, applying global water data to selected Food Production Units in Uzbekistan, Kazakhstan, Kyrgyzstan, Tajikistan, Turkmenistan, and Afghanistan. Second, the authors use concepts stemming from resilience and socio-ecological system theory to identify five transferable principles that are seen to be particularly relevant for managing transboundary water-energy-food nexus. These principles are: (1) the subsystems included/excluded from the nexus are case-specific and should be consciously scrutinized; (2) consensus is needed on what boundaries can acceptably be crossed within the nexus; (3) there is a need to understand how reducing trade-offs will modify system dependencies; (4) global stakeholders have both a responsibility and right to contribute to the shaping of the nexus; (5) combining data with global and local perspectives can help to enhance transferability and understanding of shared problems in our globalized world [22].

The analysis by Wegerich et al. [23] looks at water security in Syr Darya River Basin through case studies focusing on the water management of irrigated agriculture. Noting that current definitions for water security largely forget the supply-side of water security, the authors move on to analyze in a novel manner past and current water security approaches in Syr Darya Basin. The results indicate that the majority of water-related international activities in the basin have been either at very local level or at international level, with the importance of the meso-level and its actors (including public irrigation departments) going largely unnoticed. In conclusion, the authors call for the long-term strengthening of public administration—so-called water bureaucracy—as it is often the weakest link for water security.

Jalilov et al. [24] provide an experimental case study focusing on Amu Darya River Basin and its water-energy-agriculture nexus. Using a hydro-economic model, the authors estimate the energy- and agriculture-related economic benefits from a major water infrastructure development project, namely the Rogun hydropower plant. The results show that differing operating schemes for the plant would increase the total energy-agriculture benefits while ensuring environmental flow needed to maintain key water-related ecosystems. This emphasizes the potential for benefit sharing between the basin countries, and also indicates that nexus synergies are indeed achievable: It seems possible to build and operate a dam in a way that would simultaneously enhance energy- and food-related benefits from the current state. The authors also note that any practical application of water-energy-food nexus must be context-specific and, if necessary, also include other relevant sectors. In the case of Amu Darya, the analysis must also consider economically very important—and water-intensive—non-food agriculture such as cotton.

Moving to the Mekong Region, Keskinen et al. [26] discuss the water-energy-food nexus in the context of the Tonle Sap Lake in Cambodia. The analysis shows that that the current plans for large-scale hydropower in the Mekong will radically alter water and food security at both local and national level, and the nexus is thus a very timely, complementary concept to address the development challenges in the region. As result, for example the Mekong River Commission MRC—that has for years built its activities on IWRM concept—concluded in its most recent Basin Development Plan that addressing water-related trade-offs "requires strong IWRM understanding and water-food-energy nexus thinking" [31] (p. 42). Keskinen et al. also discuss the benefits of the nexus as an approach, noting that within their research project the nexus was particularly useful in facilitating collaboration and stakeholder engagement. The authors conclude that when compared to IWRM, the nexus is simultaneously more focused (spelling explicitly out the sectors that are to be included) and broader (expanding the discussion from mere water resources management to a general discussion about water, energy and food security). At the same time, however, there is a danger that the themes and aspects that are not explicitly included—such as livelihoods, climate change, or the environment—are simply left out or get secondary weight.

Matthews and Motta [28] provide a detailed analysis of the actual drivers for the Mekong's hydropower boom, focusing on the critical role that Chinese state-owned enterprises have for hydropower investments. Using a political economy analysis, the authors clearly show the powerful political and economic forces behind hydropower development. Importantly, the analysis also indicates how such forces create narratives that downplay the water-energy-food nexus interconnections and position hydropower as a win-win for both upstream and downstream states—in contrast to scientific studies that demonstrate the substantial trade-offs related to hydropower development. This finding emphasizes the political dimension of the nexus, and reminds us that the way the nexus is defined and used often differ greatly depending on the actors and their ultimate interests.

The Special Issue also includes five articles that have more theoretical view on the nexus. Belinskij [27] provides a refreshingly different view on the entire approach, looking at how the international water law supports the nexus approach in the context of international river basins. Focusing on two international conventions (UN Watercourses Convention and ECE Water Convention) and using the Mekong as a case study, the analysis shows that international water law provides a useful platform for the cooperation between states and different sectors. While international water law supports the transboundary nexus approach by offering an institutional framework, it does not offer concrete specifications for its procedural elements: agreeing on this remains the responsibility of riparian states. The author concludes that the most difficult part of the nexus cooperation in transboundary river basins is the reconciliation of different water uses in situations where there is not enough water to meet all the competing needs. In such situations, the general provisions in international water law on equitable and reasonable utilization and on the minimization of harmful transboundary effects become central.

Endo et al. [30] provide an insightful methodological view for the nexus, drawing on a comparative analysis of an array of different methods to look at the water-energy-food linkages. Covering both qualitative methods (questionnaire surveys, ontology engineering, and integrated maps) and quantitative methods (physical models, benefit-cost analysis, integrated indices, and optimization management models), the authors consider the pros and cons of each method with the help of set of local case studies from Japan and the Philippines. The authors also discuss the different stages and scales within the nexus analysis process that the methods can provide most contribution to, noting the importance of both temporal and spatial scale. The article, thus, reminds us how the nexus as an approach connects to already existing methods, and how the nexus linkages can be examined through a variety of different methods.

De Strasser et al. [25] discuss a methodology to assess the water-energy-food-ecosystem nexus in transboundary river basins. Drawing on groundbreaking work by the Task Force on the Water-Food-Energy-Ecosystems Nexus under the UNECE Water Convention, the authors introduce the Transboundary River Basin Nexus Approach (TRBNA) methodology and present findings from its application in three transboundary river basins: the Alazani/Ganykh, the Sava and the Syr Darya. The TRBNA methodology includes a six-step process, consisting of following steps: (1) Socio-economic and geographical context; (2) Identification of key sectors and key actors; (3) Analysis of key sectors; (4) Intersectoral issues; (5) Nexus dialogue; and (6) Solutions and benefits. Through their in-depth description of the TRBNA methodology, the authors provide a unique view of an actual nexus assessment process implemented within the political context of an intergovernmental organization. Methodologically, the authors conclude that the characteristic that makes the nexus approach innovative is the shift from a sector- or resource-centric perspective to a multi-centric one: nexus takes into account the links and dynamics between resource systems to harmonize their outlook and management. Noting that the nexus is often compared to IWRM, the authors also note that the traditional integrated approaches typically have limited analytical scope and often do not consider re-enforcing stresses or indirect links.

The article by Pittock et al. [29] introduces an influence model that can be used to systematically assess the water, energy and food nexus in large rivers. With the help of a series of influence diagrams, the authors model the hydropower-food supply nexus in the Lower Mekong River Basin. The diagrams show in visually powerful manner that the nexus in the Mekong is part of a complex system where an intervention in one aspect—for example construction of hydropower dam—will have cascading consequences for a wide range of other sectors in the socio-ecological system. In this way, the influence diagrams illustrate vividly the need for strategic management interventions, focusing on the key points of leverage where such interventions have the biggest potential to change the outcomes and to minimize negative impacts. The authors also note that the use of influence models facilitates systematic nexus analysis by identifying links across sectors, enabling discourse among actors on trade-offs and synergies across water, energy, food and other sectors.

The Special Issue is concluded with an article by Keskinen et al. [9], discussing the applicability of water-energy-food nexus in transboundary contexts based on novel comparative analysis as well as on key findings from the other articles of the Special Issue. The authors start by providing a set of definitions for the nexus, recognizing three perspectives that see the nexus as an analytical tool, governance framework and as an emerging discourse. The analysis of three Asian regions—Central Asia, South Asia and the Mekong Region—and their related transboundary river basins builds on comparative nexus triangles that aim to visualize the key nexus linkages in all these regions. Following, the authors discuss the implications that the nexus approach has for transboundary context and vice versa.

3. Key Findings and Conclusions

Together, the articles constituting the Special Issue produce several important findings. In terms of the geographic focus of this Special Issue, the articles illustrate clearly the crucial linkages that water has with both energy and food production in different regions of Asia. Equally important are the linkages between energy and food security: Irrigated agriculture is a heavy user of energy (hence benefiting from enhanced energy security), while several forms of energy production—particularly hydropower and bioenergy—have direct impacts on food security.

The articles also show how strongly both energy and food production impact the river systems, with large-scale hydropower development being a particularly powerful—and contested—driver in a number of river basins. Such changes also translate into tensions between different water uses, both within and between the countries sharing the same river basin. Indeed, the upstream-downstream tensions over the water use—commonly including both energy (hydropower) and food (agriculture and/or fish)—in transboundary river basins of Asia seem to be rather a rule than an exception. This underlines the political dimension of the nexus (see also [32,33]), but also emphasizes the need to look beyond the river basin and water use per se, towards broader regional linkages and benefits related to water, energy and food. The articles also note that water-energy-food linkages in the study regions evolve over time, and that intensifying water and resource scarcities are likely to reinforce the already grand challenges related to such linkages.

Interesting methodological findings emerge also in relation to the nexus approach itself. First of all, it becomes clear that there are varying interpretations of the nexus, with different authors emphasizing different dimensions and also extending its scope from mere water-energy-food to include for example non-food crops and ecosystems. As noted by Keskinen et al. [9] in this issue, three complementary perspectives can therefore be established for the nexus, viewing it as an analytical approach, governance framework and emerging discourse. They also note that while the need for integration in relation to the nexus generally focuses on the linkages between three nexus sectors, it seems as important to understand the cross-level linkages between these different nexus perspectives and related disciplines [9]. Several articles also discuss the different scales related to the nexus. As concluded by Endo et al. [30] in this issue, the consideration of both vertical and horizontal scales and dimensions is needed to reduce trade-offs and to optimize the water-energy-food linkages.

Although the importance of water-energy-food linkages is clear, actual empirical evidence on the water-energy-food nexus approach remains relatively thin. This is visible also in this Special Issue, with the majority of the articles actually focusing more on the nexus as a theme (water-energy-food) than as an actual approach for analysis and/or planning and policy-making. On the other hand, several articles also discuss the methodological aspects of the nexus, including the article describing the pioneering work done by the UNECE and its partners on a thorough nexus assessment process in different transboundary river basins [25], as well as a useful methodological review by Endo et al. [30].

The articles also portray the nexus as a clearly water-centered concept—partly due to the defined scope (large Asian river basins) of the Special Issue. Such water-centrism has been, however, noted also by other authors, already starting from the Bonn2011 Conference that placed "available water resources" in the center of its triangle diagram depicting the water, energy and food security nexus [1] (p. 16). In this way, it becomes natural to compare nexus with other water management approaches, most importantly IWRM. It also makes it natural to focus the attention to river basin as a political and geographical unit, and water use and allocation as key elements of the nexus.

We should, thus, acknowledge upfront that one of the reasons for the very existence of the nexus is the need to find new ways to cope with increasing water scarcity and deterioration of aquatic ecosystems, and, hence, to address the pressures posed by food and energy production on water ecosystems. Instead of "available water resources" at the center of the Bonn2011 diagram, "aquatic ecosystems" feels in many ways more appropriate focus (see also [25,34]). This would emphasize the importance of environmental impacts (and align the nexus better with common paradigms, such as ecosystem approach) and also remove the possible misinterpretation where water is seen merely as a resource to be allocated to and shared between agricultural and energy producers.

We argue that the water and resource-centrism is also the biggest challenge of the nexus. As illustrated by number of articles in this Special Issue, the ultimate success of the nexus depends from its ability to guarantee a buy-in also from the actors of energy and food sectors. We believe that such buy-in is best ensured if those actors feel that the nexus process provides them with something useful and that they are treated as equal partners, not just 'externalities' causing pressures on water.

Ultimately, however, the discussion about the nexus approach is not that relevant. What really matters is not the approach itself but the theoretical and practical processes it helps to establish and outcomes it helps to create (see also [35]). For us the key opportunity provided by the nexus is two-fold: its clear definition of the themes included (setting focus), and its emphasis on the linkages and feedback loops between those themes (not only on the themes per se). If the nexus in this way helps us to climb a step higher—from the sectoral valleys towards the mountain of sustainability—and take a more systemic view on those themes, then it is likely to live up to its expectations.

Acknowledgments: This Special Issue brought together a unique set of experts. Thank you Ilkhom Soliev, Kai Wegerich, Jusipbek Kazbekov, Joseph Guillaume, Matti Kummu, Stephanie Eisner, Daniel Van Rooijen, Nozilakhon Mukhamedova, Shokhrukh-Mirzo Jalilov, Paradis Someth, Aura Salmivaara, Antti Belinskij, Aiko Endo, Kimberly Burnett, Pedcris M. Orencio, Terukazu Kumazawa, Christopher A. Wada, Akira Ishii, Izumi Tsurita, Makoto Taniguch, Nathanial Matthews, Stew Motta, Lucia de Strasser, Annukka Lipponen, Mark Howells, Stephen Stec, Christian Bréthaut, Mirja Kattelus, Miina Porkka, Timo Räsänen, Jamie Pittock, David Dumaresq and Andrea Bassi for sharing your views related to the water-energy-food nexus and large Asian river basins. Special thanks for the editorial team of the Water journal for your collaboration throughout the editorial process. Thank you also for the numerous reviewers, who shared their time, expertise and critical views to improve the articles as well as this Editorial. Finally, thank you to our colleagues at Aalto University's Water & Development Research Group—particularly Shokhrukh-Mirzo Jalilov and Joseph Guillaume—for your support throughout this process. The work for the Special Issue was enabled by the funding from the Academy of Finland (#269901 NexusAsia), while the work for this editorial was done under Strategic Research Council (SRC) funded Winland project.

Conflicts of Interest: The authors declare no conflict of interest.

References

1. Hoff, H. *Understanding the Nexus*; Background Paper for the Bonn2011 Conference: The Water, Energy and Food Security Nexus; Stockholm Environment Institute: Stockholm, Sweden, 2011.

2. Future Earth. *Future Earth 2025 Vision*; International Council for Science: Paris, France, 2014.

3. Food and Agriculture Organisation of the United Nations (FAO). *An Innovative Accounting Framework for the Food-Energy-Water Nexus—Application of the MuSIASEM Approach to Three Case Studies*; FAO: Rome, Italy, 2013.

4. Flammini, A.; Puri, M.; Pluschke, L.; Dubois, O. *Walking the Nexus Talk: Assessing the Water-Energy-Food Nexus in the Context of the Sustainable Energy for All Initiative*; Food and Agriculture Organisation of the United Nations (FAO): Rome, Italy, 2014.

5. European Commission. Horizon 2020 Topic: Integrated Approaches to Food Security, Low-Carbon Energy, Sustainable Water Management and Climate Change Mitigation. Available online: http://ec.europa.eu/research/participants/portal/desktop/en/opportunities/h2020/topics/water-2b-2015.html (accessed on 10 August 2016).

6. *Policy Recommendations—Bonn Conference 2011: The Water, Energy and Food Security Nexus—Solutions for the Green Economy*; Bonn2011 Conference: Bonn, Germany, 2011.

7. Brandi, C.; Richerzhagen, C.; Stepping, K. *Post 2015: Why is the Water-Energy-Land Nexus Important for the Future Development Agenda?*; German Development Institute (DIE): Bonn, Germany, 2013.

8. Olsson, G. Water, energy and food interactions—Challenges and opportunities. *Front. Environ. Sci. Eng.* **2013**, *7*, 787–793. [CrossRef]

9. Keskinen, M.; Guillaume, J.; Kattelus, M.; Porkka, M.; Räsänen, T.; Varis, O. The Water-Energy-Food Nexus and the Transboundary Context: Insights from Large Asian Rivers. *Water* **2016**, *8*, 193. [CrossRef]
10. Kummu, M.; Gerten, D.; Heinke, J.; Konzmann, M.; Varis, O. Climate-driven interannual variability of water scarcity in food production potential: A global analysis. *Hydrol. Earth Syst. Sci.* **2014**, *18*, 447–461. [CrossRef]
11. Salmivaara, A.; Porkka, M.; Kummu, M.; Keskinen, M.; Guillaume, J.; Varis, O. Exploring the Modifiable Areal Unit Problem in Spatial Water Assessments: A Case of Water Shortage in Monsoon Asia. *Water* **2015**, *7*, 898–917. [CrossRef]
12. World Economic Forum. *Global Risk Report 2015*, 10th ed.; World Economic Forum: Geneva, Switzerland, 2015.
13. World Economic Forum. *Global Risk Report 2016*, 11th ed.; World Economic Forum: Geneva, Switzerland, 2016.
14. United Nations. Transforming Our World: The 2030 Agenda for Sustainable Development. In *A/RES/70/1: Resolution Adopted by the General Assembly on 25 September 2015*; United Nations General Assembly, 2015. Available online: http://www.un.org/ga/search/view_doc.asp?symbol=A/RES/70/1&Lang=E (accessed on 3 October 2016).
15. Varis, O.; Enckell, K.; Keskinen, M. Integrated water resources management: Horizontal and vertical explorations and the 'water in all policies' approach. *Int. J. Water Resour. Dev.* **2014**, *30*, 433–444. [CrossRef]
16. Keskinen, M. Water resources development and impact assessment in the Mekong Basin: Which way to go? *AMBIO* **2008**, *37*, 193–198. [CrossRef]
17. Mehtonen, K.; Keskinen, M.; Varis, O. The Mekong: IWRM and Institutions. In *Managemeng of Transboundary Rivers and Lakes*; Varis, O., Tortajada, C., Biswas, A.K., Eds.; Springer: Berlin, Germany, 2008.
18. Varis, O.; Rahaman, M.M. The Aral Sea Keeps Drying out but is Central Asia short of water? In *Central Asian Waters: Social, Economic, Environmental and Governance Puzzle*; Rahaman, M.M., Varis, O., Eds.; Water & Development Publications—Helsinki University of Technology: Esbo, Finland, 2008; pp. 3–9.
19. Rahaman, M.M.; Varis, O. Integrated water management of the Brahmaputra basin: Perspectives and hope for regional development. *Nat. Resour. Forum* **2009**, *33*, 60–75. [CrossRef]
20. Varis, O.; Kummu, M.; Salmivaara, A. Ten major rivers in monsoon Asia-Pacific: An assessment of vulnerability. *Appl. Geogr.* **2012**, *32*, 441–454. [CrossRef]
21. Soliev, I.; Wegerich, K.; Kazbekov, J. The costs of benefit sharing: Historical and institutional analysis of shared water development in the ferghana valley, the Syr Darya Basin. *Water* **2015**, *7*, 2728–2752. [CrossRef]
22. Guillaume, J.; Kummu, M.; Eisner, S.; Varis, O. Transferable principles for managing the nexus: Lessons from historical global water modelling of Central Asia. *Water* **2015**, *7*, 4200–4231. [CrossRef]
23. Wegerich, K.; Van Rooijen, D.; Soliev, I.; Mukhamedova, N. Water security in the Syr Darya Basin. *Water* **2015**, *7*, 4657–4684. [CrossRef]
24. Jalilov, S.-M.; Varis, O.; Keskinen, M. Sharing Benefits in Transboundary Rivers: An Experimental Case Study of Central Asian Water-Energy-Agriculture Nexus. *Water* **2015**, *7*, 4778–4805. [CrossRef]
25. De Strasser, L.; Lipponen, A.; Howells, M.; Stec, S.; Bréthaut, C. A Methodology to Assess the Water Energy Food Ecosystems Nexus in Transboundary River Basins. *Water* **2016**, *8*, 59. [CrossRef]
26. Keskinen, M.; Someth, P.; Salmivaara, A.; Kummu, M. Water-energy-food nexus in a transboundary river basin: The case of Tonle Sap Lake, Mekong River Basin. *Water* **2015**, *7*, 5416–5436. [CrossRef]
27. Belinskij, A. Water-energy-food nexus within the framework of international water law. *Water* **2015**, *7*, 5396–5415. [CrossRef]
28. Matthews, N.; Motta, S. Chinese state-owned enterprise investment in mekong hydropower: Political and economic drivers and their implications across the water, energy, food nexus. *Water* **2015**, *7*, 6269–6284. [CrossRef]
29. Pittock, J.; Dumaresq, D.; Bassi, A. Modelling the hydropower-food nexus in large river basin. *Water* **2016**, *8*, 425. [CrossRef]
30. Endo, A.; Burnett, K.; Orencio, P.; Kumazawa, T.; Wada, C.; Ishii, A.; Tsurita, I.; Taniguchi, M. Methods of the water-energy-food nexus. *Water* **2015**, *7*, 5806–5830. [CrossRef]
31. Mekong River Commission. *Integrated Water Resources Management-Based Basin Development Strategy 2016–2020 For the Lower Mekong Basin*; Mekong River Commission: Vientiane, Laos, 2016.
32. Allouche, J.; Middleton, C.; Gyawali, D. Technical Veil, Hidden Politics: Interrogating the Power Linkages behind the Nexus. *Water Altern.* **2015**, *8*, 610–626.
33. Foran, T. Node and Regime: Interdisciplinary Analysis of Water-Energy-Food Nexus in the Mekong Region. *Water Altern.* **2015**, *8*, 655–674.

34. UNECE. Task Force on the Water-Food-Energy-Ecosystems Nexus. Available online: http://www.unece.org/env/water/task_force_nexus.html (accessed on 27 January 2016).

35. Pittock, J.; Hussey, K.; Dovers, S. *Climate, Energy and Water: Managing Trade-Offs, Seizing Opportunities*; Cambridge University Press: New York, NY, USA, 2015.

Article

The Costs of Benefit Sharing: Historical and Institutional Analysis of Shared Water Development in the Ferghana Valley, the Syr Darya Basin

Ilkhom Soliev [1,2,*], **Kai Wegerich** [2] **and Jusipbek Kazbekov** [2]

1   Chair of Landscape and Environmental Economics, Technical University of Berlin, Straße des 17. Juni 145, Berlin 10623, Germany

2   International Water Management Institute, PO Box 2075 Colombo, Sri Lanka; K.Wegerich@cgiar.org (K.W.); J.Kazbekov@cgiar.org (J.K.)

*   Correspondence: isoliev@daad-alumni.de; Tel.: +49-303-147-3333; Fax: +49-303-147-3517

Academic Editor: Marko Keskinen

Received: 9 December 2014; Accepted: 26 May 2015; Published: 9 June 2015

Abstract: Ongoing discussions on water-energy-food nexus generally lack a historical perspective and more rigorous institutional analysis. Scrutinizing a relatively mature benefit sharing approach in the context of transboundary water management, the study shows how such analysis can be implemented to facilitate understanding in an environment of high institutional and resource complexity. Similar to system perspective within nexus, benefit sharing is viewed as a positive sum approach capable of facilitating cooperation among riparian parties by shifting the focus from the quantities of water to benefits derivable from its use and allocation. While shared benefits from use and allocation are logical corollary of the most fundamental principles of international water law, there are still many controversies as to the conditions under which benefit sharing could serve best as an approach. Recently, the approach has been receiving wider attention in the literature and is increasingly applied in various basins to enhance negotiations. However, relatively little attention has been paid to the costs associated with benefit sharing, particularly in the long run. The study provides a number of concerns that have been likely overlooked in the literature and examines the approach in the case of the Ferghana Valley shared by Kyrgyzstan, Tajikistan and Uzbekistan utilizing data for the period from 1917 to 2013. Institutional analysis traces back the origins of property rights of the transboundary infrastructure, shows cooperative activities and fierce negotiations on various governance levels. The research discusses implications of the findings for the nexus debate and unveils at least four types of costs associated with benefit sharing: (1) Costs related to equity of sharing (horizontal and vertical); (2) Costs to the environment; (3) Transaction costs and risks of losing water control; and (4) Costs as a result of likely misuse of issue linkages.

Keywords: transboundary water cooperation; equity; environment; water governance; issue linkage; institutions; Central Asia

---

1. Introduction

In order to promote cooperation over shared water resources, it is important to highlight the potential for cooperation including the broadest range of possible projects and benefits, options and choices available to riparian parties. In doing so, institutional analysis can be helpful to identify both the accepted norms, traditions, rules, principles and the modes of cooperation [1–3] which could generate greatest net as well as individual benefits [4–11]. This study reviews the benefit sharing approach in the context of international water management from institutional economic, social, environmental as well as power relations perspectives. The major advantage of benefit sharing is its capacity to facilitate cooperation among riparian parties by redirecting the focus from quantities of water to benefits derivable through its use and allocation and therefore turning the zero sum game into a positive sum interaction [4–11].

The article looks into historical data to derive lessons for potential application of benefit sharing in case of the Ferghana Valley, located in the upstream of the Syr Darya Basin and shared by Kyrgyzstan, Tajikistan and Uzbekistan. The Valley is rich in transboundary water resources along with shared infrastructure and because of the unity within one country in the past (until 1991 the republics were soviet socialist republics (SSRs), part of the Union of Soviet Socialistic Republics (USSR), the republics have a long history of relationship of initiating, implementing and maintaining the existing infrastructure on various governance levels. We are mindful that the benefit sharing approach was proposed for promoting cooperation among independent states, whereas the analysis in this article covers a period prior to independence. This is done to allow deriving lessons for the countries in the long run, at the same time possibly adding value to the research in application of the approach to riparians, which are part of a federal structure as it was in case of the Soviet Union or are countries in transition.

Although debates on benefit sharing are not as young as those on water-energy-food nexus (e.g., [12,13]), both seem to lack a rigorous historical and institutional perspective. This is at the very core of our manuscript and the analytical approach presented here attempts to fill this gap and expand understanding of the role of institutional settings in shaping the scope and effect of management decisions while viewing these decisions as a process.

The article continues with providing an overview on benefit sharing, which is followed by a background and methodology section. The analysis of the data has shown that there were five distinctive periods, each with a significant shift in the way benefits from the shared water resources were shared influenced by development of different formal and informal institutions (property rights, autonomy in decision-making, sharing criteria, changes and interaction in governance institutions, interests and priorities on different levels). While the prevailing approach has been to look at developments as before and after independence, findings of our research reveal the value of taking a more detailed look. The results section is therefore structured into these five distinctive periods. Further, the discussion section elaborates on major findings and attempts to systematize them. In the final section key conclusions are provided on implications of the research on broader scholarship of managing shared water resources as well as on possible constructive changes specifically in the Central Asian context.

2. Benefit Sharing—An Overview

In managing shared water resources, benefit sharing has been increasingly proposed as an approach to move from unilateral to cooperative actions by showing greater benefits of doing so. The approach not only redirects attention from volumes of water to benefits related to water, but also from pre-existing tensions or disagreements to new developments and arrangements. However, for sustainability of positive sum, it is central to ensure that the redirection of attention does not result in ignoring or worsening of problems, overweighing benefits in the long run. To understand the power of the benefit sharing approach to make cooperation more attractive one has to clarify: (1) *What benefits are there?* (2) *How can they be shared?* (3) *What are the costs of achieving shared benefits?*

Several studies define and categorize benefits and benefit sharing as follows.

Sadoff and Grey [4] determined four categories of benefits associated with cooperation as environmental (Type 1), with increasing benefits to the river; economic (Type 2), with increasing benefits from the river; political (Type 3), with reducing costs because of the river; and catalytic (Type 4), with increasing benefits beyond the river. The main critique on the typology is its practicality [10,14–17] as well as weakness in prioritization or identification of entry points. The latter is addressed by Phillips [8] whose methodology (Transboundary Waters Opportunity (TWO) Analysis) helps to see areas of priority when brainstormed by riparians. Overall, most scholars agree on the typology [4] as it covers the whole spectrum and allows distinguishing directions for cooperation.

Further, Sadoff and Grey [5] (p.3) define "benefit sharing" as "*any action designed to change the allocation of costs and benefits associated with cooperation*". The term "any action" can be interpreted as hindering but also enabling factor of the definition, since it broadens the spectrum of processes beyond the water sector [8–11,17,18]. Sadoff and Grey [5] acknowledge the fundamental principles of international water law—equitable and reasonable use—first established in the 1966 Helsinki Rules and then codified in the 1997 United Nations (UN) Convention on the Law of the Non-navigable Uses of International Watercourses. However, they propose the benefit sharing approach as an alternative. Dombrowsky [19] disproved it as an alternative approach showing the importance of underlying property rights if mutual benefits to be achieved and suggested that the approach could be rather complementary in certain cases. This is captured by a more specific definition suggested by Phillips and Woodhouse cited in [20] (p. 1): " . . . *as the process where riparians cooperate in optimising and equitably dividing the goods, products and services connected directly or indirectly to the watercourse, or arising from the use of its waters.*"

Later, Sadoff *et al.* [21] (pp. 28–29) explaining "*fair sharing of benefits*" refer to Article 6 of the 1997 UN Convention, which enumerates seven non-weighted guiding principles. Theoretically, this seems to translate the already existing dilemma of equitable distribution in the traditional (water volume based) approach into the benefit sharing approach. From practical perspective, Sadoff *et al.* [21] suggest learning from the actual practices derived from existing international treaties related to management of shared water resources as a starting point of negotiations referring to the database of transboundary agreements developed by Wolf [22]. However, the authors admit that "*the benefits derived from water development have generally not been shared equitably*" [21] (p. 29). The approach seems to be rather future oriented focusing on *ex ante* conceptualization of possible options to facilitate cooperation.

More broadly, the idea of benefit sharing [4,5] seems to replicate the mutual gains approach of the negotiation research introduced earlier [23]. However, one should acknowledge that both strongly relate to and based on the utilitarian concepts of the game theory and welfare economics, particularly to the problems looking for a Pareto improvement. However, unlike the game theoretic concepts, literature on both benefit sharing and mutual gains go beyond computing possibilities and show enthusiasm calling for creativity in problem solving, thinking beyond quantities, issues at the table, sectors involved, and assumptions. While encouragement for cooperation is supported by all means here, the question arises whether the increased emphasis to cooperate and achieve "yes" in a negotiation might overshadow or even cause some possible crucial negative consequences. Especially in a complex environment of shared water resources, broadening the basket and bringing in other, often as complex, issues, thus merging two or more complex resource systems, might easily lead to increased transaction costs by creating even a greater number of potentially conflicting interactions in a longer period.

The original mutual gains approach [23] addresses such questions as risks and circumstances under which one should not agree to a deal. In contrast, the studies testing the applicability of the mutual gains as well as benefit sharing in managing shared water resources seem to lack this holistic view. In fact, one of few available studies specifically on mutual gains in international rivers by Grzybowski *et al.* [24] promotes the benefits of the approach (also see: Special Issue "Getting to Yes" in United States–Canadian Water Disputes ed. by Sewell and Utton in 1986 [25]). That study, with a strong international law perspective, provides the case of the Columbia River Basin as one of the successful cases. Although, unlike Sadoff and Grey [5] and similar to Dombrowsky [19], Grzybowski *et al.* [24] argue that the mutual gains approach is complementary to the fundamental principles of international water law, *i.e.*, equitable and reasonable use, prevention of significant harm and obligation to cooperate. However, another paper, with as strong legal perspective [26], views benefit sharing as an artificial substitute to the traditional water sharing approach and concludes that in the long run the Columbia River Treaty could be questioned both on the grounds of equity of sharing and the costs to the environment.

Furthermore, focusing not only on the benefits but also on the costs of the benefit sharing approach, Dombrowsky [19] reveals a number of essential pre-conditions for benefit sharing to be successful. These include clear property rights and enforcement mechanisms, both of which are often problematic, as well as compensatory pay-off structures. However, Dombrowsky [19] seems to look into options to cooperate mostly during the negotiation process, with little emphasis on implementation and assuming that the coordination as well as operation and maintenance come at no cost.

Philips [8] (p. 14) specifically focusing on a practical application with the TWO Analysis mentions *"it [TWO Analysis] also assists markedly in defusing any pre-existing tendencies of riparians in relation to conflict"*. Defusing pre-existing tendencies of riparians in relation to conflict is indeed an advantage of benefit sharing, but it might also be its disadvantage if a riparian has to give up on a critical matter in order to gain immediate (however important those can be) benefits. Hence, what appear to be missing are possible longer-term implications. As Tarlock and Wouters [26] (p. 524) reason, focusing on benefits might result in *"unequal bargaining among states; the premature "sale" of future use opportunities; and the increased risk of aquatic ecosystem degradation"*. Riparians might be tempted by what can appear as short-term benefits and agree to arrangements that can pre-define or limit the range of decisions in a longer term.

Another study by Dombrowsky *et al.* [27] seems to acknowledge the problem of implementation in a different context, findings of which support the mentioned concerns [26]. Already looking at projects in preparation stages, they provide an example of how, due to *"unforeseen effects"* or because *"some things did not work as it was planned"*, the project-affected population became less satisfied with fairness of compensations provided for resettlement [27] (p. 1096). Concerns of the authors over implications of benefit sharing internationally and locally are timely, but long-term implementation still remains unexplored.

Overall, the long-term problems related to benefit sharing could be summarized as (1) inequitable allocation of benefits (internationally and locally, respectively; thereinafter, horizontal and vertical, respectively) as well as (2) likely underestimation of costs to the environment and related implications which are often not immediate [26]. Tarlock and Wouters [26] by benefit sharing refer to monetary compensation in return for a compromise in a shared river basin development (hydropower dams, the Case of Columbia River Treaty between the United States and Canada) or allocation (barter agreements, the Case of the Aral Sea Basin). What is not addressed is another form of benefit sharing—issue linkages. Even though issue linkage can be seen as an in-kind form of compensation, there seem to be two possible problems specifically related to issue linkages: (1) Increased transaction costs and more difficult control over implementation of the agreed terms; and (2) Possible use of issue linkages by a more advantaged party to impose its solution on other issues [9].

Similarly, Hensengerth *et al.* [11] conceptualizing benefit sharing on dams in transboundary rivers and analyzing five dams highlighted that *"the neglect of negative social and environmental concern may lead to conflict and lengthy renegotiations at a later stage"*. They also touched upon the importance of *"a history of cooperation between basin states and of institutionalized cooperation"* as a factor influencing benefit sharing [11] (p. 27). The paper attempts to expand this framework by systematic identification of the costs of benefit sharing as an approach in the long run as well as further exploring the idea that taking these costs into account is important to make cooperation more sustainable, including in river basins with history of cooperation and institutions to build on.

3. Background and Methodology

3.1. Study Area

While the central part of the Ferghana Valley lies mainly within the territory of Uzbekistan, the surrounding mountainous slopes are mostly part of Kyrgyzstan and Tajikistan (Figure 1). More specifically, the Ferghana Valley covers the territories of 7 administrative units (provinces): parts of Batken, Jalalabad and Osh Provinces of Kyrgyzstan, Sogd Province of Tajikistan as well as the entire territories of Andijan, Ferghana and Namangan Provinces of Uzbekistan. The 7 provinces have a total area of 124,000 km^2 and a population of about 14 million people, which is more than 20% of the whole population of Central Asia.

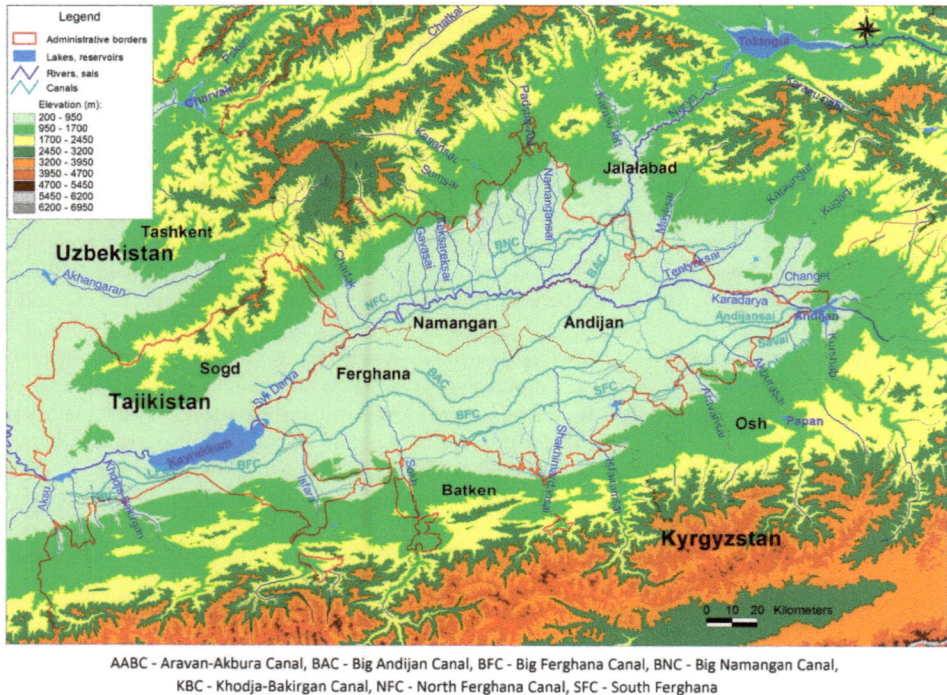

AABC - Aravan-Akbura Canal, BAC - Big Andijan Canal, BFC - Big Ferghana Canal, BNC - Big Namangan Canal, KBC - Khodja-Bakirgan Canal, NFC - North Ferghana Canal, SFC - South Ferghana

Figure 1. Topography, transboundary water resources and infrastructure in the Ferghana Valley (map by Alexander Platonov, 2015; courtesy of the International Water Management Institute).

The transboundary water resources of the valley consist of the Syr Darya, with an annual average flow of 37 billion cubic meters (BCM), formed from the confluence of the Naryn (13.8 BCM) and Karadarya (3.9 BCM), both of which originate in the mountains of Kyrgyzstan [28]. The flow of the Naryn River is regulated by the Toktogul Reservoir (14 BCM active storage capacity), located upstream in the territory of Kyrgyzstan, and the flow of the Karadarya by the Andijan Reservoir (1.75 BCM active storage capacity), which is on the border between Osh Province of Kyrgyzstan and Andijan Province of Uzbekistan. When exiting the Ferghana Valley, the Syr Darya is regulated by the Kayrakkum Reservoir (2.6 BCM active storage capacity), located in the territory of Tajikistan. Within the valley, there are also about 20 Small Transboundary Tributaries (STTs) with significant combined contribution to the flow of the main stem of 7.8 BCM [29]. Often these STTs have their own smaller reservoirs [30].

According to the Scientific Information Center of the Interstate Commission for Water Coordination (SIC ICWC) [31], the total irrigated area under command of irrigation canals in the Valley is 1.3 million ha (no data provided for Batken province). The breakdown on population, territories and irrigated lands by the countries and their associated provinces are presented in Table 1. The main economic activities are agriculture and livestock. The main crops are cotton, wheat, maize, orchards, tobacco, rice and vegetables in irrigated farming [31,32].

Table 1. Brief information on the Ferghana Valley, upstream of the Syr Darya Basin.

Country	Province	Population, Inhabitants	Population Density, Inhabitants/km^2	Territory, km^2	Irrigated Lands Data for 2010 [31], Thousand ha
Kyrgyzstan (KG)	Batken	469,700 Data for 2012 [33]	27.6	17,000 [34]	no data
	Jalalabad	1,099,200 [35]	31.6	33,700 [35]	125.6
	Osh	1,199,900 [36]	41.1	29,200 [36]	126.8
Sub-total (KG)		2,768,800	34.7	79,900	252.4
Tajikistan (TJ)	Sogd	2,349,000 Data for the period 2000–2010 [37]	93.2	25,200 Data for the period 2000–2010 [37]	178.0
Sub-total (TJ)		2,349,000	93.2	25,200	178.0
Uzbekistan (UZ)	Andijan	2,805,500 As of 1 January 2014 [38]	668.0	4,200 [39]	269.5
	Ferghana	3,386,500 As of 1 January 2014 [38]	498.0	6,800 [39]	357.7
	Namangan	2,504,100 As of 1 January 2014 [38]	316.0	7,900 [39]	282.1
Sub-total (UZ)		8,696,100	460.1	18,900	909.3
Total		13,813,900	111.4	124,000	1,339.7

3.2. Data

The data were gathered through archival research during several projects of the International Water Management Institute between 2010 and present (see acknowledgment). The specific geographical focus is on the relationship between Osh Province of Kyrgyzstan and Andijan and Ferghana Provinces of Uzbekistan, however, developments in the neighboring provinces and republics are also studied to illustrate wider issues. Since we look at historical data, it should be noted that the current Jalalabad Province (established in 1939) was part of Osh Province between 1959 and 1990 [35], whereas Batken Province was established only in 1999, which, until then, had been part of Osh Province as well [34]. Similarly, Andijan and Namangan Provinces were established in 1941 and Namangan was part of Ferghana and Andijan Provinces between 1960 and 1967 [40,41].

The data mainly represent interactions between the republics signed or prepared to manage the shared land and water resources and other related matters as well as higher level (regional) laws, decrees, agreements, declarations, *etc.*, reflected in 203 pieces of various documents covering the period between 1917 and 2013 (please see Tables S1 and S2). To refer to a specific document from Tables S1 and S2, the following acronyms are used in parenthesis [S1:N], where N is the corresponding number of the document as listed in the supplementary table (in this example, Table S1). In addition, the data with main characteristics of transboundary infrastructure were derived from the earlier studies of Wegerich *et al.* [28] for the smaller infrastructure (Table S3) as well as from the above documents and other sources for the larger infrastructure (Table 2).

3.3. Analytical Approach

The case study is based on in-depth qualitative analysis of the documents particularly from benefit sharing perspective: according to the types of benefits considered (Type 1, 2, 3 and 4) [4] and the ways sharing was envisioned, benefit-sharing mechanisms applied (compensations: monetary or in kind, issue linkages: outside or within water sector, across different basins), location of the object(s), property rights associated with the object(s), implementation of the agreed terms when relevant, and other information to see the connection and reference between the documents. Both direct costs of the developments and arrangements (such as cost of construction) and indirect costs of benefit sharing as an approach are analyzed.

The specific focus during the historical analysis was given to institutional changes. To be able to distinguish between different levels of institutions as well as to understand their level of development from temporal perspective it is referred to Williamson's [3] framework of institutional analysis:

Informal institutions such as customs, traditions, norms—Level 1; Formal institutions defining the rules such as autonomy in decision-making and property rights—Level 2; Governance institutions such as formation of main principles and organizations—Level 3; and Institutions for resource efficiency such as incentives to continuously improve marginal benefits—Level 4. As a result of the analysis, five distinctive periods of benefit sharing were distinguished where significant shift in establishment of these institutions took place. The results of the analysis form the respective five sub-sections of the following section.

4. Results

4.1. From 1917 to 1953: Border Delimitation and Irrigation Development

During this period under Stalin's strong hand, benefit sharing between the republics was imposed by the central planning government in Moscow; there was no negotiation and benefits from projects involving riparians were shared *de facto*. The republics had only a symbolic autonomy in decision-making. However, the period marks developments, which would have crucial impacts on the types of benefits and the way those benefits would be shared later.

First, a complete nationalization of lands in 1917 [S1:1] was followed by border delimitation (till 1936) forming the new republics decision-making bodies, which eventually would become the present independent states. Due to the complexity of the landscape, varying economic potential, and mixed ethnicities across the valley, the sides had contesting claims and many border questions were left open [30,42–47].

Second, the extensive irrigation development placed emphasis on cotton independence of the USSR. The studies [28,30,31,42–48] indicate that the entire institutional setting was aimed at two types of benefits [4]. The increased agricultural production is assumed to have contributed to the region's economy directly (Type 2). The water infrastructure development was in line with the Soviets' agenda to restore social and political stability in the region by increasing employment and attempts to redirect the attention from political life to implementation of the projects. The combined effect can be classified as benefits beyond the water resources, Type 4.

Third, constructed irrigation canals created the foundations for property rights on the shared water infrastructure. The infrastructure was constructed in areas that were easier to irrigate (within the valley). Since water flows were mainly utilized by downstream collective farms (*kolkhozes* and *sovkhozes*) and districts, the majority of the projects with shared command area were operated by the authorities in the Uzbek SSR, even though some were located upstream within the territories of the Kyrgyz or Tajik SSR (Table S3). This is the root of why some of the infrastructure with shared benefits within territories of Kyrgyzstan (and Tajikistan) today belong to Uzbekistan and occasionally *vice versa*.

Through 11 shared projects, the republics regulated the water resources with a command area of 57,542 ha (including 10,300 ha in the territory of the Kyrgyz SSR) (Table S3). In 3 out of 6 cases, the Kyrgyz SSR did not have any land irrigated despite the headwork/infrastructure location was in the Kyrgyz SSR. In addition to irrigation, pastures of the republics were re-distributed for long-term use. The data from 1946 indicate that the Uzbek SSR was the main recipient of pasturelands (4 million ha), while the Kyrgyz SSRs was the main provider of pasturelands (1.1 million ha), with a minor input from the Tajik SSR (71 thousand ha). This was connected to the greater number of Livestock Units (LSU) in the Uzbek part of the Ferghana Valley than in the Kyrgyz part: 0.3 million LSU and 0.2 million LSU, respectively.

The costs of construction of the shared infrastructure were financed through the budget of the Uzbek SSR, although the other republics had benefits too. In addition, during this period, a significant movement of labor force took place: first, forced migration before World War Two, second, massive resettlement during and after World War Two, which included highly qualified specialists from Russia and western parts of the USSR to Central Asia, especially Uzbekistan [42]. Thus, even without detailed data on the extent and proportions, it is evident the costs borne in providing the labor force for the construction and ameliorative works were colossal. In addition, the documents within this period do not prioritize environmental preservation or prevention of possible negative impact of the developments on available water quality and quantity.

4.2. From 1953 to 1970: Negotiation and Mega Projects to Boost Water Supply

The year 1953 marked the end of the Stalin period. Although the new leadership of the Soviet government continued with further policies to increase agricultural output, there were the following important differences influencing various aspects of benefit sharing.

First, the republics gradually started to gain autonomy in decision-making. Negotiations over the shares on several projects were held directly between the republics and explicitly documented within the Protocols. For example, after the start of the works on the Toktogul Reservoir, the Kyrgyz SSR claimed and secured compensation for the lands allocated for it through negotiations on the Andijan Reservoir [S1:30] (more details follow). At the same time, the share of the Kyrgyz SSR in the allocated pasturelands increased significantly too, amounting to 834 thousand ha (328% increase compared to 1946) without decreasing the areas allocated to the other republics [S1:8]. Later, the autonomy increased with the 1968 Union-Wide Law on Land, which called for direct dispute resolution between the republics [S1:34].

Second, in 1953–1970, negotiations and construction works of several larger projects were initiated, which led to a sharp increase in issue linkages and closed the basin in the long run (Table 2).

The table shows issue linkages of increased complexities both within and outside the basin. For the Kyrgyz SSR, who provided lands for the construction of the Toktogul Reservoir, in 1961, Moscow's idea was to compensate the lands by giving expansion rights (15,000 ha) and water for it in the Burgandy Massive through regulation of the Sokh River [S1:87]. However, in the 1962 negotiations of the Andijan Reservoir, the Kyrgyz SSR sought compensation directly from the Uzbek SSR by requesting construction of the Left-Shore Kampyr-Ravat Canal (LSKR) to the Burgandy Massive to irrigate additional 12,000 ha [S1:18]. The Uzbek SSR agreed to 8000 ha and that in addition to the LSKR Canal, the design of the Sokh Reservoir would take into account feeding these 8000 ha [S1:30].

The outcome of the period was that (1) the parties on all levels (regional, national, meso and local) were expecting significantly higher water supplies in the long term and therefore boost in irrigation expansion and (2) the agreed plans were rather ambitious and as was claimed in several cases, would exceed the capacities of the republics to implement the projects within agreed timeframes. In 1965 the Osh province Water Management Department (WMD) proposed to expedite the construction of the Toktogul, Andijan, Papan, Sokh, and Tortgul Reservoirs as water supply was not higher than 50% of water demand in the right shore tributaries of the Karadarya [S1:26]. The ambitious plans resulted in delays: transfer of land for the construction and their compensation were delayed due to administrative, technical and financial constraints [S1:56]. Some projects had delays for several decades, being only partially implemented (the Sokh Reservoir) or not implemented at all (the LSKR Canal). This had unfavorable implications for both sides. The Kyrgyz SSR was left without its expected increase in water supplies from these projects who prepared additional lands in advance [30]. Hence, incentives to look for compensation from other sources were created. The Uzbek SSR would, on the other hand, have to compensate for possible losses related to the latter and would have a weaker bargaining power in future negotiations with the Kyrgyz SSR (more details in the later periods).

Table 2. Projects with shared benefits in the Ferghana Valley initiated/constructed between 1953 and 1970.

Project	Negotiation	Commissioning	Irrigation Benefits: Command Area, thousand ha (and/or Share in Water Allocation, %)			Other Benefits
			Uzbek SSR	Kyrgyz SSR	Tajik SSR	
Kayrakkum Reservoir with the active capacity of 1.7 (BCM) on the Syr Darya River [49]	Late 1940s–1950s	1956	At the exit of the Ferghana Valley, benefiting the downstream of the Valley and contributing to 185.3 thousand ha of the Tajik irrigated lands in the Syr Darya Basin [49]. Six thousand hectares in the Arka Massive of the Kyrgyz SSR through pump-stations in the Tajik SSR			No initial data. "For the period of 1990–1998, the Kairakkum hydroelectric power station annually generated about 323 million kWh on average in the growing season" [49] (p. 115).
Toktogul Reservoir (14 BCM of active capacity) on the Naryn River	no data (assumed in late 1950s)	1974	Built for long-term regulation of the Naryn flow. Water supply increase for 918 thousand ha, expansion by 400 thousand ha in the Syr Darya River Basin (exact shares of the republics were not possible to calculate) [50]			Hydropower (4.1 billion kWh a year) initially it was agreed that the flow released as a result of hydropower generation is allocated at the ratio of 85.5% for the Uzbek SSR and 14.5% for the Kyrgyz SSR.
Left-shore Naryn Canal (18 m^3/s) and Druzhba pump-station	1960s	1969–1970	5.2	3.5	not applicable (n/a) due to its geographic location	–
Tortgul Reservoir on the Isfara STT (0.09 BCM)	1960s	1971	1.6, (8% water) [51] (p. 23).	9.23 (37% water)	21.3 (55% water)	Kyrgyz SSR and Tajik SSR share water from canal Machai (2 km upper than water intake to Tortgul Reservoir) on proportion of 80% and 20%, respectively [51] (p. 24).
Papan Reservoir (0.24 BCM of active capacity) on the Akburasai STT	1960s	1985 [51:109]	26.6 [52]	10	n/a	1.5 m^3/s for domestic use of Osh city
Sokh Reservoir (0.32 BCM of active capacity) on the Sokh STT [51:83]	1960s	Not completed	45.2	18.2	n/a	Compensation for lands provided for the construction of the Toktogul Reservoir. The reservoir would increase its irrigated lands in the Burgandy Massive by 22,000 ha (with 0.2 BCM from the reservoir) in the Kyrgyz SSR and increase water supply for the existing irrigated lands in the Uzbek SSR.
Karkidon Reservoir (0.22 BCM)	1961	1968	87% water	13% water	n/a	–
Andijan Reservoir (1.75 BCM of active capacity) on the Karadarya	1962	1978	247.1	49.6	n/a	Unlimited expansion upstream of the reservoir for the Kyrgyz SSR, hydropower release from the Nurek Reservoir (on the Amu Darya Basin) 85.5% for the Uzbek SSR, 14.5% for the Kyrgyz SSR.
Left-shore Kampyr-Ravat (LSKR) Canal	1965	Not constructed	15.9	8	n/a	Project not implemented.
Right-shore Kampyr-Ravat (RSKR) Canal	1965	1970s	14.57	–	n/a	–
Kasansai Reservoir (0.3 BCM) on the Kasansai STT [51:83]	1967 (second phase)	1972	28.8	1.3	n/a	–

The analyzed documents show the continued focus on the economic benefits, *i.e.*, increased water supply and right to expand irrigated agriculture as a result of joint infrastructure development. The costs of the smaller infrastructure were still covered through the budget of the Uzbek SSR (Table S3). There is lack of data on the detailed allocation of the costs of the Toktogul Reservoir. The construction of the Andijan, Karkidon, and Kasansai Reservoirs, Left-Shore Naryn Canal, as well as of not completed Sokh Reservoir and not implemented LSKR canal were the responsibilities of the Uzbek SSR while the Kyrgyz SSR was responsible to contribute with provision of lands for construction. While both monetary compensation, including payments to compensate losses related to population resettlement, and non-monetary compensation mechanisms were practiced within this period, the costs to the environment were still not considered.

4.3. 1970s: Competition, Allocation Criteria and Counter Hegemony

In 1970, the future of benefit sharing was significantly influenced by two important developments. The 1970 Order [S1:40] from Moscow allocated increased investments for further land reclamation as well as regulation and re-allocation of the runoff of the rivers for the next 15 years but pointed out the projects would be approved on a case by case basis. This meant official competition for the right to use land and water resources between the republics. On the other hand, the 1970 Union-Wide Law on Water [S1:41] formalized the basin approach under which so called "Schemes" of complex use should have been developed for each river basin.

The initial version of the Syr Darya Scheme developed in the beginning of the 1970s [S1:42] (p. 5) explained the principle land and water allocation criteria as:

- Proximity of the lands to the source of irrigation;
- Higher productivity of the lands, lower demand for irrigation, less investments and time;
- Preference for the lands in more southern latitudes suitable for more valuable sorts of cotton;
- Proximity of the lands to the reserve contingents (labor, infrastructure);
- Needs of the republics in connection with the Union's interests.

The idea was to locate the lands based on the above criteria that would then receive a proportional share of water based on the area, crop pattern and other features. This is how the water allocation criteria tied to the irrigated area started to develop.

The irrigated area in the Valley in 1970 was 1058 thousand ha [S1:42], 720.9 thousand ha (68%) of which was in the Uzbek part [44]. The data in Table S3 show, there was a significant decrease in the number and scope of the shared infrastructure constructed. The new infrastructure was added due to the construction of the canals in early 1970s linked to the Dustlik pump-station which itself had been constructed in 1969. This means that almost no irrigation infrastructure (except the Jiyda canal in 1974 with the capacity to irrigate only 905 ha) was agreed between the riparians on the STTs in this period. Three other projects were the dams with flood control function. The focus shifted from the smaller infrastructure (Table S3) to the implementation measures of the larger infrastructure (Table 2). While a number of projects were completed in the 1970s, the LSKR Canal and Sokh Reservoir for upstream expansion had long delays. The Kyrgyz SSR referred to the agreements reached with the Uzbek SSR on the Andijan Reservoir as an example to persuade Moscow in providing more expansion rights [S1:47], however, Moscow dismissed such requests. Perhaps, the dismissal put the Kyrgyz SSR in the position to raise numerous claims both regarding the irrigation expansion and pasture use unlike in the previous periods. The Kyrgyz SSR had a number of unilateral projects with the potential to irrigate an additional 137,260 ha prepared for implementation within the Ferghana Valley with 66,260 ha being directly connected to shared water resources, *i.e.*, Kayrakkum Reservoir, Khodja-Bakirgan STT, Sokh STT, RSKR Canal, LSKR Canal, and Aravansai STT [S1:58]. In 1974, the Kyrgyz SSR requested Moscow to return the pasturelands used by the other republics within the territory of the Kyrgyz SSR [S1:59].

While there is evidence of monetary (Andijan Reservoir, Karkidon Reservoir) and non-monetary compensation (several cases of land compensation), the Kyrgyz SSR also requested the Uzbek SSR to be connected to gas pipelines as a subsidy (0.5 BCM annually), documenting the first explicit quantitative expression of issue linkages outside the water sector during negotiations [S1:60]. The downstream Uzbek SSR as well as the Kazakh SSR, unlike the Kyrgyz SSR, was to bear the environmental costs as a result of massive expansion. A rapid drop in the level of the Aral Sea and a sharp increase in salinization was expected [S1:42]. There was an estimated 9000-ton loss in fishery from the Aral Sea annually. The impact and the need for diversion of Siberian rivers to the basin was highlighted on the highest level [S1:40], with first design works to be completed in 1971–1975. However, there is no evidence that any design documentation was prepared by that time.

4.4. 1980s: Attempts to Clarify and Solve Conflicting Issues

By 1980, most of the larger infrastructure had been completed and there was a need for new sharing arrangements taking into account all the changes. The following four significant developments were found which shaped the new period of benefit sharing in the 1980s: (1) Increased complexity of issue linkages; (2) Amplified autonomy in decision making and negotiation; (3) Further expansion and basin closure; and (4) Increased cooperation and lost tracks of linked issues previously.

First, the complexity of issue linkages increased to its maximum: while the newer versions of the Schemes connected the infrastructure and developments in the entire Syr Darya Basin in more detail, a new Protocol from 1980 [S1:64] connected all of the STTs in the Ferghana Valley as one package. In addition to the linkages between and across the basins, the non-monetary compensation in the form of land transfer and exchange was discussed and applied more often whereas monetary compensation was no longer observed.

Second, autonomy in decision-making and negotiation amplified further. For example, there is evidence when the Kyrgyz SSR officially contested the decisions approved by Moscow regarding the ways the water shares in the 1980 Protocol were calculated [S1:65]. The design institute argued the main allocation principle was followed [S1:66]. Moscow's purpose to maximize cotton production in the basin had been well established by this period as the Scheme for the basin was in its final stages and discussions were on details rather than on principles. Hence, Moscow gave even more space to the republics for negotiations on the details, as the main purpose with its direct economic benefits for Moscow was more or less secured. On the other hand, the intensifying socio-economic crisis in the USSR during the late 1970s and 1980s [53] was not favorable for Moscow to continue with its active coordination and oversight. In any case, the Kyrgyz SSR kept demanding more water. After the arrangement to share the STTs as one package in 1980 [S1:64], the Kyrgyz SSR, in 7 cases out of 9, including 5 cases where the terms had been implemented, requested to increase its share due to the optimization of water use in the Uzbek part [S1:80].

Third, both the increasing costs to the environment due to basin closure (as the water was utilized to its fullest) as well as the increasing pressure from the Kyrgyz SSR to re-consider allocations implied increased costs for the Uzbek SSR. As of 1 January 1981, the Ferghana Valley had 1227.30 thousand ha of irrigated lands: 255.5 thousand ha (21%) in the Kyrgyz SSR, 124.8 thousand ha (10%) in the Tajik SSR and 847.0 thousand ha (69%) in the Uzbek SSR. The expansion maximum was estimated at 1341.6 thousand ha, which would also change the ratio to 24% (+3%), 10% and 66% (−3%), respectfully [S1:83]. The number of constructed pump-stations in the Uzbek SSR increased rapidly in this period to compensate water to the lands affected by the upstream expansion [54]. Although the lift was unsustainable in the long run due to its high operation and maintenance costs [54], keeping the irrigated lands was important for preventing high social costs at least in a short run and keeping the shares of water tied to the areas of land by the Scheme in a longer run.

Fourth, there was an increased cooperation on the Sokh Reservoir and the Sokh STT, although the construction of infrastructure with shared benefits further slowed down in the 1980s. There were only two shared canals constructed with combined capacity to irrigate 890 ha in the Uzbek part of the Valley (Table S3). The other three projects were flood-controlling dams. In case of the Sokh Reservoir construction, the Uzbek SSR was responsible for the costs, the construction works began and intensified, but there were still delays to address resettlement issues of the affected population [S1:88]. In case of the Sokh STT, in 1989, the Kyrgyz SSR secured a significant increase in the share from the STT of more than additional 0.2 BCM to irrigate the Burgandy Massive [S1:92]. Expansion in the Burgandy Massive was initially agreed as part of compensation for the lands provided by the Kyrgyz SSR for the Toktogul Reservoir (see the period 1953–1970). The agreement was to irrigate the massive through intakes from the Andijan Reservoir and the Sokh Reservoir. The share from the Andijan Reservoir was 0.2 BCM to be delivered with the LSKR Canal. Although the increased share from the Sokh STT in 1989 exceeded this previously agreed limit, within the same Protocol where this agreement was reached, it was agreed to pursue the projects of the LSKR Canal and the Sokh Reservoir further.

4.5. From 1991 to 2013: Independence and Response to New Old Challenges

From institutional perspective to benefit sharing, the most important distinction of this period is that the republics found themselves between the highest level of autonomy in decision making (sovereignty) by far on one hand, and the highest level of physical (inter-)dependence (shared resources, infrastructure and issue linkages) on the other hand. Irrigation expansion exceeded the planned levels of basin closure, a report from 1991 indicates that the irrigated area in the Ferghana Valley by 1988 was 1382 thousand ha: 290 thousand ha (21%) in Kyrgyzstan, 919 thousand ha (66%) in Uzbekistan, and 173 thousand ha (13%) in Tajikistan [55].

It should be noted, that to date there is abundance of literature on analysis of reforms, problems and opportunities on all possible levels and numerous case studies explaining the situation and possible steps ahead after independence. We do not intend to go through those all but rather maintain our focus on the gap—institutional changes and developments influencing the new period of benefit sharing as well as costs and benefits thereof.

With independence of the states in 1991, the benefit sharing from the existing infrastructure, arrangements and agreements did not stop. In fact, the 1992 Almaty Agreement confirmed the will of all five Central Asian states to adhere to the existing pattern and principles as well as acting regulations of water allocation from interstate resources [S2:1]. This was reinforced within other agreements and declarations later (Table S2). However, implementation of these agreements in a longer run faced a number of challenges.

First, financial difficulties: With problems on how to restore economic and social stability, while the infrastructure built during the Soviet Union was getting outdated and in need of increased investments, the problem was now how to balance between the required more rational use with less finances and meeting the demand for water which became even more crucial for the national economies than before. With the 1998 Syr Darya Framework agreement [S2:7] focusing on the releases from the Toktogul Reservoir, Kyrgyzstan managed to successfully agree with Kazakhstan and Uzbekistan on the compensation mechanisms, which linked water releases with hydropower and fossil fuels between the countries. Tajikistan joined the agreement in 1999. However, due to implementation problems, the Framework Agreement was not renewed after its first five years cycle [29].

Second, although environmental protection received more attention on the regional level agreements (Table S2), implementation of those did not reflect much in the analyzed lower level documents (Table S1), where economic benefits remained dominant. Most of the cooperation on the meso level was mainly related to maintenance issues—to reconstruct, renovate existing reservoirs (Andijan Reservoir, Papan Reservoir), irrigation and drainage networks. Additional difficulties were observed due to the lengthy clearance processes for crossing the national borders often resulting in delays or indefinite halt of planned maintenance activities. After independence, three shared transboundary projects were constructed (Table S3). While one of them is on the existing canal (Madaniyat-2 pump-station) the other two are flood control infrastructure, hence, all was constructed only to support the existing infrastructure.

Third, no specific interstate organization or framework has been created with focus on managing the shared STTs and their infrastructure. Thus, for the actors on the lower levels in the

Ferghana Valley, the institutional arrangement was that the sides were supposed to continue their relationship based on the previous agreements and practice. This implies that there are the following agreements/institutional arrangements in place.

- From transboundary perspective, the latest agreement in place was the 1980 Protocol [S1:64]. However, already during the Soviet period, the sides had disagreements on a number of the agreed terms within the Protocol as described in the analysis of the previous period. A Report from the Kyrgyz side in 2012 [S1:183] mentions the 1989 Protocol [S1:92] as an agreement in place for the Sokh STT. A Report from the Uzbek side of the same year [S1:184] informs that in 2001 an oral agreement was reached to share 3 STTs on a 50/50 basis. However, it is not evident whether it was a one-time agreement to address the drought year. The sides address issues on an ad hoc basis; they exchange requests in case of emergencies such as floods and for annual agreement of decadal allocation from the shared water resources. In addition, with lost linkages behind the LSKR Canal and the Sokh Reservoir, these two continue to be a topic of complaints.
- From a meso level perspective, Uzbekistan has partly shifted from water management according to administrative boundaries (provinces and districts) towards management based on hydrographic/hydrological boundaries of basins and irrigation systems. This was done on the main canals of the Valley (BFC, BAC, SFC). However, the other canals and STTs are left with the WMDs of the provinces [56]. In Kyrgyzstan, in addition to the shift to basin principle, as it was mentioned, Osh province was reorganized into three provinces while the process of restructuring water management in Tajikistan is still in progress [31].

As an outcome of the above challenges and mismatch of institutional arrangements, the incentives of the countries increased to secure more water within their national boundaries, especially since the 1998 Framework Agreement was no longer implemented [29]. With operational change of the Toktogul Reservoir by Kyrgyzstan to meet its energy demand, mid and downstream countries had to find pragmatic solutions increasing internal storage capacities in Uzbekistan and Kazakhstan, re-arranging agreements on certain parts of the Valley as in case of the Isfayramsai, Shakhimardansai and Sokh STTs between Kyrgyzstan and Uzbekistan or the Khodja-Bakirgan STT between Kyrgyzstan and Tajikistan, or attempting to be independent from transboundary infrastructure as the case of Tajikistan on the BFC [29,57].

5. Discussion

Going back to the discussions on water-energy-food nexus, it seems that benefit sharing, as a positive and result-oriented negotiating approach, could be useful to bring about the needed changes and transition, specifically in managing shared waters. It could serve as a much-needed instrument for what Hoff [12] describes as "stimulating development through economic incentives" (p.37). The historical and institutional analysis, as provided here, seems to offer practical lessons for reconciliation of long-term and global objectives (such as ecosystem stewardship and equity goals) with shorter-term economic benefits, identified as one of the main challenges in the nexus debate [12]. Further, the case study also shows how the isolated focus (e.g., on the Toktogul Reservoir and larger rivers) might have reduced the system efficiency in the long run [12]. Overall, it seems that nexus, which thus far has largely lacked the historical perspective and has not fully viewed management decisions (whether on water, energy, food or their inter-linkages) as a process, could almost entirely borrow the presented analytical approach for assessing evolving institutional settings shaping the scope and effect of the management decisions.

Carrying on with more specific case study findings and looking particularly from benefit sharing point of view, it becomes evident that from one period to the other the benefit sharing increased and incorporated more benefits to the both riparian states (Table 3). Notably, if to follow the typology [4] (Type 1—Environmental; Type 2—Economic; Type 3—Political; and Type 4—Catalytic benefits), the Type 2 benefits remained dominant throughout the entire analyzed period (one should note that here the costs and benefits are deliberately not provided in any explicit way; it is questionable whether issues with this level of complexity and over such long period of time would allow quantifying costs and benefits with any accuracy at all). This highlights the concerns for sustainability of water resources and ecosystems, also discussed in the nexus literature where water is seen as a source or at least as a central factor of economic growth [12,13].

Overall, taking a historical/dynamic or comparative approach highlights that there is a clear gap of how to show differences, particularly since most of the agreements are within Type 2. In addition to the direct costs of benefit sharing development or arrangement (such as construction costs), the analysis pointed to four other possible concerns in the long run, which we term as indirect costs of benefit sharing. In turn, looking at the nature of the lessons on long-term costs, one can state these costs do not necessarily have to limit to benefit sharing, but could be similarly taken into account in the discussions of the nexus approach [12,13].

5.1. Costs Related to Equity of Sharing

Here we are proposing to include "equity" within these particular types/categories so that it would be possible to highlight an increase, stagnation or decrease of these particular types. It could cover the concern pointed out by Tarlock and Wouters [26] regarding transboundary (horizontal) equity in allocation of benefits and it might work well with the social concern of Hensengerth *et al.* [11], which addresses the equity of development vertically. It supports findings of the study by Dombrowsky *et al.* [27] specifically focusing on this aspect of benefit sharing.

The case study brought forward that for the transboundary infrastructure within the Ferghana Valley, property rights and therefore long term sustainability of operation and maintenance of infrastructure are key. Furthermore, while the benefits generated through the infrastructure were shared, the obligation (costs) of operating and maintaining the infrastructure were and are still (except occasionally) not shared. This point highlights the additional need for clearly emphasizing not only benefits but also costs. Looking only at the sharing of benefits might show, that benefit sharing is not equitable.

Besides, in cases when the decisions on forced labor were made solely for the purpose of constructing and operating the infrastructure (1917–1953) internalization of these costs would change the ratio of costs and benefits. Another example, increased unilateral ambitions of the Kyrgyz SSR starting in 1970s emerged because it appears that the Kyrgyz SSR was unsatisfied with the equity of sharing due to the delayed and non-implemented projects. At the same time, the Kyrgyz SSR often argued that the Uzbek SSR increased its water supply levels through unilateral optimization works and therefore requested to re-consider shares to achieve proportional supply levels. This seems to have created a strong disincentive for increasing efficiency as well as incentives for misrepresenting data. In general, such an approach, penalizing a good manager, seems to be a result of serious mismatch between the allocation criteria and improving efficiency.

In addition, looking at Williamson's concept [3], it appears that although there have been tremendous changes regarding the water scarcity situation and the external environment (financial overflow 1960s and 1970s, withdrawal of Moscow and basin closure in the 1980s, independence and financial collapse in 1990), which have triggered adaptation in negotiations and changes of water agreements, so far these changes have not altered the official property rights situation. Besides, the region presents a possibly unique, or at least, very rare case of property rights where a country's infrastructure is located beyond its national boundaries. Further studies are necessary to clearly determine in which case property rights and therefore the obligation to operate and maintain have been altered and the consequences thereof.

Table 3. Summary of the periods.

Periods	Benefits	Benefit Sharing	Mechanisms	Institutions Established
From 1917 to 1953	Types 2 and 4	Increased through boost in smaller infrastructure, pasture exchange	Existed only technically (not voluntarily), founded the shared infrastructure	Central government	Republican borders, property rights on land and infrastructure (Level 2 institutions)
From 1953 to 1970	Type 2	Increased through boost in larger infrastructure, pasture exchange	Emerged with the initiation of larger shared projects, autonomous bilateral negotiation, specific shares of each republics	Monetary and non-monetary compensation, issue linkages within and outside (pastures) water sector, across basins (Nurek Reservoir)	Autonomous negotiations, irrevocable commitments (Toktogul, Andijan and other projects) for revocable ones (Sokh Reservoir, LSKR Canal)
1970s	Type 2	Increased through basin scheme development to use the basin resources to their fullest	Existed and challenged by further autonomy of the republics, increased claims (counter-hegemony) of the Kyrgyz SSR	Monetary and non-monetary compensation, issue linkages within and outside (pastures, gas pipelines) water sector	Proportional water allocation tied to irrigated areas (Level 3 institutions), competition for expansion
From 1980 to 1991	Type 2	Increased through basin closure, rise in pump-stations	Strengthened by further autonomy and official disputation of the Moscow's decisions	Non-monetary compensation, issue linkages within (linking all STIs together) and outside water sector	Governance institutions (Level 3 institutions): managing through sub-basin allocations
From 1991 to 2013	Types 2 and 1	Partly maintained through operation and maintenance of existing infrastructure, enhancement of flood control	Encouraged and tested on regional level but failed (1998 Framework Agreement), practiced on meso level (linked infrastructure and financial incentives), being replaced by national solutions	Issue linkages within and outside water sector (framework of compensations linking water releases, hydropower generation and fossil fuels)	Level 1 (traditions, customs, norms) and Level 2 institutions (above) carried over, Level 3 partly valid, Level 4 (allocative and resource efficiency) attempted by national reforms

5.2. Costs to the Environment

Environmental concern highlighted in the literature [11,26] proved to be absolutely valid throughout the analyzed periods. Given the scales of the developments, integration of "the costs to the water resources" (or "negative benefits to the river") would likely reduce the net economic benefits (Type 2). Even though there are a number of intergovernmental agreements after independence on a national level calling for cooperation in the area of environment and rational use of natural resources the data indicate that the parties focusing on benefits (irrigation expansion) on lower levels have only occasionally considered rising water tables where in fact the focus was on potential economic damage. Institutionalization of a water allocation principle that did not prioritize environmental flow appears to be the main factor in this respect.

5.3. Transaction Costs and Risks of Losing Water Control

Development of uneconomic lift irrigation to secure benefits from water sharing arrangements showed how focusing on benefits might lead to higher costs in the long run especially in a case of multiple interconnected issue linkages.

Similarly, the analysis showed that although there was a clear issue linkage in the beginning (regarding LSKR Canal and the Sokh Reservoir), the two uncompleted infrastructures appeared in different contexts. Furthermore, today's cooperation appears to be based on a tit-for-tat approach because of the multiple integrated infrastructures. Hence, there is a dynamic of issue linkages within the context of Ferghana Valley. Therefore the original issue linkages (documented in agreements) appear to be in constant flux and utilized as bargaining positions whenever necessary.

Because of the interdependence on transboundary infrastructure cooperation appears to be the most viable option taking a more holistic approach for all infrastructure. It is a likely reason why many projects in the Valley with isolated focus did not succeed as expected. Bigger donors such as the World Bank, Asian Development Bank and United States Agency for International Development focused on the larger rivers without going into details of the lower level inter-dependencies [47,58]. The initiatives of the Deutsche Gesellschaft für Internationale Zusammenarbeit (GIZ) on the Isfara and Khodja-Bakirgan STTs focused on signing bi-lateral agreements, which led to exclusion of Uzbekistan from the Isfara STT [47,51,58]. The projects of the Swiss Agency for Development and Cooperation (SDC) on the Shakhimardansai and Khodja-Bakirgan STTs, although focused on bottom up cooperation, basically did not succeed due to a weak link up with higher frameworks [47,58].

One should note that all that transboundary tributaries, where the previous agreement was challenged, are within the same 'newly created' administrative unit (Batken Province), similarly, the small reservoir (Kasansai), which appears to have the most problems regarding cooperation [30] is also located in a "newly created" administrative unit (Jalalabad Province). This puts into question whether decentralization as practiced by Kyrgyzstan has decreased cooperation, since it decreased the possibility of issue linkage. Similarly, the water reforms in Uzbekistan (the partly implemented hydrographization [56]) might have negative effects on cooperation, since it reduced the bargaining positions of the former players (Andijan and Ferghana Provinces). In this respect, it might be important to highlight that the practice of honoring past agreements (national level) might be put into question, particularly if lower levels are tasked with the implementation and these lower levels cease to exist or have reduced bargaining power. Having stated this, one could also question whether the national level in Kyrgyzstan has control over the meso level administrative units [59].

5.4. Costs Resulting from Misuse of Issue Linkages

The issue linkages, on one hand, have helped to achieve cooperation and conclude multiple agreements. On the other hand, it created a number of linkages between asymmetric issues. The Toktogul was linked during the Andijan Reservoir negotiations to compensate the lands under the Toktogul by expansion rights in the Burgandy Massive. The Burgandy Massive was linked to the

LSKR Canal and Sokh Reservoir. While the Toktogul and Andijan Reservoirs became irrevocable commitments the LSKR Canal and Sokh Reservoir were revoked and never completed. The significant increase in the share from the Sokh STT, which boosted irrigation in the Burgandy Massive for Kyrgyzstan, did not stop them from continuing or even reconsidering the claims on the LSKR Canal and Sokh Reservoir. Hence, both the scope and symmetry of issues to be linked are important to be able to follow through and implement the agreements in a longer period.

Similarly, what seems to be not explored enough from benefit sharing perspective is the focus beyond the river, which entails the brokering (including financial incentives and issue linkages) as well as arbitration role of third parties, in this case of Moscow. As the analysis suggests, the interests and influence of third parties might completely re-design the structure of both benefits and sharing.

6. Conclusions

Countries need dialogue and coordinated actions to address dynamic challenges and to shift towards more holistic views in managing shared water resources. While the water-energy-food nexus is the most recent way to promote more holistic views, it seems to largely lack both historical and institutional perspectives: this study has emphasized the importance of such perspectives. Our research indicated evolution and implications of institutional settings for shaping management decisions and revealed multiple factors limiting as well as enabling cooperation in a highly complex environment. The focus on benefit sharing as an approach demonstrated that new arrangements and developments with shared benefits and mutual gains provide a good platform for the needed dialogue. Yet, the research findings also brought to attention possible indirect costs associated with benefit sharing in the long run, which might have been overlooked in the literature. It seems incorporations of these costs could contribute to making cooperation and dialogue more constructive and informed and therefore new arrangements more stable.

The case study has identified five different periods of development in the relationship related to management of the shared water resources in the Ferghana Valley between 1917 till present between Kyrgyzstan and Uzbekistan. A particular focus has been placed on what can be learned from benefit sharing perspective. From the earlier Soviet period under the Stalin's strong regime when the property rights on land and more importantly on shared infrastructure were established, the analysis showed that the institutional transformation between the republics took place already in the period from 1953 to 1970 in time of heroic engineering projects targeting cotton independence of the USSR. However contradictory, already then the republics got to negotiate whether to construct, what to construct and how to share benefits. A very strong top down administration started to transform into a bottom up hierarchy. In the 1970s, the republics gained even more autonomy when Kyrgyzstan claimed its major expansion and return of pasturelands. Ambitious plans to boost the water supply resulted in increased expectations leading to new water shortages. Later in the 1980s, the official disputation of the decisions approved by Moscow became acceptable; Uzbekistan had to compensate the loss caused by Kyrgyz expansion in the previous decade. Finally, the period of independence continued with what was left from the Soviets but with significantly less financing, which led to both some cooperative and some national solutions.

Along the entire analyzed period, institutions that are still, at least partly, valid were established. In addition to the property rights, proportional allocation principle is still referred as the central principle for allocation of water. The principle is biased to the criteria of the time it was developed. That is partly why the governance institutions do not function effectively. In addition, the principle itself is contradictory to increasing efficiency, as it requires reconsideration of the allocation with any disproportional change in water supply, which in turn contradicts with the closure of the basin and fixed shares. Without taking into consideration these concerns, benefit sharing might become prone to inequity both horizontally and vertically, failure to internalize environmental costs, loss of water control due to the scope of issue linkages as well as vulnerability in implementation due to asymmetrical commitments.

Separation of the issues on border crossing due to the security concerns from the water and land management sectors is indeed one of the constraints for successful cooperation because of the nature of property rights for infrastructure located beyond the national boundaries. In this regard, a similar case of the Tuyamuyun Reservoir with the pump-stations on the Amu Darya River shared by Turkmenistan and Uzbekistan could be studied for possible lessons. An additional framework agreement on passing the borders at least for operations and maintenance purposes would reduce the *ad hoc* nature of the issues and bring more stability to the existing cooperation. The case of the Chu and Talas Rivers seems to be relevant for further comparative studies from issue linkages perspective as well as to learn more successful agreements of maintenance sharing.

Overall, the situation is extremely complex: geographically, infrastructure-wise as well as institutionally. However, it is necessary for the complexity to be taken into account in the development of appropriate policy. Simplification of issues might have actually led to the decline in cooperation, since the later arrangements in the Syr Darya, as well as Amu Darya and larger Aral Sea basins, were mainly brokered by donors, which did not engage comprehensively with the big picture. One lesson from the historical complexity is the desire for each state to have independence in water management—with each nation focusing on its own water resources. However, the possible gains from further dialogue and cooperation are clear.

Supplementary Materials: Supplementary materials can be found at http://www.mdpi.com/2073-4441/7/6/2728/s1.

Acknowledgments: The data analyzed in this article were gathered during the IWMI's Irrigation Bureaucracy project in Central Asia, the Integrated Water Resource Management—Ferghana Valley project funded by the Swiss Agency for Development and Cooperation, the Water Security project funded by the Ministry for Foreign Affairs of Finland and the Water Cooperation in the Ferghana Valley work package funded by the Consultative Group on International Agricultural Research (CGIAR-wide) Research Program on Water, Land and Ecosystems. Funding for the doctoral studies from the German Academic Exchange Service (DAAD), within which this research was carried out, as well as from the IWMI's Irrigation Bureaucracy project in Central Asia is gratefully acknowledged. We are grateful to Volkmar Hartje, Head of Chair of Landscape and Environmental Economics at Technical University of Berlin, Germany, for his valuable advice and insights on various aspects of benefit sharing and Alexander Platonov, GIS and Remote Sensing Specialist at the IMWI Central Asia, for developing the map. We thank three anonymous reviewers and the editors of the Special Issue, whose constructive comments helped to improve the quality of the paper.

Author Contributions: Ilkhom Soliev developed the initial and final versions of the framework, analyzed the data, organized the systematic discussion of the costs and led the drafting process of the study incorporating the contributions from co-authors as well as from the reviewers and editors; Kai Wegerich proposed the initial idea of testing the approach in case of the Ferghana Valley, contributed by interim editing of the paper, drafting the initial version of the discussion section, structure of the study and raising critical questions on the approach; Jusipbek Kazbekov contributed by providing his expertise and insights on the study area and historical-institutional arrangements as well as clarifications in understanding complications and connections of the collected data, facilitated the development of the updated map.

Conflicts of Interest: The authors declare no conflict of interest.

References

1. North, D.C. *Institutions, Institutional Change, and Economic Performance*; Cambridge University Press: New York, NY, USA, 1990.

2. Ostrom, E. *Governing the Commons: The Evolution of Institutions for Collective Action*; Cambridge University Press: New York, NY, USA, 1990.

3. Williamson, O.E. Transaction cost economics: How it works; Where it is headed. *Economist* **1998**, *146*, 23–58. [CrossRef]

4. Sadoff, C.W.; Grey, D. Beyond the river: The benefits of cooperation on international rivers. *Water Policy* **2002**, *4*, 389–403. [CrossRef]

5. Sadoff, C.W.; Grey, D. Cooperation on international rivers: A continuum for securing and sharing benefits. *Water Int.* **2005**, *30*, 420–427. [CrossRef]

6. Phillips, D.; Daoudy, M.; McCaffrey, S.; Öjendal, J.; Turton, A. *Trans-Boundary Water Cooperation as A Tool for Conflict Prevention and Broader Benefit Sharing*; Ministry of Foreign Affairs of Sweden: Stockholm, Sweden, 2006.

7. Phillips, D.; Allan, J.A.; Claassen, M.; Granit, J.; Jägerskog, A.; Kistin, E.; Patrick, M.; Turton, A. *The TWO Analysis: Introducing a Methodology for the Transboundary Waters Opportunity*; Report No. 23; SIWI: Stockholm, Sweden, 2008.

8. Phillips, D. *The Transboundary Water Analysis as a Tool for RBOs*; Report No. CSIR/NRE/WR/ER/2009/0124/B; SADC Water Division under contract to GTZ: Gaborone, Botswana, 2009.

9. Dombrowsky, I. *Conflict, Cooperation and Institutions in International Water Management: An Economic Analysis*; Edward Elgar Publishing Limited: Cheltenham, UK, 2007.

10. Qaddumi, H. *Practical Approaches to Transboundary Water Benefit Sharing*; Overseas Development Institute: London, UK, 2008.

11. Hensengerth, O.; Dombrowsky, I.; Scheumann, W. *Benefit-Sharing in Dam Projects on Shared Rivers*; Deutsches Institut für Entwicklungspolitik: Bonn, Germany, 2012.

12. Hoff, H. Understanding the Nexus. In Proceedings of the Bonn 2011 Conference: The Water, Energy and Food Security Nexus; Stockholm Environment Institute: Stockholm, Sweden, 2011.

13. Allouche, J.; Middleton, C.; Gyawali, D. *Water and the Nexus, Nexus Nirvana or Nexus Nullity? A Dynamic Approach to Security and Sustainability in the Water-Energy-Food Nexus*; STEPS Working Paper No. 63; STEPS Center: Brighton, UK, 2014.

14. Tafesse, T. *The Nile Question: Hydropolitics, Legal Wrangling, Modus Vivendi and Perspectives*; Lit: Münster, Germany, 2001.

15. Nicol, A. The dynamics of river basin cooperation: The Nile and the Okavango basins. In *Transboundary Rivers, Sovereignty and Development: Hydrological Drivers in the Okavango River Basin*; Turton, A., Ashton, P., Cloete, E., Eds.; African Water Issues Research Unit: Johannesburg, South Africa, 2003; pp. 167–186.

16. Zeitoun, M. Hydro-hegemony theory—A framework for analysis of water-related conflicts. In Proceedings of the First International Workshop on Hydro-hegemony, King's College, London, UK, 21–22 May 2005.

17. Turton, A. A South African perspective on a possible benefit-sharing approach for transboundary waters in the SADC region. *Water Altern.* **2008**, *1*, 180–200.

18. Klaphake, A. Cooperation on international rivers from an economic perspective: The concept of benefit-sharing. In *Transboundary Water Management in Africa: Challenges for Development Cooperation*; Scheumann, W., Neubert, S., Eds.; German Development Institute: Bonn, Germany, 2006; pp. 103–173.

19. Dombrowsky, I. Revisiting the potential for benefit sharing in the management of trans-boundary rivers. *Water Policy* **2009**, *11*, 125–140. [CrossRef]

20. Southern African Development Community (SADC). SADC Concept Paper on Benefit Sharing and Transboundary Water Management and Development. 2010. Available online: http://www.orangesenqurak.org/UserFiles/File/SADC/SADC%20concept%20paper_benefit%20sharing.pdf (accessed on 4 June 2015).

21. Sadoff, C.W.; Greiber, T.; Smith, M.; Bergkamp, G. *Share: Managing Water across Boundaries*; IUCN: Gland, Switzerland, 2008.

22. Wolf, A.T. Criteria for equitable allocations: The heart of international water conflict. *Nat. Resour. Forum* **1999**, *23*, 3–30. [CrossRef]

23. Fisher, R.; Ury, W. *Getting to Yes: Negotiating Agreement Without Giving in*; Penguin: New York, NY, USA, 1981.

24. Grzybowski, A.; McCaffrey, S.C.; Pailey, R.K. Beyond international water law: Successfully negotiating mutual gains agreements for international watercourses. In Symposium Issue, Proceedings of the Conference "Critical Intersections for Energy & Water Law: Exploring New Challenges and Opportunities", Calgary, AB, Canada, 20–21 May 2009.

25. Sewell, D.; Utton, A. Special issue on A. US—Canada Transboundary Resource Issues. *Nat. Resour. J.* **1986**, *26*, 2.

26. Tarlock, A.D.; Wouters, P. Are shared benefits of international waters an equitable apportionment? *Colo. J. Int. Environ. Law Policy* **2007**, *18*, 523–536.

27. Dombrowsky, I.; Bastian, J.; Daeschle, D.; Heisig, S.; Peters, J.; Vosseler, C. International and local benefit sharing in hydropower projects on shared rivers: The Ruzzi III and Rusumo Falls. *Water Policy* **2014**, *16*, 1087–1103. [CrossRef]

28. Wegerich, K.; Kazbekov, J.; Kabilov, F.; Mukhamedova, N. Meso-Level cooperation on transboundary tributaries and infrastructure in the Ferghana Valley. *Int. J. Water Resour. Dev.* **2012**, *28*, 525–543. [CrossRef]

29. Wegerich, K.; Kazbekov, J.; Lautze, J.; Platonov, A.; Yakubov, M. From monocentric ideal to polycentric pragmatism in the Syr Darya: Searching for second best approaches. *Int. J. Sustain. Soc.* **2012**, *4*, 113–130. [CrossRef]

30. Pak, M.; Wegerich, K. Competition and benefit sharing in the Ferghana Valley: Soviet negotiations on transboundary small reservoir construction. *Cent. Asian Aff.* **2014**, *1*, 225–246. [CrossRef]

31. Dukhovny, V.A.; Sokolov, V.; Galustyan, A.; Djalalov, A.A.; Mirzaev, N.N.; Horst, M.G.; Stulina, G.V.; Muminov, S.; Ergashev, I.; Kholikov, A.; *et al. Report on Comprehensive Hydrographic Study of the Ferghana Valley*; SIC ICWC: Tashkent, Uzbekistan, 2011.

32. Musabaeva, A.; Moldosheva, A. *The Ferghana Valley: Current Challenges*; United Nations Development Fund for Women (UNIFEM): Bishkek, Kyrgyzstan, 2005.

33. Osmonaliev, A.; Bayjumanov, D.; Kasymbekov, B.; Tekeeva, L.; Isaliev, K.; Koychumanova, K.; Plesovskih, R.; Turdubaeva, C. *Statistical Review of Agriculture of Kyrgyz Republic for 2008–2012*; National Statistics Committee of Kyrgyz Republic: Bishkek, Kyrgyzstan, 2013.

34. Kyrgyz Information Portal. Batken Province. Available online: http://www.welcome.kg/ru/kyrgyzstan/region/fretrr/ (accessed on 19 November 2014). (In Russian)

35. Kyrgyz Information Portal. Jalalabad Province. Available online: http://www.welcome.kg/ru/kyrgyzstan/region/dffer/ (accessed on 19 November 2014). (In Russian)

36. Kyrgyz Information Portal. Osh Province. Available online: http://www.welcome.kg/ru/kyrgyzstan/region/xaaaa/ (accessed on 19 November 2014). (In Russian)

37. Hasanova, G.; Shokirov, S.; Asoev, A.; Norov, K.; Silemunshoev, N.; Gukasova, T.; Abdulloev, M.; Kulov, A.; Turaev, B.; Jdanova, L. *Demographic Yearbook of the Republic of Tajikistan*; Agency under the President of the Republic of Tajikistan on statistics: Dushanbe, Tajikistan, 2013.

38. State Statistical Committee of the Republic of Uzbekistan. *Statistics of Permanent population*; State Statistical Committee of the Republic of Uzbekistan: Tashkent, Uzbekistan, 2014. (In Uzbek)

39. Provinces of Uzbekistan. Available online: http://uzbekembassy.es/index.php/ru/perfil-de-uzbekistan-7/regiones (accessed on 19 November 2014).

40. Official web-site of the Namangan province administration. Available online: http://www.namangan.uz/index.php/uz/namangan-viloyati/viloyat-tarixi (accessed on 10 March 2015). (In Uzbek)

41. Historical and Genealogical Dictionary Directory. Available online: http://www.defree.ru/publications/p01/p90.htm (accessed on 20 February 2015). (In Russian)

42. Polian, P.M. *Against Their Will: The History and Geography of Forced Migrations in the USSR*; Central European University Press: Budapest, Hungary, 2004.

43. Weinthal, E. State making and environmental cooperation: Linking domestic and international politics in Central Asia. In *Global Environmental Accord: Strategies for Sustainability and Institutional Innovation*; MIT Press: Cambridge, MA, USA, 2002.

44. Thurman, M. Modes of Organization in Central Asian Irrigation: The Ferghana Valley, 1876 to Present. Ph.D. Thesis, University of Indiana, Bloomington, IN, USA, 1999.

45. Abashin, S.; Abdullaev, K.; Abdullaev, R.; Koichiev, A. Soviet rule and the delineation of borders in the Ferghana Valley, 1917–1930. In *Ferghana Valley: The Heart of Central Asia*; Starr, F., Beshimov, B., Bobokulov, I., Shozimov, P., Eds.; M.E. Sharpe, Inc.: New York, NY, USA, 2011; pp. 94–118.

46. Rahimov, M.; Urazaeva, G. *Central Asian Nations and Border Issues*; Central Asian Series; Defence Academy of the United Kingdom, Conflict Studies Research Centre: London, UK, 2005.

47. Bichsel, C.; Mukhabbatov, K.; Sherfedinov, L. Land, water and ecology. In *Ferghana Valley: The Heart of Central Asia*; Starr, F., Beshimov, B., Bobokulov, I., Shozimov, P., Eds.; M.E. Sharpe, Inc.: New York, NY, USA, 2011; pp. 253–277.

48. Benjaminovich, Z.; Tersitskiy, D. *Irrigation of Uzbekistan II*; Fan: Tashkent, Uzbekistan, 1975.

49. Khamidov, M.; Leshanskiy, A. Review of the Proposal of Constructing an Operation Model for Kairakkum Reservoir. In *Final Report Example Allocations of Operating and Maintenance Costs of Interstate Water Control Facilities Employing the Use-of-Facilities Method*; Hutchens, A., Ed.; U.S. Agency for International Development: Washington, DC, USA, 1999; pp. 110–120.

50. Khamidov, M. Experience of coordinated water resources use of the syrdarya basin states. In Presented in Advanced Research Workshop Socio-Economic Stability and Water Productivity: Implications of Food and Water security in the Central Asian Region, Tashkent, Uzbekistan, 18–20 March 2008; Available online: http://www.icwc-aral.uz/workshop_march08/pdf/khamidov_en.pdf (accessed on 28 November 2014).

51. Djaylobaev, N.; Sakhvaeva, E.; Matushkina, O.; Chernikova, T.; Mendikulova, Z.; Neronova, T.; Aytbaev, B.; Mamadiev, K.; Shukurov, J.; Ibraimov, D.; *et al. Basin Plan for the Isfara River, Batken District, Kyrgyz Republic*; GIZ: Bishkek, Kyrgyzstan, 2014. (In Russian)

52. Mirzaev, N.N. Application of the IWRM principles to the Akburasai river basin. In *Problems of Ecology and Use of Water Land Resources in the Countries of Eastern Europe, Caucasus and Central Asia*; Dukhovny, V.A., Ed.; SIC ICWC: Tashkent, Uzbekistan, 2010; pp. 167–176.

53. Shozimov, P.; Beshimov, B.; Yunusova, K. The Ferghana Valley during perestroika, 1985–1991. In *Ferghana Valley: The Heart of Central Asia*; Starr, F., Beshimov, B., Bobokulov, I., Shozimov, P., Eds.; M.E. Sharpe, Inc.: New York, NY, USA, 2011; pp. 178–204.

54. Wegerich, K. Unpacking the disconnect of hydraulic mission and loss of water control: Three decades of external and internal changes and their implication for water management for the irrigation bureaucracy in Ferghana province/Uzbekistan. forthcoming.

55. *Analysis of Contemporary Conditions in the Aral Sea Basin and Use of Land-Water Resources of the Uzbek SSR*; Ministry of Melioration and Water Resources of the Uzbek SSR: Tashkent, Uzbekistan, 1991. (In Russian)

56. Wegerich, K. Shifting to hydrological/hydrographic boundaries: A comparative assessment of national policy implementation in the Zerafshan and Ferghana Valleys. *Int. J. Water Resour. Dev.* **2015**, *31*, 88–105. [CrossRef]

57. Pak, M.; Wegerich, K.; Kazbekov, J. Re-Examining conflict and cooperation in Central Asia: A case study from the Isfara River, Ferghana Valley. *Int. J. Water Resour. Dev.* **2014**, *30*, 230–245. [CrossRef]

58. Strategy and Project Activities to Support Improved Regional Water Management in Central Asia, prepared by the United Nations Development Programme (UNDP). July 2004. Available online: http://waterwiki.net/images/5/53/UNDP-CA-Reg_Water_proposal_and_strategy_for_EU_clean.doc (accessed on 10 March 2015).

59. Czerniecka, K. *The State at Its Borders: The Internal Dimensions of Kyrgyzstan's Border Security*; Central Asia Security Policy Brief No. 4; OSCE Academy in Bishkek, GCSP: Bishkek, Kyrgyzstan; Geneva, Switzerland, 2011.

Article

Transferable Principles for Managing the Nexus: Lessons from Historical Global Water Modelling of Central Asia

Joseph H. A. Guillaume [1],*, Matti Kummu [1], Stephanie Eisner [2] and Olli Varis [1]

[1] Water & Development Research Group, Aalto University, Tietotie 1E, Espoo 02150, Finland;
     matti.kummu@aalto.fi (M.K.); olli.varis@aalto.fi (O.V.)
[2] Center for Environmental Systems Research, University of Kassel, Kassel 34117, Germany;
     eisner@usf.uni-kassel.de
*   Correspondence: joseph.guillaume@aalto.fi; Tel.: +358-50-407-2906

Academic Editor: Miklas Scholz
Received: 26 May 2015; Accepted: 24 July 2015; Published: 31 July 2015

Abstract: The complex relationships within the water-energy-food security nexus tend to be place-specific, increasing the importance of identifying transferable principles to facilitate implementation of a nexus approach. This paper aims to contribute transferable principles by using global model data and concepts to illustrate and analyze the water history of Central Asia. This approach builds on extensive literature about Central Asia and global change as well as recent advances in global water modeling. Decadal water availability and sectorial water consumption time series are presented for the whole 20th century, along with monthly changes in discharge attributable to human influences. Concepts from resilience and socio-ecological system theory are used to interpret the results and identify five principles relevant to managing the transboundary nexus: (1) the subsystems included/excluded from the nexus are case-specific and should be consciously scrutinized; (2) consensus is needed on what boundaries can acceptably be crossed within the nexus; (3) there is a need to understand how reducing trade-offs will modify system dependencies; (4) global stakeholders have both a responsibility and right to contribute to the shaping of the nexus; (5) combining data with global and local perspectives can help to enhance transferability and understanding of shared problems in our globalized world.

Keywords: Central Asia; nexus; WaterGAP; resilience

1. Introduction

A nexus approach is one that "reduces trade-offs and builds synergies across sectors" [1], notably between the management of water, energy, and food security. Practical implementation of a nexus approach involves understanding inter-relationships between sectors and seeking suitable governance and management options [2]. Those relationships are naturally quite case-specific. From a water perspective, interactions will be very different in regions with snow or glacial melt, monsoon, groundwater, mountains, and deserts. Interactions involving the energy sector might differ depending on the importance and availability of fossil fuels, biomass, hydropower, nuclear, wind, tidal energy, *etc.* Relationships with food notably vary depending on energy intensiveness, labor- and land-productivity [3], as well as the importance of trade [4], irrigation [5], non-food agriculture (e.g., cotton) [2], and non-agricultural food production (e.g., freshwater or marine fisheries) [6]. The nexus is readily acknowledged as unique in every place. This is even more noticeably the case in transboundary nexus situations, where negotiations are often focused on a particular issue, for example, on a specific transboundary watercourse and its specific competing uses.

Learning about the management of the nexus globally therefore depends on our ability to identify and share transferable principles that appear to apply across scales around the world. One approach is to develop a theory about processes, including transferable methodologies (e.g., [7,8]). Another is to use case studies to contribute to the growing checklist-style body of knowledge about relationships to look out for and corresponding governance arrangements (e.g., [2,9,10]). The key aim in each case is to cross scales, linking the local implementation-scale situation to a broader global understanding.

Considering Central Asia as a case study, there is already a solid foundation of research on which to build, including the use of the nexus approach and global data. Water in Central Asia has been extensively discussed, particularly focusing on the history and future of the Aral Sea basin. Notable publications include several books (e.g., [11,12]), policy reports [13,14], as well as special issues on "Water and Security in Central Asia: Solving the Rubik's cube" [15], "Water in Central Asia—Perspectives under global change" [16], and (this year) "Sustainable Water Management in Central Asia" [17].

This existing work has included nexus approaches. This has included analyses of historical and ongoing multilateral cooperation and benefit sharing, for example, at basin scale with regard to energy and water [18] and at regional scale with regard to water, energy, food, and agriculture [2,19]. Other analyses focused on international influences on management of the nexus in Central Asia and the need for capacity-building [20], economic analyses of energy supply and irrigation [18], and the need for economic reform as a means of addressing nexus challenges [21]. Stucki and Sojamo [22] notably give a useful introductory overview of the "nouns and numbers" of the nexus in Central Asia through definitions, indicators, and data. The special issue that this article is a part of focuses on transboundary nexus issues in Asian river basins, including other case studies focused on Central Asia (e.g., [23,24]).

Global data has also been used in multiple ways in assessments of Central Asia. Varis and Kummu [25] produced vulnerability profiles of basins. Porkka *et al.* [26] investigated the effect of changes in import and export on water stress and shortage based on data in the year 2000. Aus der Beek *et al.* [27] applied a global model (WaterGAP3) to differentiate the impact of climate and water use on flows in the Aral Sea basin. Malsy *et al.* [28] used the same model to analyze the potential impact of climate change on Central Asian water resources.

More generally, recent developments in global water modeling and global change literature have provided new means of looking at local issues. Global water models, such as in the Water and Global Change (WATCH) project [29,30], provide widely applicable datasets for water availability and multi-sectorial water consumption. Their growing maturity, particularly in estimation of water use (e.g., [31]), has led to broad application across many scales (e.g., [4,5,26–28,32]). Simultaneously, literature on global change, socio-ecological systems, and resilience has developed descriptive and normative theory regarding planetary boundaries [33] specifically, and has taken a resilience-based perspective to water sustainability [34] more generally. These concepts have aims similar to the nexus [1,35], and can be expected to yield useful principles regarding its management.

This paper aims to contribute transferable principles that could be applicable across scales around the world. The approach taken is to begin with global model data and concepts and apply them to a local case study. This helps to identify possible generalizable implications for the implementation of a nexus approach. We look at the nexus through a water lens, where food and energy are users of the resource [7]. Specifically, we use output from the WaterGAP global water model, based on a new 100-year input dataset of irrigated areas [36], to illustrate the history of the nexus in Central Asia from 1900 to 2000. This history is described through a series of spatiotemporal assessments of blue water availability and consumption combining powerful visualizations and their interpretation using literature about Central Asia. Central Asia has been selected for its particular history and wealth of prior analyses. Discussion of the analysis then makes use of globally applicable concepts from resilience theory and socio-ecological systems, notably boundaries and system dependencies, in order to uncover transferable principles for implementation of a nexus approach. This paper does not aim to provide specific recommendations regarding management of the nexus in Central Asia.

The structure of this article follows. Section 2 (Method and Data) describes the global water model used and the assessments performed. Section 3 (Results and Interpretation) presents the results and their interpretation in the form of a history of the nexus in Central Asia. Section 4 (Discussion) delves deeper into concepts underlying this history in order to identify implications for management of the nexus, as well as possible extensions to this analysis. Section 5 (Conclusion) summarizes key conclusions about the history of the nexus in Central Asia as well as transferable principles for management of the nexus.

2. Method and Data

2.1. Global, Spatially Distributed Estimates of Water Availability and Consumption

The analysis uses monthly time series of blue water availability and consumption for the 20th century, produced using a preliminary version of WaterGAP2.2 [37]. Blue water availability is calculated by a daily water balance model for each 0.5° grid cell based on meteorological forcing and landscape factors. Blue water corresponds to "liquid water in rivers and aquifers", as opposed to green water, which refers to "naturally infiltrated rain, attached to soil particles and accessible to roots" [38]. Blue water is of particular interest in Central Asia, notably due to the highly publicized effects of irrigation water use on the Aral Sea, as discussed further in Section 3.

Spatially explicit estimates of water consumption are provided by the WaterGAP water use models for five sectors: households and small businesses (domestic sector), thermal power plant cooling (electricity sector), irrigated agriculture, livestock farming, and manufacturing industries [31]. Water consumption refers to water that is "evaporated or incorporated into products" [37]. Key drivers include water use intensities, consumptive water use coefficients, and structural or technological change factors.

Monthly time series of blue water availability are simulated in two model setups: (1) including human interference on the natural regime through water abstraction and reservoir operation, and (2) under naturalized conditions, *i.e.*, excluding any human interference. The model setup was nearly identical to the WATCH project [30], but it differs in a few important aspects. While the WATCH precipitation and temperature forcing datasets were used [29], irrigation water use is additionally based on a new dataset of irrigated land from 1900–2005 compiled from a variety of subnational sources [36]. Secondly, spin-up of the model allows data to be used from the year 1901, whereas the WATCH dataset treats the 1901–1905 period as model spin-up, such that these cannot be used for analysis.

The data used naturally has limitations (discussed throughout the article), and needs to be interpreted with care. The data itself has therefore not been made available, and our conclusions regarding the nexus and Central Asia have been additionally supported with references to previous research. An interested reader can nevertheless find similar publically available datasets from the WATCH or Inter-sectoral Impact Model Intercomparison Project (ISI-MIP) projects [30,39].

2.2. Construction of Historical Assessments in Central Asia

Water availability and consumptions were aggregated to Food Production Units (FPUs), which are a combination of river basin and economic regions used by a number of previous studies [4,40–42]. Other units of analysis may yield some variation in quantitative results [32]. FPUs are a hydro-politically relevant scale of analysis [40], and the conclusions drawn with this unit of analysis were found to be consistent with existing literature about Central Asia.

This paper focuses on a subset of FPUs within Uzbekistan, Kazakhstan, Kyrgyzstan, Tajikistan, Turkmenistan, and Afghanistan (Figure 1). FPUs that are part of larger river basins (e.g., the Volga and Ob) have been omitted, as have some small, sparsely populated endorheic basins with little water use. The study area is shown in Figure 1, including mean water availability, country borders, basin borders, key rivers, and boundaries of the FPUs.

33

For the majority of the analysis, the monthly data have been aggregated to obtain decadal water availability and consumption time series. This allows a focus on broad temporal patterns, and reduces the impact of error at shorter time scales. The use of decadal scales lessens the importance of inter-annual storage variation, such that inaccuracies in reservoir operations, snowmelt, and groundwater have less effect.

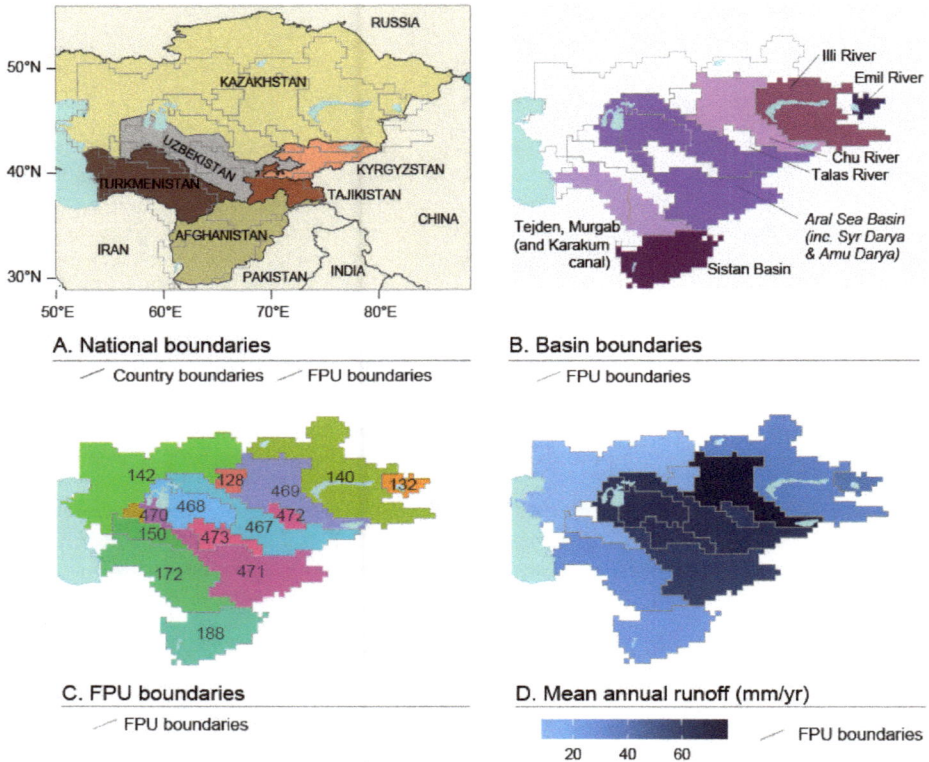

Figure 1. Map of study area: (**A**) national boundaries; (**B**) basin boundaries; (**C**) FPU boundaries; (**D**) mean decadal water availability for each food production unit (FPU); overlain with a selection of lakes [43,44].

FPUs were split into clusters based on the similarity of their decadal water availability time series. Hierarchical clustering (the *hclust* function in R [45]) was used to identify four clusters using positive Kendall correlations as measure of similarity.

In interpreting the results, we treat consumptive uses as different pathways by which water becomes inaccessible to humans, notably through evaporation by which water returns to the atmosphere. In this paper, we focus on blue water in particular. If there were no human uses, then blue water would eventually evaporate from rivers, lakes, or from the sea. Given that the catchments in Central Asia are endorheic, this means that, for example, any water that reaches the Aral Sea or Lake Balkhash will ultimately evaporate from there. This environmental water is available to aquatic ecosystems, and is therefore key to maintaining healthy waterways and ecosystem services. Reflecting the importance of sharing water between society and ecosystems [46], we include total environmental water as a consumptive use, and calculate it as the difference between availability and total human consumption of blue water.

Per capita consumption and decadal stress and shortage were also calculated. The latter two are common indicators of water scarcity [47–49], respectively calculated as the ratio of human water consumption and availability and per capita availability of water. Their interpretation is discussed in Section 3.3 when they are used. Population data for each decade was obtained from the spatially explicit History Database of the Global Environment (HYDE) dataset [50].

To complement the high-level decadal view, we also calculated the monthly difference between natural discharge and human-influenced discharge, *i.e.*, with flow regulation and water abstraction. Errors may be more significant when using global scale modeling of shorter time scales. Models may be able to estimate relative changes even if absolute values are not quite right. We therefore avoid examining availability and consumption data directly and instead focus on the change in seasonal flows that has occurred due to human water use.

3. Results and Interpretation: History of the Nexus in Central Asia through a Global Water Lens

This section illustrates the 20th century history of the water-energy-food nexus in Central Asia through a water lens. Assessment of decadal blue water availability is followed by decadal water consumption and human-induced seasonal changes in discharge.

3.1. Decadal Water Availability

Physical water availability is a key factor influencing the water-energy-food security nexus. Figure 1 shows mean decadal blue water availability in the study region while Figure 2B shows time series of decadal blue water availability. Decadal water availability is an indicator of the size of the resource and its variation in the long-term. Across Central Asia, most basins are endorheic and transboundary. Mean water available is relatively low (Figure 1). There is still significant variation spatially, such that the FPUs can be split into clusters according to the similarity of the variation over time (Figure 2A). Summaries of these clusters follow.

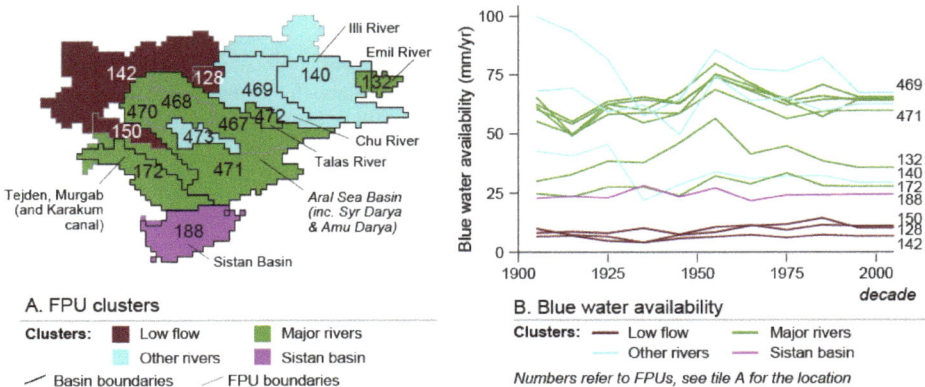

Figure 2. Decadal water availability: (**A**) Map of food production units (FPUs) showing clusters based on correlation of their time series, along with major basins; (**B**) water availability time series for each FPU, colored according to the same clusters.

In general, the "major rivers" cluster is characterized by high water availability, driven by glacial and snow melt from the Pamir and Tien Shan mountains to the east [51]. This is a dominant feature of the region, as the arid downstream areas depend on the continental climate of these headwaters and the relative reliability of glacier flows [52]. The time series in Figure 2 does, however, show lower water availability at the start of the century, a peak in the middle of the century, followed by a return to lower levels, consistent with the strong role of observed climatic variation [51] and a cold, wet period in the 1950s [53]. There is, however, no clear trend of water availability, consistent with studies showing that climate change effects on runoff vary substantially between catchments, though increased temperature, evaporation, and glacier melt are generally accepted to be playing a major role [51]. Note that the model used here does not explicitly represent glacier melt, though it is still considered to have acceptable performance in Central Asia [27,28], at least for the purpose of this paper.

This "major rivers" cluster most notably includes the Aral Sea drainage basin (FPU 468, see Figures 1 and 2), including the Amu Darya (FPU 470, 471) and Syr Darya main stream (FPU 467). The Talas river (FPU 472) has similar flow patterns. It also descends from the mountains in Kyrgyzstan and disappears in the desert in Kazakhstan [54]. FPU 132 and 172 were identified as having relatively similar temporal patterns but lower water availability. FPU 132 includes the catchment of the saline Lake Alakul, including the Emin (Emel or Emil) river, which flows from the mountains in China into eastern Kazakhstan [55]. FPU 172 includes the Tedjen (or Hari) and Murghab rivers flowing from the mountains of Afghanistan. While the model does not include the Kara-kum canal, which contributes significantly to irrigation in Turkmenistan by carrying water from the Amu Darya [56], the data still sufficiently demonstrates key features of the region's water consumption history.

The "other rivers" cluster spans very different river systems, some of which are characterized by sometimes quite high but highly variable water availability (Figure 2). The cluster includes important transboundary river basins such as the Chu river (FPU 469) and the Illi river, which flows into Lake Balkash (FPU 140). The same FPUs include smaller basins such as the endorheic lake Issyk Kul and Sary su River (FPU 469) and the Ramsar-listed Tengiz-Korgalzhyn Lake System (FPU 140), as well as other smaller endorheic basins. FPU 473 provides another contrast, with the anthropogenic Aydar-Arnasay system of lakes in Uzbekistan and nearby irrigation areas [57], as well as the city of Zarafshan in the Kyzylkum desert. The area makes use of water from both the Syr Darya and Amu Darya rivers.

In the "low flows" cluster, FPU 150 includes extensive irrigation areas in Turkmenistan and Uzbekistan near the Amu Darya. FPUs 142 and 128 cover large parts of Kazakhstan. While FPU 142 does include some endorheic rivers (including the Emba River, which leads to the Caspian Sea), water availability is generally much lower than in the other regions. The climate is predominantly cold, arid desert and steppe.

In the cluster "Sistan basin", FPU 188 has been singled out as uncorrelated with others. The largest river in the basin is the Helmand (Hirmand), which is notably fed by snowmelt from mountains to the northeast of the region, and is used for irrigation in both Afghanistan and Iran [58]. Blue water availability is also relatively low.

3.2. Decadal Blue Water Consumption

From a nexus point of view, the key is how the available water is used. In this section, we focus on different consumptive uses of water (Figure 3). As discussed in Section 2.2, consumptive uses can be considered as different pathways by which water becomes inaccessible to us, notably through evaporation by which water returns to the atmosphere. This includes both human uses and water that is available to aquatic ecosystems (referred to as environmental water). When water is consumed (evaporated or transpired), it returns to the atmosphere and, hence, the global water circulation system. The analysis of van der Ent *et al.* [59] indicates that in Central Asia, a large proportion of water available comes from its consumption (*i.e.*, evaporation) on the European and Asian continents to the west, and when it is consumed (*i.e.*, evaporates) in the study area, the majority will return as precipitation on the Asian continent to the east. The water is not lost, but can no longer be easily reused in its previous location.

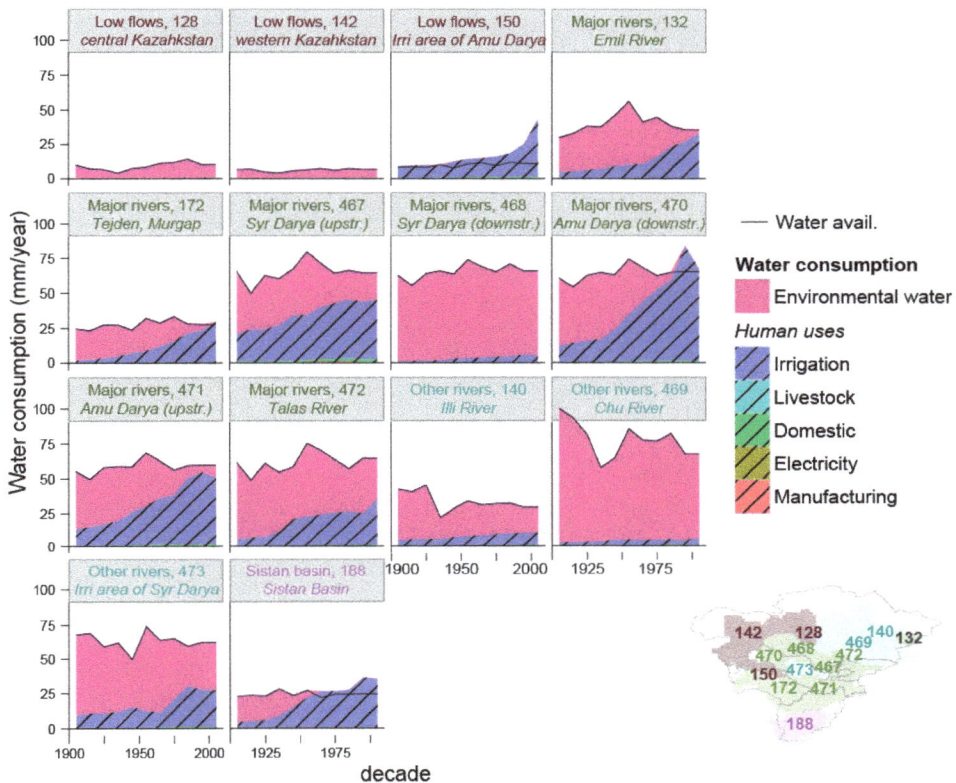

Figure 3. Total blue water use split by end-use of water (*i.e.*, pathway in which it evaporates) for each food production unit (FPU) for 1905–2005, overlain with total water availability. Environmental water is defined here as the water that is available to aquatic ecosystems, calculated as the difference between availability and total human use [46]. Water use sectors other than irrigation and environmental water are so small that they are not visible on this figure.

3.2.1. Partitioning between Human Use and Environmental Water

In our findings (Figure 3), two consumptive uses (*i.e.*, evaporation pathways) stand out: environmental water and irrigation. Other uses are very small by comparison, e.g., at the bottom of the plot for FPU 467. At the start of the 20th century, the majority of water was environmental water in all but one FPU. This means that the water evaporates in a pathway that provides environmental flows [60] and maintains natural ecosystem services.

In some FPUs, environmental water has not changed much. This is predominantly the case in FPUs with low water availability (FPUs 128, 142). In these areas, it is possible that there is not enough water to maintain a large water-dependent population and economy. In FPU 142, oil and gas production is instead particularly prominent, with associated problems with water quality [61]. FPUs 468 and 469 have also seen only relatively small decreases in environmental water. An exception is FPU 150, which is fed by irrigation channels from the Amu Darya [27].

In many other FPUs, environmental water has diminished over time, and human use (notably irrigation) has increased. As the population increases and water-consuming economic activity grows, the use of water becomes determined by humans rather than natural processes [62]. Existing ecosystems have evolved to suit local water availability, such that when tolerable human blue water consumption is exceeded, it can lead to changes in the way the socio-ecological system operates [63] and a "high probability of (possibly abrupt) water-induced changes with large detrimental impacts on human societies" [64].

In some FPUs, human use has even exceeded available water (FPUs 150, 470, 188). In the context of the model, this likely indicates the use of long-term stored water, such as the non-sustainable use of groundwater and unaccounted physical water transfers.

The case of the Aral Sea is a high-profile example of the effect of increasing human water use (and diminishing environmental water) [65]. The Amu Darya (FPUs 471 and 470) and Syr Darya (FPUs 467, 468) are the main rivers feeding the Aral Sea. Irrigation expansion (visible in Figure 3) reduced inflows, resulting in reduced lake area and increased salinity, loss of fish species, desertification, dust storms, and climate change along the shoreline [65]. These changes to the ecological system led to changes in the associated social system, including the collapse of fishing industries, high unemployment, the loss of irrigated land to salinization, poorer diets, and health problems [65].

3.2.2. Distribution of Human Water Consumption per Capita

Human water use is not distributed equally per capita. Societies have different needs and wants and, hence, different water footprints. Figure 4 shows per capita human water use over time. Some FPUs have much larger per capita consumption than others, which reflects intensive use of water within the economy rather than high consumption by individuals. As already noted, the human use of water is dominated by irrigation. The large changes in per capita consumption (e.g., in FPU 468) can therefore be explained by the expansion and contraction of irrigation water in relation to the population. Some declines (e.g., FPUs 140 and 469) are, however, more likely to be explained by improvements in the efficiency of water use or increases in the population. The population grew throughout the 20th century for all but a few smaller FPUs in Central Asia (results not shown). If population increases and water-dependent industries do not grow, then per capita consumption falls.

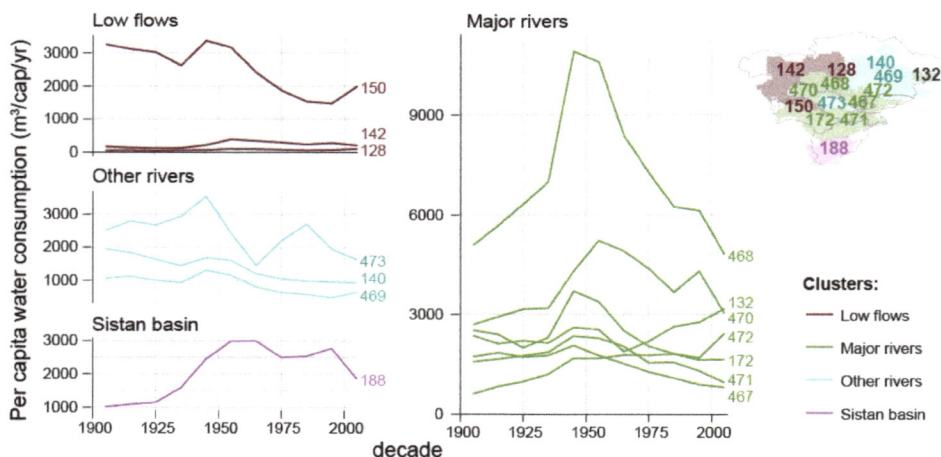

Figure 4. Human blue water consumption per capita.

3.2.3. Distribution of Water by Activity

A core issue of the water-energy-food nexus is the trade-off between water-dependent activities. Time series of consumption for each sector are shown in Figure 5. Before interpreting this figure, it is important to remember that the nexus exists within a broader context. When a water resource is not stressed, all industries take as much as they want (at the increasing expense of environmental water). The environment acts as a buffer, mitigating conflict between uses. Similarly, when the population is small, there is less competition for water than when the population is large [48]. Only when water becomes scarce does meeting all activities' water requirements become an issue. This is the case in some basins in Central Asia, which raises the need to understand the requirements and impacts of particular consumptive pathways.

3.2.4. Irrigation

Irrigation is the predominant human use of water. The importance of irrigation globally and locally is well known [5]. From a water management point of view, it is therefore a key point of interaction in the nexus between water, food, and energy. The history of drivers of irrigation area and intensity in Central Asia has been a focus of research attention, particularly in the Aral Sea basin [56,66]. Irrigation is essential because of the aridity of the area. For water-intensive crops such as cotton, there is insufficient green water (soil moisture derived directly from rainfall), so the growth of plants is water-limited. The Soviet Union played a major role in the expansion of irrigation. Large water projects intended first to allow the Soviet Union to become self-sufficient in cotton, and later enable export earnings [56]. Rather than just food security, non-food agriculture therefore plays a large role in the nexus in Central Asia.

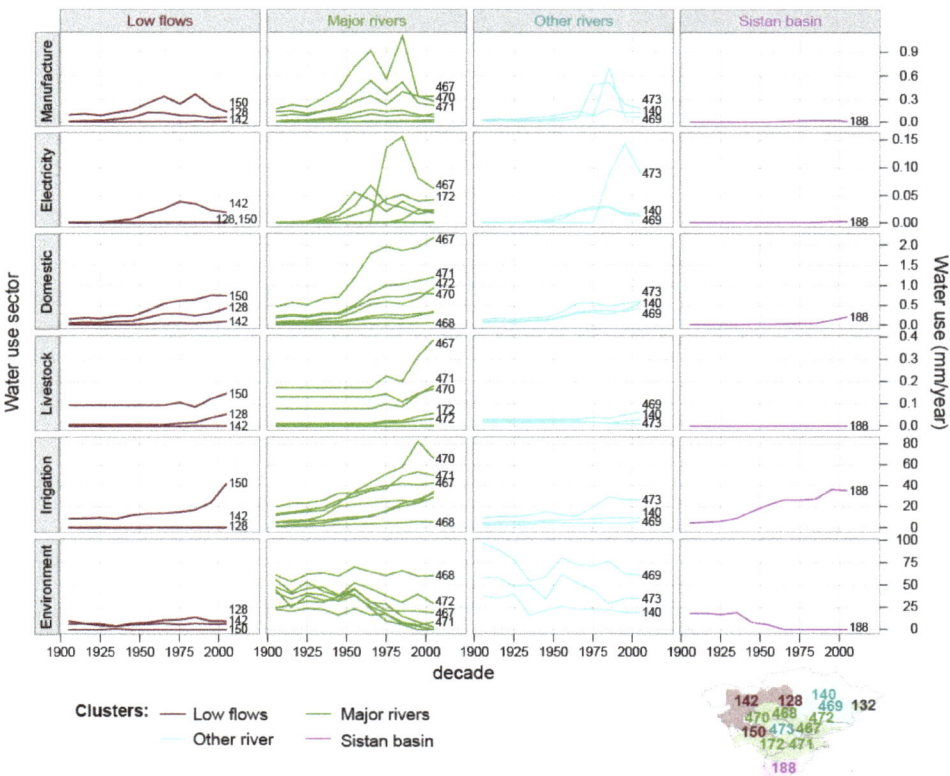

Figure 5. Temporal evolution of water uses for each food production unit (FPU). Each row represents a specific water use sector. Scales differ between rows.

Until recently, irrigation demand had been rising in most FPUs. This had the effect of shifting evaporation from downstream areas to cropland areas [67], often in locations where intensive cropping would not otherwise have been possible. In addition, there was a decrease in productive water use through an undesirable "vapor shift" [68]. Previously, the water evaporated through an environmental water pathway, meaning that all the water was available to aquatic ecosystems, which in turn provided useful ecosystem services. Afterward, only part of that water contributed to valuable crop transpiration. A significant portion was instead wasted through non-productive evaporation from dams, open irrigation canals (officially 28% of the water in Turkmenistan [69]), and the low productivity of the irrigation of crops grown in poor-quality soils [21]. Excess irrigation flowing to groundwater was not productive either, as the rising water table often contributed to salinization [69]. This situation was aggravated by increases in the irrigation rate rather than just the irrigation area [53]. While this increased the water available to crops and, hence, cash value, it decreased the total productive water use when taking into account the opportunity costs of other industries and ecosystem services. A narrow focus on agricultural productivity came at the expense of potential broader societal benefits.

Irrigation water use and total human water use did reach a peak or stabilize in several FPUs. In some cases, such as the Amu Darya (FPU 470, 471) this does coincide with reaching the limits of water availability. Aus der Beek [27] cites two main reasons of this change: improvements in irrigation efficiency and the conversion of cotton fields to food crop fields. The latter has been caused by falls in the price of cotton, the dissolution of the Soviet Union, and the resulting need to increase food self-sufficiency [70].

The fall in irrigation therefore should probably not be interpreted as a response to increasing water stress. In order to continue the expansion of agricultural land, water was seen as an input to be acquired, rather than a limit on development. Within basins, water was prioritized for agricultural use at the expense of downstream flows. As part of what has been called a "hydraulic mission" [22,71,72], dams were built to store water over time. Between basins, water was transferred through a large number of canals. Though it was later abandoned, some preliminary work had even been done on reversing the flow of rivers in Siberia and European Russia, to flow towards Central Asia, away from the Arctic Ocean [73].

There has been a (perhaps temporary) interruption in this tendency to use physical transfers of water to overcome water availability limits. Firstly, the dissolution of the Soviet Union resulted in decreased or stabilizing irrigation water use, as noted above. Secondly, such water transfers would have now required transboundary negotiations between countries, which is still a difficult political issue today [74]. Finally, it became more widely accepted that water does need to be left for ecosystem services (including intrinsic cultural value) [73].

3.2.5. Energy Production

It is widely known that the provision of energy for heating and water for irrigation have long been intertwined in Central Asia, first through barter arrangements of fuel-for-water and more recently through trade-offs between maximizing hydroelectricity production and irrigation [19]. The impacts are however more closely related to water storage, withdrawals, and water quality impacts rather than consumption, driven by the specific role of water in hydropower (which dominates in Kyrgyzstan and Tajikistan), gas (Turkmenistan and Uzbekistan), or oil and coal (Kazakhstan) [22].

The "electricity" water use in Figure 5 corresponds to the use of cooling water in thermoelectric power plants [75]. This is, in most cases, the smallest of the consumptive water uses in Central Asia. It is estimated based on power production for specific technological plant types and cooling systems. Its trend over time is therefore likely explained by changes in demand for thermo-electric power, corresponding to the expansion and contraction of the economy, and increased reliance on hydro-electric power [74]. While consumptive use is small, fossil fuels and thermoelectric power plants are associated with a number of problems with water, soil, and air pollution [61].

In this paper, water consumption of hydropower is not estimated. In principle, the storage of water in dams results in higher evaporation [76]. However, in Central Asia, evaporation is much larger in the downstream, arid zone (as high as 2250 mm) than in the upstream, mountainous zone (as low as 500 mm) [56], such that evapotranspiration from downstream uses is much more important than evaporation due to upstream storage of water for hydropower. Additionally, it is difficult to attribute evaporation from these dams to a particular sector. The dams are used for multiple purposes, supplying irrigation water, producing hydropower, and regulating floods [24,74]. There is no reason to attribute consumption to one sector over another from a mass conservation point of view, though it has been suggested that consumption could be allocated based on the ecosystem service benefits produced by each sector [77]. It is therefore of greater interest to focus on the non-consumptive impacts of hydropower, as discussed in Section 4.4.

3.2.6. Other Activities

The trends in other water uses in Figure 5 are best understood through their underlying model assumptions [78]. Domestic water use is estimated based on population and per capita water use intensity. National water use intensity is assumed to depend on income (GDP per capita), consumptive-use-coefficients, and technological change rates. For downscaling, additionally, rural *vs.* urban setting, and access to safe drinking water are taken into account. All FPUs show increases in domestic use over time consistent with increasing population and increasing water use intensities.

Livestock water consumption is estimated based on 10 types of livestock and water consumption per head and per year [78]. Increased livestock water consumption in a number of FPUs therefore primarily reflects increased livestock numbers at a decadal scale, though numbers might have varied from year to year. Note that variations in livestock numbers prior to 1960 are not known, and are therefore kept fixed at the 1960 level.

National time series of manufacturing water use are estimated based on the gross value added (GVA) economic measure, technological change rates, and consumptive-use-coefficients [31]. Country-scale estimates are allocated to grid cells according to the distribution of the urban population. Consistent with Central Asia's turbulent history, manufacturing water use has varied significantly in a number of FPUs, with a dip in the 1970s, followed by a temporary increase in the 1980s, and a fall since the 1990s with the dissolution of the Soviet Union. Manufacturing water consumption is small relative to agriculture, but generally results in the production of higher-value goods. As noted by Stucki and Sojamo [22], manufacturing provides a larger contribution to GDP than agriculture while using far less water.

3.3. Decadal Water Scarcity

The preceding sections discussed the role of decadal water availability and consumption within the history of the nexus in Central Asia. These issues are closely related to the concept of water scarcity, in particular through the water stress and shortage indicators. Figure 6 shows the trajectories of each FPU over time in terms of these two indicators.

Water stress is measured as the ratio of human water consumption and availability. A stress level of 100% means that all available water is used for human purposes. As discussed above, ecosystem services may already be affected at lower stress levels. Falkenmark and Lindh [49] suggest that water supply becomes a limiting factor of economic development when human use exceeds 20% of the available blue water. A stress level greater than 100% indicates the use of long-term stored water, such as the non-sustainable use of groundwater or unaccounted physical water transfers.

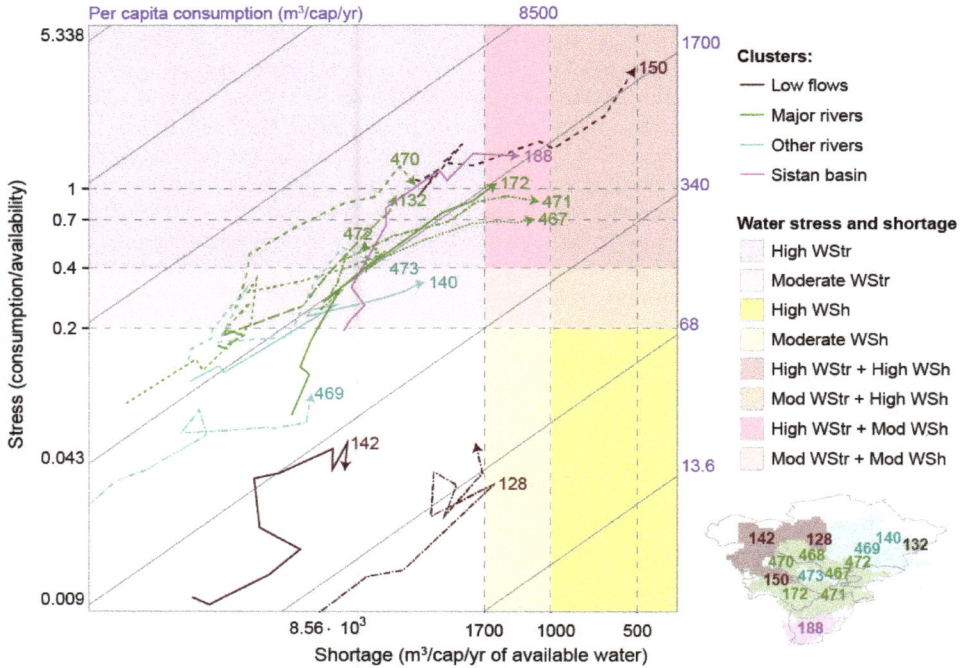

Figure 6. Water scarcity trajectories of each FPU for the 20th century, showing change in shortage, stress, and per capita consumption over time. The diagonal lines refer to per capita consumption isolines. Background colors show existence of moderate and high water stress (consumption respectively >20% and >40% of available water) and moderate and high water shortage (<1700 m^3/cap/year and <1000 m^3/cap/year), as used in Porkka *et al.* [26].

Water shortage is measured by the per capita availability of water. It is also referred to as water crowding, in which case it is interpreted as an indicator of competition for water [48]. A water shortage of 500 m^3 per capita per year corresponds to 2000 people sharing one gigaliter of water per year [47].

A point in the trajectories in Figure 6 represents the average stress and shortage of an FPU for a particular decade, calculated using the data presented in previous sections. Consistent with the preceding results, the overall shape of the trajectories from the bottom-left to the top-right indicates an increase in both stress and shortage over time as human water use and population increases. Per capita consumption is shown with a diagonal grid. The change in position of the trajectories relative to the diagonal grid shows an increase and, in some cases, later decrease consistent with the expansion of irrigation and the dissolution of the Soviet Union. Boundaries indicating multiple levels of water stress and shortage are shown using dashed lines and shading in the background. The trajectories for all the major rivers cross into regions of the plot, corresponding to some level of water stress, and some are also in regions corresponding to water shortage. Other FPUs appear to be on a trajectory toward stress and shortage as the population increases. In low flow FPUs, small changes have a large effect, as seen by large relative changes in stress or shortage between points.

These trajectories are suggestive of FPUs' attitude to change and to the crossing of boundaries. One might expect that as an FPU approaches a stress limit, efforts would be made to curb water consumption. Similarly, as an FPU reaches a shortage limit, migration or population control would stabilize water shortage. Instead, what we see is an apparent disregard for boundaries. As discussed in the previous sections, increasing water scarcity has instead been met by physical transfers to increase water availability and engineering works to reduce the impacts of stress. The literature on Central Asia even suggests that the dissolution of the Soviet Union is in fact responsible for the FPUs that do appear to have reached a stress ceiling (FPUs 471, 467, 470), not the response to decreasing water availability. In any case, when stabilization did eventuate, significant changes had already occurred, particularly in the case of the Aral Sea.

Naturally, there were efforts to return or remain within the boundaries and, hence, avoid or mitigate associated impacts [56]. While some plans were halted, such as the reversal of the Northern rivers [73], other changes simply happened too fast and too strongly to avoid crossing at least some boundaries. From a societal point of view, changes could not be controlled or regulated due to problems with the economy, the loss of expertise, the breakdown of cooperation, and social unrest [56]. Unfortunately, in the face of such a fast rate of change, the transformations of natural ecosystems have been quite significant [79]. Humans have become very successful at making changes—expanding our population and impact—but are not always very good at regulating those changes, at least when it comes to water consumption.

3.4. Impact of Human Use on Seasonal Discharge

Previous sections focused on decadal change, which does not show human impacts at shorter time scales. Figure 7 shows the monthly change in discharge between human and natural scenarios, summed for each FPU, and averaged across each decade, obtained using two model setups as described in Section 2.1. Relative changes are expressed as a percentage of natural discharge. This emphasizes changes in environmentally important low flows rather than high flows [60].

As expected, the analysis shows an overall decrease in discharge due to withdrawals for irrigation, which becomes more substantial over time. What is important here is the distribution across months. Most FPUs show greater withdrawals during the irrigation season, which varies in time and by location. For example, in the Syr Darya (FPU 467), decreases in discharge are most prominent from March to July in 1901–1910 and extend from March to November in 1991–2000, with decreases larger than 50% in some months. Historically, some FPUs have seen increases in discharge at other times, corresponding to a seasonal shift in the timing of flows due to reservoir operation [80]. While the model may not fully capture historical reservoir operations, this shift corresponds to the combined effect of multiple uses, most notably irrigation and hydropower generation. The need for irrigation in months with naturally low flows may result in higher discharges, particularly upstream of irrigation areas. More importantly, from the point of view of the nexus, Tajikistan and Kyrgyzstan are highly dependent on hydropower for winter heating, resulting in releases of stored water that could have been used for irrigation [2,74]. This is notably the focal point of conflict between Uzbekistan and Tajikistan over the Rogun Dam [74,81].

The conflict over seasonal flows raises the broader issue of temporal scales of water availability and use. Fundamentally, different users may have water needs that are continuous, seasonal, or episodic. Water needs may also be essential to survival ("obligate"), or optional ("facultative") [82]. Basic human needs must be satisfied (nearly) continuously, but some ecosystems can persist even with occasional floods. In Central Asia, the rise in irrigation, the seasonal shift in discharge, and the conflict with winter hydropower suggest that irrigation and hydropower would be "obligate" needs, as a result of quotas and the lack of economic alternatives [70]. However, the subsequent reduction in water use shows that irrigation and hydropower for winter heating are in fact essentially "facultative" and seasonal. It is technically possible to obtain food, foreign currency, and energy by other means, for example, by emphasizing trade, the development of urban economies, and knowledge-intensive industries [21]. Seasonal irrigation and hydropower can also vary each year depending on water availability and in response to food and energy crises, as is the case with current *ad hoc* annual agreements [74]. In the Soviet era, food security did not depend on local irrigation and winter heating did not depend solely on hydropower [10]. Admittedly, those arrangements also led to increased water consumption and, hence, the loss of the Aral Sea, and have since been destabilized by the dissolution of the Soviet Union. The history of Central Asia does however show that we are not truly locked into a particular path. There is the capacity for change to accommodate multiple, evolving objectives, which is, of course, beneficial in pursuing a nexus approach.

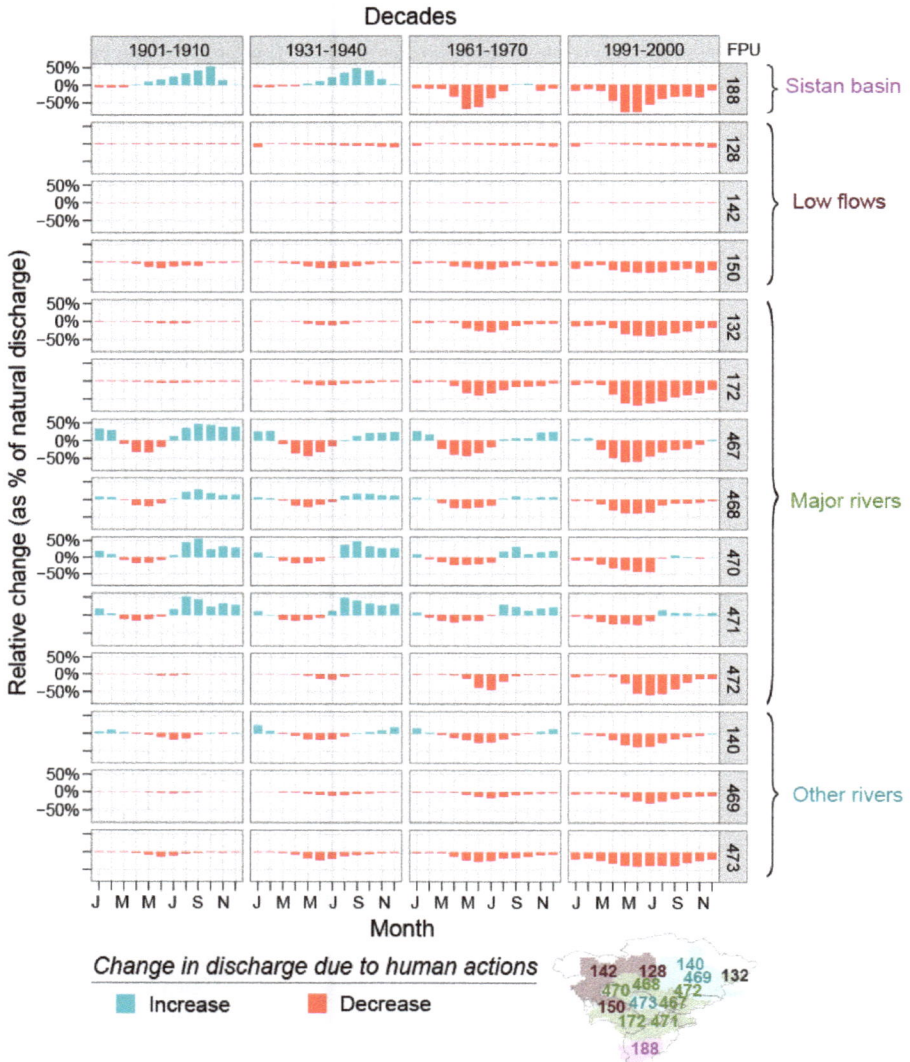

Figure 7. Monthly change in discharge due to human influence, relative to natural discharge, averaged for decades from 1905–2000. Estimated discharge under a natural scenario is subtracted from estimated discharge including human influences, *i.e.*, with flow regulation and water abstraction.

4. Discussion: Implications for the Water-Energy-Food Security Nexus

The intention of this paper was to derive generalizable conclusions regarding the implementation of the nexus by taking a global perspective when illustrating the history of the nexus in Central Asia. We do not aim to make specific recommendations for Central Asia. The use of a global water model put the focus on the basic principles underlying water availability, multi-sector consumption, and their historical changes.

To further help derive generalizable conclusions, we now draw on key concepts from literature on global change, socio-ecological systems, and resilience theory. A resilient socio-ecological system usually operates within a stable regime, compensating for minor disturbances [34,83]. However, if changes to a system are too great, a boundary is crossed, resulting in either the sudden or gradual transformation of the system such that it operates in a new (potentially undesirable) regime. A prominent example is the concept of "planetary boundaries" [33,84], which argues that the Earth system is currently in a regime particularly suited for human life, and that crossing certain boundaries will cause a transformation outside the "safe operating space for humanity". Boundaries may be social as well as natural, as in the analogous concept of a "just operating space for humanity", which defines the minimum requirements for social well-being [85,86]. The discussion based on these concepts will cover five key points:

- Firstly, changes in the system do not necessarily occur within the typical nexus sectors of water, energy, and food security. Key changes in Central Asia instead relate to non-food irrigation and the loss of ecosystems. This raises the question: what subsystems should be considered within the nexus?
- Secondly, trade-offs within the nexus (e.g., between water, energy, and food security objectives) are to some extent inevitable (e.g., the use of water by irrigation will always reduce flows). This means that is important to understand where the boundary is located that determines whether changes are or are not permissible (e.g., a small reduction in flows to the Aral Sea *vs.* loss of livelihoods). The question becomes: what boundaries can acceptably be crossed within the nexus?
- Thirdly, crossing boundaries implies that the system is transformed, often resulting in new dependencies being introduced (e.g., on water releases from transboundary reservoirs in other upstream countries). New trade-offs can therefore be created in the process of reducing existing trade-offs within the nexus (e.g., providing water for local food needs *vs.* cotton export). There is a need to ask: how will managing the nexus change system dependencies?
- Fourthly, these trade-offs, boundaries, and dependencies were historically already influenced by global relationships (e.g., trade of cotton), and this trend is only strengthening as a result of globalization. We discuss: what say should global stakeholders have in managing the water-energy-food security nexus?
- Finally, given that this paper has focused on the use of a global model, we address the question: what role does global data play?

4.1. What Subsystems Should Be Considered within the Nexus?

In the case of Central Asia, key changes in system regimes occurred in non-food industries and ecosystems, neither of which fit within the water, energy, and food security nexus, if taken literally. The increase in the irrigated area was at first related to cotton production, and resulted in the decline of the Aral Sea ecosystems and, hence, the broader socio-ecological system, with the associated loss of livelihoods as well as health impacts. Key issues in a case study do not necessarily relate solely to the security of water, energy, and food supply. If too narrow a scope is used, there is a risk that economic values of energy and food production may be over-emphasized at the expense of other values. This appears to have occurred during the Soviet Era, where comprehensive arrangements were made between republics for the provision of food, energy, and water, as well as export income through cotton. Strictly speaking, a nexus approach was already implemented, but significant problems still emerged because the protection of ecosystems and cultural values was at least partly neglected. Admittedly, this problem has already been recognized by nexus researchers. For example, Hoff [1] explicitly notes that "While not part of most water security definitions yet, availability of and access to water for other human and ecosystem uses is also very important from a nexus perspective" (p. 11). The message is clear: in implementing a nexus approach, it is essential to treat the relevant subsystems (or sectors) as case-specific and to consciously scrutinize what is included in the nexus and what is excluded from it.

4.2. What Boundaries Can Acceptably Be Crossed within the Nexus?

The trade-off between human and environmental water uses highlights the point that some trade-offs within the nexus are inevitable due to the principle of conservation of mass. Water simply cannot be present at two places at the same time. A change in consumptive use (*i.e.*, evaporation pathways) will always result in some degree of change to the system. Even when water is not consumed but returned to the system, a change in the timing of flows will also have some impact elsewhere in the system. Water trade-offs can, of course, be diminished in size by making smaller changes to the water system, or identifying synergies to create more value from the same-sized change. There is, however, no such thing as a change without impact, and there will always be some degree of trade-off between the sectors of the nexus no matter what synergies are found.

The inevitability of trade-offs does not necessarily imply conflict. We noted that environmental water in particular provides a buffer that delays competition for water (see Figure 3 and Section 3.2.1). However, trade-offs are often problematic when they cross key boundaries, such as upper limits and minimum requirements. Crossing boundaries transforms the identity of the system, modifying the processes that dominate system behavior and the functions that the system provides, as summarized in Table 1. While these boundaries would be difficult to quantify, one could easily see that these types of transformations would be contested and therefore a source of conflict. For example, the irrigated production of cotton is still a point of contention (see transformation for Figure 5 and Section 3.2.4), particularly in light of its effect on the Aral Sea (Figure 3 and Section 3.2.1).

Crossing boundaries (*i.e.*, transformation) may, however, also be necessary. Literature on socio-ecological systems (SES) talks about resilience as "the capacity of a SES to continually change and adapt yet remain within critical thresholds" [87]. At the same time, it highlights that resilient societies need to be able to transform themselves when their conditions become untenable. In particular, transformation at smaller scales helps support the survival of the broader socio-ecological system in the long-term [87]. In Central Asia, physical transfers of irrigation water allow greater human use of land and greater economic wealth (see transformation for Figure 2 and Section 3.2). Achieving certain changes to seasonal discharge associated with hydropower allows greater energy security (Figure 7 and Section 3.4).

The implication is that a key challenge of implementing the nexus is to resolve differences regarding the desired identity of the system: which boundaries are acceptable to cross, and which are not; what changes to processes and functions are permitted, and which are forbidden. The idea is to eliminate what is categorically unacceptable in order to delimit a narrowed negotiation space within which to search for solutions, as has been recommended elsewhere [88,89]. This is naturally of particular importance in a transboundary context, where different boundaries are imposed not just by the water, energy, and food sectors, but also by the interests of different governments and cultures [90]. Negotiations also need to recognize that it is possible that no feasible solution satisfies all of a given set of socio-economic and ecological boundaries, meaning that some boundaries may, in fact, be incompatible. Negotiating a consensus on boundaries raises important issues of power and justice. While these issues are out of scope of the present discussion, they are a key concern in rights-based approaches to development, which similarly emphasize boundaries that should not be crossed, and have been argued to be vital in implementing the water-energy-food security nexus [35,91,92].

Table 1. Key functions and processes illustrated by each assessment in Section 3, and transformations that could occur after boundaries are crossed. Icons refer to Figures 2–7.

Assessment	Relevant Figure	Function	Process to Achieve Function	Transformation after Boundary Is Crossed
Water Availability		Availability of blue water for socio-ecological systems	Partitioning of water flow by location within global water system	Unsustainable reliance on stored water, e.g., fossil groundwater, dependence on energy-intensive physical, or virtual water transfers
Human *vs.* Environmental Water Consumption		Provision of goods and services by ecosystem and humans	Partitioning of water between human use and environmental water (evaporation pathway)	Reduction in ecosystem services, shift towards dependence on energy-intensive human services
Human per Capita Consumption		Provision of individual human needs and wants, including food and water	Partitioning of water (and its benefits) within population	In extreme cases, starvation, malnutrition, hunger, but also poverty, inequality, and social unrest

Table 1. *Cont.*

Assessment	Relevant Figure	Function	Process to Achieve Function	Transformation after Boundary Is Crossed
Water by Activity		Provision of human goods and services underpinning human quality of life globally (including energy)	Partitioning of water by activity (reflecting values and power relationships)	Shift in dominant economic power, employment, cultural identity of population, and their diversity
Water Scarcity		Adaptation to cope with external drivers and internal changing needs and wants	Feedback between sub-systems, including water users, governance, and broader environment	Inability of society to adapt to changes; competitive advantage to actors that are better at learning
Impact of Human Use on Seasonal Discharge		Maintenance of flows and water availability at operational time-scales	Partitioning of water flows in time	Impacts of other transformations depend on their need for or aversion to variability and peak flows, e.g., flooding and timing of irrigation season

4.3. How Will Managing the Nexus Change System Dependencies?

A defining feature of system regimes is the relationship between its subsystems, including its social and ecological subsystems at different scales and in different locations. Dependencies of subsystems can impede growth by acting as a limiting factor, for example, the dependence of agriculture on water. Their interruption can cause failures, like the loss of cooperation when centralized decision-making ceased in the Soviet Union. Dependencies also relate to power, with countries engaging in both cooperation and conflict over the control of downstream flows, countered by the need of those countries for foreign currency, food, or fossil fuels [90]. Changes in dependencies are therefore a fundamental part of transformations between system regimes that occur when boundaries are crossed.

Table 2 summarizes a set of alternative dependencies that can occur in different system regimes. Each pair of dependencies forms a continuum defining the strength of each dependency. At various times in its history, Central Asia has operated according to system regimes at different points in these continuums. We do not aim here to identify specific regimes, as this is discussed elsewhere (e.g., [23,71]). We aim only to draw attention to the alternative paths that have been and could have been followed.

Table 2. Examples of pairs of alternative dependencies, drawn from this analysis of Central Asia. System regime may rely to varying degrees on different dependencies. Avoiding or reducing one dependency may introduce another.

Dependence on . . .	*vs.* Dependence on . . .
Dams and Diversions	**Naturally Available Water Supply**
Centralized decision making, e.g., within Soviet Union	Capabilities and interactions of separate nation states, *i.e.*, decentralized
Goodwill and trust with other countries sharing water resource	Maintaining control over water resource, e.g., coercion or hegemony
Ongoing maintenance of large-scale infrastructure, e.g., dams and canals	Sufficiency of small-scale user-maintained schemes
Demand for exports, e.g., price of cotton	Self-sufficiency of closed local economy
Livelihoods through jobs, incl. export industries	Subsistence farming
Food imports	Food self-sufficiency or sovereignty
External expertise, e.g., construction, maintenance of complex irrigation infrastructure	Local expertise, e.g., local solutions with local training
Engineering solutions to maintain ecosystems	Continuing naturally required inflows
Economic strength to pay for system operation	Limitations of solutions with low ongoing monetary commitments
Institutional capacity to understand and act on complex interactions	Suitability of system regimes without complex interactions

There does appear to be a trend toward increasing interdependence [23] and away from self-sufficiency and direct dependence on immediate surroundings, consistent with the general concept of globalization [93]. While a key focus of the nexus is to reduce trade-offs between sectors, such reductions can instead create new dependencies between systems that could previously operate (more or less) independently. In other words, it creates new trade-offs elsewhere. In Central Asia, massive engineering projects can be seen as an attempt to reduce trade-offs between users and between activities. Overcoming the limits of natural water availability through dams and pipelines means there is more water to share between human uses, and less risk of conflict. However, the expanded irrigated area is now dependent on that increased water availability and, hence, on the maintenance of extensive infrastructure as well as cooperative transboundary relationships [23]. In this case, increased interdependence may be to some extent unavoidable. Dependency on natural water availability cannot be maintained if a population grows too large, for example.

The trend towards interdependence is continuing in other ways. Rather than reducing irrigated areas, artificial barriers were created to maintain the level of the north Aral Sea with lower flows [65].

The health of the north Aral Sea is now dependent on the continued maintenance of those barriers. Maintaining (and achieving) low irrigation water losses depends on the maintenance and improvement of irrigation infrastructure and practices. While these are useful contributions to solving the current problems, they each create the need to commit to future financing, technology, skills, and labor to maintain these supporting systems. This was notably visible after the dissolution of the Soviet Union, as lack of time and finances reduced maintenance [56,70] and, hence, reduced productivity of irrigation water use. More generally, some relationships within the nexus have only recently become relevant. The conflict between irrigation and hydropower only exists because reservoirs successfully relieved trade-offs between competing irrigators and trade-offs between competing electricity users. We have ended up creating a dependency on the "institutional capacity to understand and act on the complex interactions" [7].

Similarly, the history of irrigation in Central Asia can be described in terms of interdependencies outside the region. Its political situation as part of the Soviet Union led to the dramatic expansion of irrigation. In particular, cotton was considered a valuable export crop. Even after the dissolution of the Soviet Union, cotton export remained politically important even when the contribution of the GDP of agriculture was smaller than that of the industry and services [22]. According to Porkka *et al.* [26], cotton contributes to 62% of the total agricultural blue water consumption in Central Asia, and eliminating virtual water flows (including cotton export) would reduce water scarcity for 47% of the population and completely eliminate it for 3%. Trade obviously plays a role in achieving food and energy needs, but the end result is that in Central Asia, the nexus is also "transboundary" in the sense that its history and future are closely tied to global trade relationships and agreements and the desire for foreign currency income.

There are, however, also examples of shifts towards self-reliance, and the dependencies that implies. Notably, the shift from cotton to food crops is a result of Central Asian countries trying to achieve food sovereignty [71], and the emphasis of upstream countries on hydropower aims to make them independent of imported fossil fuels [20]. We do not aim to predict or recommend what dependencies will be in place in the future, or whether self-reliance or interdependence is to be preferred. It is, however, clearly important, when trying to reduce trade-offs within the nexus, to understand how system dependencies will be altered.

4.4. What Say Should Global Stakeholders Have in Managing the Water-Energy-Food Security Nexus?

In light of Central Asia's global interdependencies, all people globally can be considered to have a stake in the boundaries to be crossed, what functions may be lost, or which dependencies might be added. This view appears to be shared by the governments involved in the International Fund for Saving the Aral Sea, who state that "international organizations, bi-lateral aid agencies and foreign governments have stepped up to cooperate" [94]. We have a role to play not just in defining the desired identity of the system, but also in supporting local people and organizations to make the appropriate transformations [20]. This is probably typical of most places in our increasingly globalized world. Folke *et al.* [87] boldly state that "society must seriously consider ways to foster resilience of smaller more manageable SESs [socio-ecological systems] that contribute to Earth System resilience and to explore options for deliberate transformation of SESs that threaten Earth System resilience."

Planetary boundaries are one definition of what it means to threaten Earth System resilience [33,84]. While there are difficulties in downscaling these boundaries, Central Asia, and the Aral Sea basin in particular, would probably be considered to be using more than its fair share of the freshwater planetary boundaries. The degradation of the biosphere it has caused is precisely the kind of impact that the planetary boundaries are trying to avoid. The planetary boundaries are expected to interact [84]. In this case, the loss of the Aral Sea is known to have resulted in impacts on "biosphere integrity" through the extinction of fish populations [55]. The increase in irrigation is naturally associated with agricultural expansion and, hence, "land-system change", alteration of "biogeochemical flows", and potentially the introduction of "novel entities", as well as an increase in "atmospheric aerosol loading" through increased dust storms arising from salinization and desertification [65]. At the same time, climate change is also obviously related to energy issues in Central Asia, such that the greater emphasis on hydropower rather than the barter of fossil fuels could be seen as a positive development. For most of these planetary boundaries, it is unclear how they translate into local boundaries, particularly in ambiguous cases where the land-system change could be seen as greening the desert and, hence, avoiding the reduction of forest cover elsewhere. It is, however, clear that the history of the nexus in Central Asia (and elsewhere) is considered to be intertwined with global sustainability.

4.5. What Role Does Global Data Play? Contributions and Limitations of the Analysis

This paper specifically makes use of global modeled data. While it is likely not as reliable as local models or measured data, it helps to appropriately set the scene. It draws attention to underlying, broadly applicable, and fundamental issues rather than more widely discussed details. Starting from a global perspective provides a broader context and makes it easier to relate case study observations to global trends and ideas. *Vice versa*, the use of global data to focus on local history adds value to the global model by providing a link to a wealth of existing place-based studies and a corresponding depth of analysis. From a practical perspective, using global data enhances the comparability of results across regions, even where data is sparse.

Other global data could provide further insights, including extending the boundaries and new dependencies identified in the previous sections. This paper focused on blue water, given the importance of irrigation in Central Asia. Green water (available in soils directly from precipitation) obviously also plays an important role in agriculture, generally [38]. Historical analysis of virtual water flows would also be beneficial [26]. The model used here treated physical transfers of water and the operation of reservoirs in a rather basic way, and glacier melt in-flows and evaporation from dams are not explicitly included. Improvement of these features is needed in order for these models to be used at finer spatial and temporal scales, particularly where large scale water supply schemes exist, as is the case in Central Asia [27]. This paper also focused primarily on water consumption. Future analyses could also consider water withdrawals and grey water impacts. The analysis of seasonal variation in discharge clearly shows that water withdrawals play a key role in transforming water availability over time. Increasing water use has been accompanied with significant water quality impacts, notably salinization [65]. Finally, this paper views the nexus through a water lens, interpreting historical water data in terms of its broader agriculture, food, and energy context. It would be of interest to use other global data (e.g., cited in [25]) to perform a similar exercise through an agricultural, food security, or energy security lens.

5. Conclusions

This paper used spatiotemporal data from a global model to illustrate the 20th century history of the nexus in Central Asia through a water lens. In the region, low natural water availability constrains potential water-dependent activities. Human consumptive use increased throughout the 20th century, primarily driven by irrigation at the expense of environmental water (e.g., the Aral Sea), resulting in high per capita water consumption. The evolution of different sectors over time follows economic development and population growth. Overall, water use and scarcity show a rapid upwards trend. However, in some areas, total water use stabilized or declined after the dissolution of the Soviet Union, likely driven by a shift in political and economic circumstances rather than as a direct response to increasing water scarcity. While the focus was on decadal water consumption, changes in monthly discharge showed evidence of the well-known effects of hydropower and irrigation. Despite its limitations, the use of global data in this paper played a key role in setting the case study within a global context and emphasizing generalizable lessons. This paper does not aim to provide specific recommendations regarding management of the nexus in Central Asia.

Accordingly, discussion of the results used the Central Asian case study to identify five transferable principles related to the nexus. This case study-based synthesis is a novel contribution of this paper, though these principles naturally build on solid foundations from existing literature, notably from socio-ecological systems and resilience theory. The first principle is that *the subsystems included/excluded from the nexus are case-specific and should be consciously scrutinized*. Discussion of the nexus in Central Asia would notably be incomplete without consideration of irrigated cotton production and ecological impacts, even though they are not strictly part of water, energy, and food security.

The second principle is that *it is important to reach an understanding of what boundaries can acceptably be crossed within the nexus*. Some trade-offs within the nexus are inevitable. Due to conservation of mass, changes in water use and storage will always result in some degree of change to the system. Consistent with the concept of planetary boundaries, we note that it is the crossing of boundaries that is of key importance, as it results in the transformation of the structure, functions, and identity of the system. These transformations notably include the loss of functions (e.g., ecosystem services) and the creation of dependencies requiring ongoing commitments. The third principle is, therefore, that *it is important to understand how reducing trade-offs will modify system dependencies*.

In Central Asia (like elsewhere in the world), history shows that limits of water availability have been treated as a boundary to be crossed rather than respected, as part of what has been called a hydraulic mission. Most visibly, infrastructure projects have reduced the conflict between some human water uses by increasing localized water availability. This has come at the expense of ecosystems and industries dependent on the Aral Sea, and an increased dependence on upstream transboundary reservoirs and a commitment to the ongoing long-term maintenance of infrastructure. This trend towards increased interdependence is manifested also in engineering solutions to the Aral Sea crisis and the continuing influence of global trade (especially cotton). On the other hand, the development of Central Asia may be considered to have disproportionally contributed to several planetary boundaries, contributing to cumulative global impacts through the broader Earth system as well as through socio-economic connections. These dependencies and impacts respectively mean, as a fourth principle, that *global stakeholders have both a responsibility and a right to contribute to the shaping of the nexus*. In this context, we have found a fifth principle useful, namely that *use of global data combined with existing place-based studies can help to provide a global perspective*, enhancing the transferability and understanding of shared problems in our globalized world.

These principles can contribute to the successful implementation of nexus approaches to understand complex interactions, reduce trade-offs, and build synergies. The process of developing integrated solutions to water, energy, and food security necessarily involves the transformation of the system. The assessment of solutions needs to understand the functions that may be lost and the dependencies that may be introduced. Evaluating these changes requires a shared understanding of which boundaries should be avoided and which can be crossed. In today's globalized world, every person is, in principle, a stakeholder in this process, contributing directly or indirectly to international decision-making. Governments, researchers, businesses, civil society, and consumers should all have their role to play, some large, some small, as discussed in other papers of this special issue. These roles should include seeking to better understand the effects of transformations, participating in judging what constitutes unacceptable changes, and reflecting on how their actions help shape transformations within the sustainable water-energy-food security nexus.

Acknowledgments: The authors would like to thank three anonymous reviewers for their comments which helped improve the paper. The authors also thank their colleagues in the Water and Development and Research Group, Aalto University for their support and useful comments. Joseph Guillaume received funding from the Academy of Finland funded project NexusAsia (Grant No. 269901) and Matti Kummu received funding from the Academy of Finland funded project SCART (Grant No. 267463).

Author Contributions: Joseph Guillaume and Matti Kummu designed the analysis. Joseph Guillaume and Stephanie Eisner performed the analysis. Joseph Guillaume, Matti Kummu, Olli Varis, and Stephanie Eisner contributed to discussion and writing and editing.

Conflicts of Interest: The authors declare no conflict of interest.

References

1. Hoff, H. *Understanding the Nexus: Background Paper for the Bonn2011 Nexus Conference: The Water, Energy and Food Security Nexus*; Stockholm Environment Institute: Stockholm, Sweden, 2011.
2. Granit, J.; Jägerskog, A.; Lindström, A.; Björklund, G.; Bullock, A.; Löfgren, R.; de Gooijer, G.; Pettigrew, S. Regional options for addressing the water, energy and food nexus in Central Asia and the Aral sea basin. *Int. J. Water Resour. Dev.* **2012**, *28*, 419–432. [CrossRef]
3. Hayami, Y.; Ruttan, V.W. *Agricultural Development: An International Perspective*; The Johns Hopkins Press: Baltimore, MD, USA; London, UK, 1971.
4. Kummu, M.; Gerten, D.; Heinke, J.; Konzmann, M.; Varis, O. Climate-driven interannual variability of water scarcity in food production potential: A global analysis. *Hydrol. Earth Syst. Sci.* **2014**, *18*, 447–461. [CrossRef]
5. Siebert, S.; Döll, P. Quantifying blue and green virtual water contents in global crop production as well as potential production losses without irrigation. *J. Hydrol.* **2010**, *384*, 198–217. [CrossRef]
6. Orr, S.; Pittock, J.; Chapagain, A.; Dumaresq, D. Dams on the Mekong river: Lost fish protein and the implications for land and water resources. *Glob. Environ. Chang.* **2012**, *22*, 925–932. [CrossRef]
7. Bazilian, M.; Rogner, H.; Howells, M.; Hermann, S.; Arent, D.; Gielen, D.; Steduto, P.; Mueller, A.; Komor, P.; Tol, R.S.J.; *et al.* Considering the energy, water and food nexus: Towards an integrated modelling approach. *Energy Policy* **2011**, *39*, 7896–7906.
8. Hussey, K.; Pittock, J. The energy-water nexus: Managing the links between energy and water for a sustainable future. *Ecol. Soc.* **2012**, *17*. [CrossRef]
9. Water in the West. *Water and Energy Nexus: A Literature Review*; Stanford University: Stanford, CA, USA, 2013.
10. Scott, C.A.; Pierce, S.A.; Pasqualetti, M.J.; Jones, A.L.; Montz, B.E.; Hoover, J.H. Policy and institutional dimensions of the water-energy nexus. *Energy Policy* **2011**, *39*, 6622–6630. [CrossRef]
11. Rahaman, M.M.; Varis, O. *Central Asian Waters: Social, Economic, Environmental and Governance Puzzle*; Helsinki University of Technology: Helsinki, Finland, 2008.
12. Dukhovny, V.A.; de Schutter, J. *Water in Central Asia: Past, Present, Future*; CRC Press/Balkema: London, UK, 2011.
13. United Nations Economic Commission for Europe (UNECE). *Strengthening Water Management and Transboundary Water Cooperation in Central Asia: The Role of Unece Environmental Conventions*; UNECE: Geneva, Switerland, 2011.
14. United Nations Environmental Programme (UNEP). *Environment and Security in the Amu Darya Basin*; UNEP: Nairobi, Kenya, 2011.
15. Stucki, V.; Wegerich, K.; Rahaman, M.M.; Varis, O. Introduction: Water and security in Central Asia—Solving a rubik's cube. *Int. J. Water Resour. Dev.* **2012**, *28*, 395–397. [CrossRef]
16. Unger-Shayesteh, K.; Vorogushyn, S.; Merz, B.; Frede, H.G. Introduction to "water in Central Asia—Perspectives under global change". *Glob. Planet. Chang.* **2013**, *110*, 1–3. [CrossRef]
17. Karthe, D.; Chalov, S.; Borchardt, D. Water resources and their management in Central Asia in the early twenty first century: Status, challenges and future prospects. *Environ. Earth Sci.* **2015**, *73*, 487–499. [CrossRef]
18. World Bank. *Water Energy Nexus in Central Asia: Improving Regional Cooperation in the Syr Darya Basin*; World Bank: Washington, DC, USA, 2004; pp. 1–59.
19. Wegerich, K. Coping with disintegration of a river-basin management system: Multi-dimensional issues in Central Asia. *Water Policy* **2004**, *6*, 335–344.
20. Abdolvand, B.; Mez, L.; Winter, K.; Mirsaeedi-Großner, S.; Schütt, B.; Rost, K.; Bar, J. The dimension of water in Central Asia: Security concerns and the long road of capacity building. *Environ. Earth Sci.* **2015**, *73*, 897–912. [CrossRef]
21. Varis, O. Resources: Curb vast water use in Central Asia. *Nature* **2014**, *514*, 27. [CrossRef] [PubMed]
22. Stucki, V.; Sojamo, S. Nouns and numbers of the water-energy-security nexus in Central Asia. *Int. J. Water Resour. Dev.* **2012**, *28*, 399–418. [CrossRef]
23. Soliev, I.; Wegerich, K.; Kazbekov, J. The costs of benefit sharing: Historical and institutional analysis of shared water development in the Ferghana valley, the Syr Darya basin. *Water* **2015**, *7*, 2728–2752. [CrossRef]
24. Jalilov, S.-M.; Keskinen, M.; Varis, O. Sharing benefits in transboundary rivers: An experimental case study of Central Asian energy-agriculture dispute. *Water*, submitted.

25. Varis, O.; Kummu, M. The major Central Asian river basins: An assessment of vulnerability. *Int. J. Water Resour. Dev.* **2012**, *28*, 433–452. [CrossRef]

26. Porkka, M.; Kummu, M.; Siebert, S.; Flörke, M. The role of virtual water flows in physical water scarcity: The case of Central Asia. *Int. J. Water Resour. Dev.* **2012**, *28*, 453–474. [CrossRef]

27. Aus der Beek, T.; Voß, F.; Flörke, M. Modelling the impact of global change on the hydrological system of the Aral sea basin. *Phys. Chem. Earth A B C* **2011**, *36*, 684–695. [CrossRef]

28. Malsy, M.; Aus der Beek, T.; Eisner, S.; Flörke, M. Climate change impacts on Central Asian water resources. *Adv. Geosci.* **2012**, *32*, 77–83. [CrossRef]

29. Weedon, G.P.; Gomes, S.; Viterbo, P.; Shuttleworth, W.J.; Blyth, E.; Österle, H.; Adam, J.C.; Bellouin, N.; Boucher, O.; Best, M. Creation of the WATCH forcing data and its use to assess global and regional reference crop evaporation over land during the twentieth century. *J. Hydrometeorol.* **2011**, *12*, 823–848. [CrossRef]

30. Haddeland, I.; Clark, D.B.; Franssen, W.; Ludwig, F.; Voß, F.; Arnell, N.W.; Bertrand, N.; Best, M.; Folwell, S.; Gerten, D.; *et al.* Multimodel estimate of the global terrestrial water balance: Setup and first results. *J. Hydrometeorol.* **2011**, *12*, 869–884.

31. Flörke, M.; Kynast, E.; Bärlund, I.; Eisner, S.; Wimmer, F.; Alcamo, J. Domestic and industrial water uses of the past 60 years as a mirror of socio-economic development: A global simulation study. *Glob. Environ. Chang.* **2013**, *23*, 144–156. [CrossRef]

32. Salmivaara, A.; Porkka, M.; Kummu, M.; Keskinen, M.; Guillaume, J.; Varis, O. Exploring the modifiable areal unit problem in spatial water assessments: A case of water shortage in monsoon Asia. *Water* **2015**, *7*, 898–917. [CrossRef]

33. Rockström, J.; Steffen, W.; Noone, K.; Persson, A.; Chapin, F.S.; Lambin, E.F.; Lenton, T.M.; Scheffer, M.; Folke, C.; Schellnhuber, H.J.; *et al.* A safe operating space for humanity. *Nature* **2009**, *461*, 472–475. [PubMed]

34. Rockström, J.; Falkenmark, M.; Allan, T.; Folke, C.; Gordon, L.; Jägerskog, A.; Kummu, M.; Lannerstad, M.; Meybeck, M.; Molden, D.; *et al.* The unfolding water drama in the anthropocene: Towards a resilience based perspective on water for global sustainability. *Ecohydrology* **2014**, *7*, 1249–1261.

35. Allouche, J.; Middleton, C.; Gyawali, D. Technical veil, hidden politics: Interrogating the power linkages behind the nexus. *Water Altern.* **2015**, *8*, 610–626.

36. Siebert, S.; Kummu, M.; Porkka, M.; Döll, P.; Ramankutty, N.; Scanlon, B.R. A global data set of the extent of irrigated land from 1900 to 2005. *Hydrol. Earth Syst. Sci.* **2015**, *19*, 1521–1545. [CrossRef]

37. Müller Schmied, H.; Eisner, S.; Franz, D.; Wattenbach, M.; Portmann, F.T.; Flörke, M.; Döll, P. Sensitivity of simulated global-scale freshwater fluxes and storages to input data, hydrological model structure, human water use and calibration. *Hydrol. Earth Syst. Sci.* **2014**, *18*, 3511–3538. [CrossRef]

38. Rockström, J.; Falkenmark, M.; Karlberg, L.; Hoff, H.; Rost, S.; Gerten, D. Future water availability for global food production: The potential of green water for increasing resilience to global change. *Water Resour. Res.* **2009**, *45*, W00A12. [CrossRef]

39. Schewe, J.; Heinke, J.; Gerten, D.; Haddeland, I.; Arnell, N.W.; Clark, D.B.; Dankers, R.; Eisner, S.; Fekete, B.M.; Colón-González, F.J.; *et al.* Multimodel assessment of water scarcity under climate change. *Proc. Natl. Acad. Sci. USA* **2014**, *111*, 3245–3250. [CrossRef] [PubMed]

40. Kummu, M.; Ward, P.J.; de Moel, H.; Varis, O. Is physical water scarcity a new phenomenon? Global assessment of water shortage over the last two millennia. *Environ. Res. Lett.* **2010**, *5*, 034006.

41. Cai, X.; Rosegrant, M.W. Global water demand and supply projections. *Water Int.* **2002**, *27*, 159–169. [CrossRef]

42. De Fraiture, C. Integrated water and food analysis at the global and basin level. An application of Watersim. *Water Resour. Manag.* **2007**, *21*, 185–198. [CrossRef]

43. Defense Mapping Agency. Digital Chart of the World, Accessed via Diva-gis Repository. Available online: http://www.diva-gis.org/Data (accessed on 14 July 2015).

44. Global Runoff Data Centre. *Major River Basins of the World*; Federal Institute of Hydrology (BfG), Global Runoff Data Centre: Koblenz, Germany, 2007.

45. R Core Team. *R: A Language and Environment for Statistical Computing*; R Foundation for Statistical Computing: Vienna, Austria, 2014.

46. Wallace, J.S.; Acreman, M.C.; Sullivan, C.A. The sharing of water between society and ecosystems: From conflict to catchment-based co-management. *Philos. Trans. Royal Soc. Lond. B Biol. Sci.* **2003**, *358*, 2011–2026. [CrossRef] [PubMed]

47. Falkenmark, M. The massive water scarcity now threatening Africa: Why isn't it being addressed? *Ambio* **1989**, *18*, 112–118.

48. Falkenmark, M. Fresh water: Time for a modified approach. *Ambio* **1986**, *15*, 192–200.

49. Falkenmark, M.; Lindh, G. *Water for a Starving World*; Westview Press: Boulder, CO, USA, 1976.

50. Klein Goldewijk, K.; Beusen, A.; Janssen, P. Long-term dynamic modeling of global population and built-up area in a spatially explicit way: Hyde 3.1. *Holocene* **2010**, *20*, 565–573. [CrossRef]

51. Unger-Shayesteh, K.; Vorogushyn, S.; Farinotti, D.; Gafurov, A.; Duethmann, D.; Mandychev, A.; Merz, B. What do we know about past changes in the water cycle of Central Asian headwaters? A review. *Glob. Planet. Chang.* **2013**, *110*, 4–25. [CrossRef]

52. Schär, C.; Vasilina, L.; Pertziger, F.; Dirren, S. Seasonal runoff forecasting using precipitation from meteorological data assimilation systems. *J. Hydrometeorol.* **2004**, *5*, 959–973. [CrossRef]

53. Belyaev, A.V. Water balance and water resources of the Aral sea basin and its man-induced changes. *GeoJournal* **1995**, *35*, 17–21. [CrossRef]

54. Wegerich, K. Passing over the conflict. The Chu Talas basin agreement as a model for Central Asia? In *Central Asian Waters: Social, Economic, Environmental and Governance Puzzle*; Helsinki University of Technology: Helsinki, Finland, 2008; pp. 117–131.

55. Aladin, N.V.; Plotnikov, I.S. Large saline lakes of former ussr: A summary review. In *Saline Lakes V*; Hurlbert, S., Ed.; Springer: Houten, The Netherlands, 1993; Volume 87, pp. 1–12.

56. O'Hara, S.L. Central asia's water resources: Contemporary and future management issues. *Int. J. Water Resour. Dev.* **2000**, *16*, 423–441. [CrossRef]

57. Rashid, K.; Nodirbek, M.; Asqar, N.; Dilafruz, K.; Jobir, S. Qualitative and quantitative assessment of water resources of Aydar Arnasay lakes system (AALS). *J. Water Resour. Prot.* **2013**, *5*, 941–952.

58. Van Beek, E.; Bozorgy, B.; Vekerdy, Z.; Meijer, K. Limits to agricultural growth in the sistan closed inland delta, Iran. *Irrig. Drainage Syst.* **2008**, *22*, 131–143. [CrossRef]

59. Van der Ent, R.J.; Savenije, H.H.G.; Schaefli, B.; Steele-Dunne, S.C. Origin and fate of atmospheric moisture over continents. *Water Resour. Res.* **2010**, *46*, W09525. [CrossRef]

60. Pastor, A.V.; Ludwig, F.; Biemans, H.; Hoff, H.; Kabat, P. Accounting for environmental flow requirements in global water assessments. *Hydrol. Earth Syst. Sci.* **2014**, *18*, 5041–5059. [CrossRef]

61. Dahl, C.; Kuralbayeva, K. Energy and the environment in Kazakhstan. *Energy Policy* **2001**, *29*, 429–440. [CrossRef]

62. Renaud, F.G.; Syvitski, J.P.M.; Sebesvari, Z.; Werners, S.E.; Kremer, H.; Kuenzer, C.; Ramesh, R.; Jeuken, A.; Friedrich, J. Tipping from the holocene to the anthropocene: How threatened are major world deltas? *Curr. Opin. Environ. Sustain.* **2013**, *5*, 644–654. [CrossRef]

63. Gordon, L.J.; Peterson, G.D.; Bennett, E.M. Agricultural modifications of hydrological flows create ecological surprises. *Trends Ecol. Evol.* **2008**, *23*, 211–219. [CrossRef] [PubMed]

64. Gerten, D.; Hoff, H.; Rockström, J.; Jägermeyr, J.; Kummu, M.; Pastor, A.V. Towards a revised planetary boundary for consumptive freshwater use: Role of environmental flow requirements. *Curr. Opin. Environ. Sustain.* **2013**, *5*, 551–558. [CrossRef]

65. Micklin, P. The aral sea disaster. *Annu. Rev. Earth Planet. Sci.* **2007**, *35*, 47–72. [CrossRef]

66. Saiko, T.A.; Zonn, I.S. Irrigation expansion and dynamics of desertification in the circum-Aral region of Central Asia. *Appl. Geogr.* **2000**, *20*, 349–367. [CrossRef]

67. Gordon, L.J.; Steffen, W.; Jönsson, B.F.; Folke, C.; Falkenmark, M.; Johannessen, Å. Human modification of global water vapor flows from the land surface. *Proc. Natl. Acad. Sci. USA* **2005**, *102*, 7612–7617. [PubMed]

68. Rockström, J. Water for food and nature in drough-prone tropics: Vapour shift in rain-fed agriculture. *Philos. Trans. R. Soc. Lond. B* **2003**, *358*, 1997–2009. [CrossRef] [PubMed]

69. O'Hara, S.L. Irrigation and land degradation: Implications for agriculture in Turkmenistan, Central Asia. *J. Arid Environ.* **1997**, *37*, 165–179. [CrossRef]

70. Abdullaev, I.; de Fraiture, C.; Giordano, M.; Yakubov, M.; Rasulov, A. Agricultural water use and trade in Uzbekistan: Situation and potential impacts of market liberalization. *Int. J. Water Resour. Dev.* **2009**, *25*, 47–63. [CrossRef]

71. Abdullaev, I.; Rakhmatullaev, S. Transformation of water management in Central Asia: From state-centric, hydraulic mission to socio-political control. *Environ. Earth Sci.* **2015**, *73*, 849–861. [CrossRef]

72. Molle, F.; Mollinga, P.P.; Wester, P. Hydraulic bureaucracies and the hydraulic mission: Flows of water, flows of power. *Water Altern.* **2009**, *2*, 328–349.

73. Yanitsky, O.N. The shift of environmental debates in Russia. *Curr. Sociol.* **2009**, *57*, 747–766. [CrossRef]

74. Libert, B.; Orolbaev, E.; Steklov, Y. Water and energy crisis in Central Asia. *China Eurasia Forum Q.* **2008**, *6*, 9–20.

75. Vassolo, S.; Döll, P. Global-scale gridded estimates of thermoelectric power and manufacturing water use. *Water Resour. Res.* **2005**, *41*, W04010. [CrossRef]

76. Mekonnen, M.M.; Hoekstra, A.Y. The blue water footprint of electricity from hydropower. *Hydrol. Earth Syst. Sci.* **2012**, *16*, 179–187. [CrossRef]

77. Zhao, D.; Liu, J. A new approach to assessing the water footprint of hydroelectric power based on allocation of water footprints among reservoir ecosystem services. *Phys. Chem. Earth A B C* **2015**, *79–82*, 40–46. [CrossRef]

78. Alcamo, J.; Döll, P.; Henrichs, T.; Kaspar, F.; Lehner, B.; Rösch, T.; Siebert, S. Development and testing of the WaterGAP 2 global model of water use and availability. *Hydrol. Sci. J.* **2003**, *48*, 317–337. [CrossRef]

79. Schlüter, M.; Khasankhanova, G.; Talskikh, V.; Taryannikova, R.; Agaltseva, N.; Joldasova, I.; Ibragimov, R.; Abdullaev, U. Enhancing resilience to water flow uncertainty by integrating environmental flows into water management in the Amudarya river, Central Asia. *Glob. Planet. Chang.* **2013**, *110*, 114–129. [CrossRef]

80. Cai, X.; McKinney, D.C.; Rosegrant, M.W. Sustainability analysis for irrigation water management in the Aral sea region. *Agric. Syst.* **2003**, *76*, 1043–1066. [CrossRef]

81. Jalilov, S.-M.; Amer, S.; Ward, F. Water, food, and energy security: An elusive search for balance in Central Asia. *Water Resour. Manag.* **2013**, *27*, 3959–3979. [CrossRef]

82. Eamus, D.; Froend, R.; Loomes, R.; Hose, G.; Murray, B. A functional methodology for determining the groundwater regime needed to maintain the health of groundwater-dependent vegetation. *Aust. J. Bot.* **2006**, *54*, 97–114. [CrossRef]

83. Walker, B.; Holling, C.S.; Carpenter, S.R.; Kinzig, A. Resilience, adaptability and transformability in social-ecological systems. *Ecol. Soc.* **2004**, *9*, 5.

84. Steffen, W.; Richardson, K.; Rockstrom, J.; Cornell, S.E.; Fetzer, I.; Bennett, E.M.; Biggs, R.; Carpenter, S.R.; de Vries, W.; de Wit, C.A.; *et al.* Planetary boundaries: Guiding human development on a changing planet. *Science* **2015**, *347*. [CrossRef]

85. Dearing, J.A.; Wang, R.; Zhang, K.; Dyke, J.G.; Haberl, H.; Hossain, M.S.; Langdon, P.G.; Lenton, T.M.; Raworth, K.; Brown, S.; *et al.* Safe and just operating spaces for regional social-ecological systems. *Glob. Environ. Chang.* **2014**, *28*, 227–238.

86. Raworth, K. A safe and just space for humanity: Can we live within the doughnut. *Oxfam Policy Pract. Clim. Chang. Resil.* **2012**, *8*, 1–26.

87. Folke, C.; Carpenter, S.R.; Walker, B.; Scheffer, M.; Chapin, T.; Rockström, J. Resilience thinking: Integrating resilience, adaptability and transformability. *Ecol. Soc.* **2010**, *15*, 20.

88. Luce, R.D.; Raiffa, H. *Games and Decisions: Introduction and Critical Survey*; Duncan: New York, NY, USA, 1985.

89. Pierce, S.A.; Sharp, J.M.; Guillaume, J.H.A.; Mace, R.E.; Eaton, D.J. Aquifer-yield continuum as a guide and typology for science-based groundwater management. *Hydrogeol. J.* **2012**, *21*, 331–340. [CrossRef]

90. Sojamo, S. Illustrating co-existing conflict and cooperation in the Aral sea basin with TWINS approach. In *Central Asian Waters: Social, Economic, Environmental and Governance Puzzle*; Helsinki University of Technology: Helsinki, Finland, 2008; pp. 75–88.

91. Gready, P. Rights-based approaches to development: What is the value-added? *Dev. Pract.* **2008**, *18*, 735–747. [CrossRef]

92. Cornwall, A.; Nyamu-Musembi, C. Putting the "rights-based approach" to development into perspective. *Third World Q.* **2004**, *25*, 1415–1437. [CrossRef]

93. Young, O.R.; Berkhout, F.; Gallopin, G.C.; Janssen, M.A.; Ostrom, E.; van der Leeuw, S. The globalization of socio-ecological systems: An agenda for scientific research. *Glob. Environ. Chang.* **2006**, *16*, 304–316. [CrossRef]

94. International Fund for Saving the Aral Sea. *From the Glaciers to the Aral Sea—Water Unites*; International Fund for Saving the Aral Sea: Tashkent, Uzbekistan, 2015.

Article

Water Security in the Syr Darya Basin

Kai Wegerich [1,*], Daniel Van Rooijen [2], Ilkhom Soliev [3] and Nozilakhon Mukhamedova [1]

[1] Leibniz Institute of Agricultural Development in Transition Economies (IAMO), Theodor-Lieser-Str. 2, Halle (Saale) 06120, Germany; nozilar@gmail.com
[2] International Water Management Institute—East Africa and Nile Basin Office, IWMI c/o ILRI, PO Box 5689, Addis Ababa 1000, Ethiopia; d.vanrooijen@cgiar.org
[3] Chair of Landscape and Environmental Economics, Technical University of Berlin, Straße des 17, Juni 145, Berlin 10623, Germany; isoliev@daad-alumni.de
* Correspondence: wegerich@iamo.de; Tel.: +49-345-2928138; Fax: +49-345-2928199

Academic Editor: Marko Keskinen
Received: 13 May 2015; Accepted: 19 August 2015; Published: 27 August 2015

Abstract: The importance of water security has gained prominence on the international water agenda, but the focus seems to be directed towards water demand. An essential element of water security is the functioning of public organizations responsible for water supply through direct and indirect security approaches. Despite this, there has been a tendency to overlook the water security strategies of these organizations as well as constraints on their operation. This paper discusses the critical role of water supply in achieving sustainable water security and presents two case studies from Central Asia on the management of water supply for irrigated agriculture. The analysis concludes that existing water supply bureaucracies need to be revitalized to effectively address key challenges in water security.

Keywords: water security; supply water security; irrigation bureaucracy; polycentric water management; transboundary; Central Asia

1. Introduction

Since the 1990s, water infrastructure security and security of water supply became important topics with regards to conflict [1] and terrorism [2,3] specifically related to water. Today, water security is even more dominant on the agenda of the international water community and three international organizations (the Global Water Partnership, the United Nations University, and the International Water Management Institute) adopted the term as a guiding framework. Current water security definitions refer to key demands or objectives of users and the ecosystem in a changing environment [4,5]. In addition to this global focus on water security, the water, energy and food nexus builds around water security objectives [6]. With the emphasis put on these objectives, more traditional approaches to water supply security, such as direct and indirect water security measures, are omitted.

An important factor of indirect water security is infrastructure. Infrastructure development for irrigated agriculture had its peak in the late 1970s (measured by World Bank lending) and was from then on a decline [7]. Today, mainly because of population pressure, new water infrastructure development is again on the agenda [8]. New large scale water infrastructure could also be an important aspect of polycentric water management within basins [9,10]. Some of the past investments in large scale water infrastructure are based on the fragmentation of former colonies and new national water security approaches, such as the construction of link-canals in the Indus Basin in Pakistan [11]. This paper contributes to the literature on indirect water security approaches in a recently fragmented basin, the Syr Darya in Central Asia. International literature on water security in the Syr Darya Basin often focuses on large transboundary infrastructure such as the Toktogul and Kayrakum reservoirs in

Kyrgyzstan and Tajikistan, respectively, as well as the planned new Kambarata 1 and 2 reservoirs in Kyrgyzstan. The prominent nature of the water-energy nexus in large water infrastructures, such as in the Syr Darya Basin, has also brought a focus on related energy security [12,13]. Hence, in the Syr Darya Basin, water and energy security focus mainly on the main river as well as its larger reservoirs. This focus ignores important aspects of historical design. The Soviet Union designed and planned water management at basin level as well as Smaller Transboundary Tributaries (STTs) and smaller infrastructure such as main canals, reservoirs or pump station schemes [14–17].

Direct water security in large scale irrigation systems has been the responsibility of irrigation bureaucracies in the past [18,19]. However, with the exception of some early experiences, Irrigation Management Transfer (IMT) became a national strategy in most developing countries in the 1980s and 1990s [20]. IMT shifts the responsibility of direct water security from the government to the users, organized in newly created Water User Associations (WUAs). While IMT and WUAs have been in the past widely promoted [21,22], more recently there have been doubts [23,24]. With the focus on IMT the lower level bureaucracy is "handed over" [25,26] and the higher level bureaucracy focuses on other functions or focuses only on the higher level like basin management [27]. Here, a case study is presented on partial IMT in one province of Uzbekistan. When focusing on water security for irrigated agriculture within Uzbekistan, so far the emphasis has been on the introduction of winter wheat (as policy to increase food security) and therefore the reduction of irrigated area under cotton [28,29] as well as creating WUAs [30,31]. The water supply organizations, the irrigation departments, have received little international attention, although they were incorporated in some donor projects.

This paper discusses both indirect and direct water supply security measures in irrigated agriculture by drawing from evidence from the Syr Darya basin and Ferghana Province, Uzbekistan. The focus on water supply, rather than on water demand security, is meant to draw attention to the way in which water management, with particular focus on irrigated agriculture, was organized. This focus on past water supply security approaches attempts to challenge the current focus of the international research community on basins and large infrastructure [12,13]. This paper also points out weaknesses in the current promotion of IMT especially at the main canal level–which shifts water supply security from the government to the water users for agricultural water uses [32,33].

The presented case study is structured into two sections. The data for the first section is based on a literature review and interviews with a key informer of the Syr Darya basin water organization (BWO) in 2014. The data presented in the second section is based on archival research of annual reports of the Ferghana Province Irrigation Department in Uzbekistan. The annual reports studied cover a period from 1978 up to 2010. Key informers of the Ferghana and Andijan Province Irrigation Departments were interviewed regarding verification of reported trends.

The paper continues with a short framework section on water security. The following case study is structured into two sections. The first section focuses on water supply security within the Syr Darya and the associated challenges faced by past and current irrigation water management strategies at the irrigation district level. The second section focuses on water security approaches within Ferghana Province and highlights changing water demands as well as the water security approaches taken so far. Within the section, large emphasis is put on the irrigation departments which after Uzbekistan's independence were not incorporated in achieving water security. Each case study is followed by a short discussion. A broader discussion follows, highlighting the possibly national as well as international reasons for not focusing on water supply organizations, which appear to have become the weakest link in water security. The conclusion stresses the need to look at poly-centric water management and a refocus on water supply organizations.

2. Water Security

As Allouche *el al.* [34] noticed "historically security has been concerned with safety and therefore can be understood as the condition of being protected from, or not exposed to, danger". Water security by the turn of the century focused on these traditional aspects. The security of larger water supply infrastructure was voiced in the debate on water wars [1], terrorism [2,3] as well as cyber-attacks [35]. While these perceived insecurities have been dismissed, they have also triggered calls for heightened security and additional systems of resilience [35,36].

More recently, the term water security gained prominence in the international literature from a different perspective. UN-water [4] defines water security as "The capacity of a population to safeguard sustainable access to adequate quantities of and acceptable quality of water for sustaining livelihoods, human well-being, and socio-economic development for ensuring protection against water-borne pollution and water-related disasters and for preserving ecosystems in a climate of peace and political stability". The definition mainly focuses on the demand side and objectives of water security. While this broad definition of water security focusses on access and is human centered ("capacity of a population"), it critically lacks reference to the supply-side of water security. Water supply is vaguely addressed and seems to extend the responsibility of water security to the wider public by making reference to 'a population'. As answer to the current challenge of water security, UN-water [4] calls for "tailored policy responses", human "capacity development" and "improved water governance". Water service providers, their challenges and strategies how to meet water demands are not directly addressed. The focus on human "capacity building" seems to neglect the human ingenuity in developing countries to cope with water insecurity. As Allouche *et al.* [34] highlights, "Missing [...] is the issue of security sought by households in the South, many of whom exist within the vast informal economy, through which they survive and cope with external circumstances".

Grey and Sadoff [5] define water security as "the availability of an acceptable quantity and quality of water for health, livelihoods, ecosystems and production, coupled with an acceptable level of water-related risks to people, environments and economies. Lautze and Manthrithilake [37] highlight that Grey and Sadoff's "broader treatment of risks strongly suggests inclusion of issues related to water for national security or independence". Sadoff *et al.* [38], when defining pathways to water security, put the emphasis on institutions, information and infrastructure. Sadoff *et al.* [38] see institutions as "formal laws, policies, regulations, and administrative organizations as well as informal networks and coalitions". According to them, institutions incorporate planning, financing, construction, operating, supplying, regulating, monitoring, enforcing and insuring. Hence, the main focus is on the public sector, and central are water supply organizations. Nevertheless, they [38] highlight the need for "a 'poly-centric' and multi-level governance system that has been described as an 'institutional tripod' involving water users, states and markets". The "institutional tripod" can be criticized from different perspectives such as diversity and power inequities of users within sectors [39–43] and competing sectors [44–46], market failure and the responsibility of markets for the water crisis [47,48] as well as states institutionalizing inequities through water rights reforms [42,49–51].

In the global debate on the water, energy and food nexus, although reference is made to water, energy and food security, the emphasis for all three is on "access" [6,47]. Different authors have highlighted that the water, energy and food nexus is under-conceptualized and that security in one is contradicting security of the other parts of the nexus [47,52,53]. Hoff [6] highlights "the emphasis on access in these definitions also implies that security is not so much about average (e.g., annual) availability of resources, but has to encompass variability and extreme situations such as droughts or price shocks, and the resilience of the poor". Hence, key would be to include in the debate the supply side of water security. Instead, Hoff [6] argues that "It is increasingly recognized that conventional supply side management is coming to an end in many cases". Nevertheless, he [6] calls for strengthening existing supply side institutions for building "new links across sectors and deal with the additional uncertainty, complexity and inertia when integrating a range of sectors and stakeholders". The assumption appears to be, that linking an undefined range of sectors and stakeholders together will by itself provide better "access". Overall, an analysis of existing water supply organizations, and their strategies to meet demands or encounter risks, is crucially missing.

Traditionally, securing water supply focused on planning and construction of large infrastructures to be able to capture and store water resources as well as satisfying urban and agricultural needs [54]. Infrastructure development was not only seen to increase indirect national water security within transboundary basins [11] but also to enable polycentric water management within basins [9,10]. Recently, due to population pressure, but also due to seasonal variability of water, a rising deficit of existing water infrastructure has been identified [8,55–57].

Looking at water supply security in irrigated agriculture, the aspect of service provision towards the users came to the forefront in the 1980s. There was a realization that the gap in maintenance of irrigation infrastructure [58] led to a deterioration of water supply services. In addition there was recognition of the failure of the irrigation bureaucracy for ensuring equity of water distribution between water users [59,60]. Both insights could be attributed to issues regarding the financial security of water supply services. However, colonial irrigation systems focused on water supply as well as demands. Water control was achieved through different components focusing on water infrastructure, the organization providing the service and water demand [61,62]. Looking at past colonial large scale irrigation systems Ertsen [18,19] highlights that water supply (infrastructure and organization) as well as demand was planned for in the British, French and Dutch irrigation systems. Because of rising political pressure, market development and also changes of land ownership and farm sizes the water control side in irrigated agriculture disintegrated [63,64]. The rising water demand within the existing irrigated area was not met with an expansion of water supply infrastructure and providing more water resources or a strengthening of the irrigation bureaucracy controlling the distribution of limited water resources. The failure to provide equitable distribution was attributed to the continuation of established control practices [65] as well as the overall low salaries of the irrigation bureaucracy and therefore the rise of corruption [66].

Similarly, in the 1990s with the fall of the iron curtain and with a focus on transitional economies, water service provision for urban areas rose high on the development agenda. Again, the focus was on maintenance of infrastructure as well as monitoring of water losses [67,68]. The failure of strengthening the supply side could be classified as financial insecurity triggering the decline in quality of water supply services.

Rising demands but also a failure to secure and increase water supply triggered the development of more resilient water supply systems, *i.e.*, cities established inter-linkages between different sources and water storage systems to cope with temporary supply shortages [69]. Similarly, for supporting irrigated agriculture, countries or even smaller administrative units (like provinces) established resilient systems to cope with international or national transboundary water supply insecurities [11,70–73]. Common in all these formal systems of resilience is a diversified access to water resources as well as less reliance on one main supply infrastructure.

Looking at the debates within the water sector, risks to water security have been identified as transboundary and inter and intra sectorial competition, water pollution, unsustainable operation and maintenance as well as reliance on a single source or supply network. Therefore, water supply security could be defined as a resilient system capable of coping with shocks, abuses and threats through direct security measures (surveillance and guards) and indirect or more passive measures through increasing maintenance and additional or alternative water supply sources, duplication of or less reliance on critical infrastructure to better cope with temporary shortages in water source availability as well as water rights or allocations to cope with competitions.

3. Water Security Approaches in the Syr Darya Basin

3.1. Geographic Background to the Syr Darya Basin

The Syr Darya rises in the Tien Shan Mountains of Kyrgyzstan and terminates in the Aral Sea in Kazakhstan. It is the longest river in Central Asia, at 3019 km, with a catchment area of 219,000 km^2. Up to the confluence with the Karadarya (also from Kyrgyzstan) the Syr Darya is called the Naryn. The Syr

Darya is shared between four riparian states, Kazakhstan, Kyrgyzstan, Tajikistan and Uzbekistan. On its way to the Aral Sea, the Naryn crosses international boundaries between Kyrgyzstan and Uzbekistan when entering the Ferghana valley and within the valley between Uzbekistan and Tajikistan as the Syr Darya. When leaving the Farghana valley, the Syr Darya enters first Uzbekistan and then crosses into Kazakhstan (Figure 1).

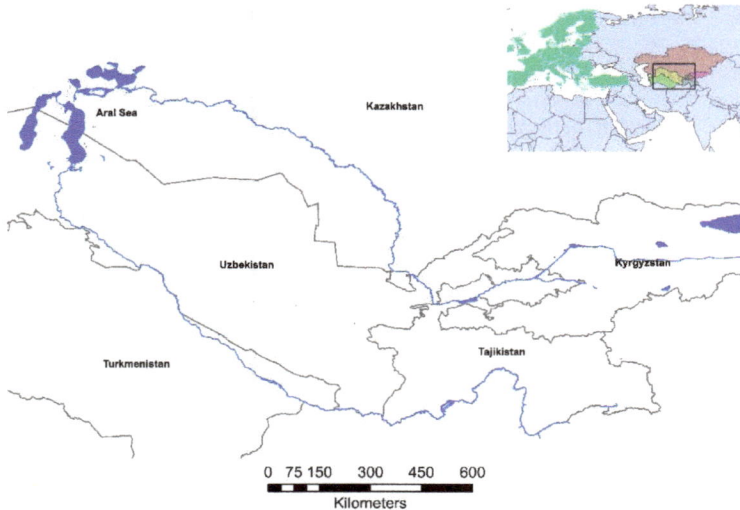

Figure 1. The Syr Darya Basin.

Due to large scale irrigation expansion, facilitated through the construction of multiple use reservoirs (Toktogul and Kayrakum), the Syr Darya basin closed in the 1980s [73,74]. The water allocation principles developed under the Schemes of Complex Use and Protection of Water Resources in the 1970s and early 1980s became the guiding principles of water allocation between the riparian states [17]. Later in 1987 the Syr Darya Basin Water Organization (BWO) was established [75]. Directly after independence, in 1992, the five Central Asian states came to an agreement to continue with these principles. However, while during the time of the Soviet Union the multiple use reservoirs operated to facilitate irrigated agriculture, after independence the operation of the reservoirs shifted mainly to winter releases for energy production to cover upstream riparian needs. The reason for the shift of operation is based on the collapse of existing compensation mechanisms. During the Soviet Union era, downstream states compensated for excess electricity produced at the reservoirs during the summer, by supplying fossil fuels and electricity during the winter.

3.2. The Common Approach to Look at Water Insecurity within the Syr Darya Basin

After independence the international emphasis on water security within the Syr Darya Basin focused on the conflicting interests of upstream hydropower production during the winter and downstream water needs for the agricultural sector during the summer. Therefore, the main emphasis was on the operation of the Toktogul reservoir and the brokering of an agreement on water and energy use in 1998. The agreement was amended to include the Kayrakum reservoir in Tajikistan in 1999. According to the agreement, purchases of energy and therefore water allocations from Toktogul are determined annually [76–78]. The implementation of the agreement has been seen as problematic in reference to water delivery to Kazakhstan [79,80] and as generally failed because of the late signing of annual bilateral agreements [78]. Overall, the primary focus of the international attempts to foster water security focused on the infrastructure controlling the main stem of the Syr Darya Basin, the Naryn, only [81]. However, the Naryn supplies about 40 percent (14.5 km^3) of the average annual flow of the Syr Darya River (37.2 km^3) only [81]. The focus on the Naryn River and the Toktogul reservoir assumed that basin management was the overarching principle. In addition, the agreement focused on national levels and did not incorporate Tajikistan as downstream water user [15].

3.3. Water Insecurity at the Meso-Level: Irrigation Districts and Within

Other research highlights that within the Syr Darya Basin, water management was organized according to "water-use regions" or "irrigation districts", which in some cases even crossed republican boundaries [82–84]. Within the Syr Darya Basin there were six irrigation districts during the Soviet era, these were: Upper Naryn, Ferghana Valley, Chirchik-Akhangaran-Keles (Chakir), Midstream, Arys-Turkestan (Artur) and Downstream (Figure 2). Three of these irrigation districts were transboundary: the Ferghana Valley irrigation district incorporated irrigated areas within the valley from Kyrgyzstan, Tajikistan and Uzbekistan; Chakir incorporated irrigated areas of Kazakhstan, Kyrgyzstan and Uzbekistan; and Mid-stream incorporated irrigated areas from Kazakhstan, Tajikistan and Uzbekistan.

Figure 2. Irrigation districts in the Syr Darya Basin.

Irrigation districts can be categorized into different groups with focus on the utilization of water sources, having access to alternative resources, and capturing winter flow (Table 1). The implication of former management according to irrigation districts is that the past system focused on poly-centric [9] and not basin-level water management and therefore crafted poly-centric water security approaches (storage and reliance on multiple sources). Therefore, the collapse of the Soviet Union and the emergence of independent states, as well as the shift of operation of the larger Toktogul reservoir in Kyrgyzstan did not create water insecurity for the whole basin, but created water insecurity for individual irrigation districts or parts of them.

The implication of looking at irrigation districts rather than the whole basin is that local water insecurity becomes more visible. Hence, after independence, irrigation districts were most water insecure if they were either dependent on one transboundary source only or if they were dependent on one transboundary infrastructure for capturing winter flows. Looking at the largest irrigator within the irrigation district only, the most potentially water insecure irrigation district would be Mid-stream. Here, the largest benefiter of the Mid-stream Kayrakum Reservoir is Uzbekistan; however, the reservoir is controlled by Tajikistan. In the case of the Ferghana Valley, although the main benefiter is Uzbekistan and the main reservoir is controlled by Kyrgyzstan, having access to alternative sources (Karadarya and small tributaries) as well as smaller reservoirs (Andijan as well as on small tributaries) could be interpreted as being in a less water insecure situation.

Table 1. Features of irrigation districts (Source: adapted from [84,85].

Irrigation District	Source (km³)	Storage (km³)	Republic	Irrigated Land (1000 ha)	Total Water Use (km³/year)
Upper Naryn	Naryn (14.5)	–	Kyrgyz SSR	130.3	–
Ferghana Valley	Naryn (14.5)	Toktogul: Total Storage (TS)-19.4 Active Storage (AS)-14.0	Uzbek SSR	409.8	4.69
			Tajik SSR	97.7	1.36
			Kyrgyz SSR	22.5	0.74
	Karadarya (3.9); Small Transboundary Tributaries (STT) (total 7.8)	Andijan: TS-1.9; AS-1.8; Some smaller transboundary reservoirs	Uzbek SSR	471.7	5.75
			Kyrgyz SSR	293.7	3.21
			Tajik SSR	30.5	0.23
Chakir	Chirchik (7.8); Akhangaran (0.7); Keles (0.3)	Charvak: TS-2.0; AS-1.6	Uzbek SSR	347.2	3.43
			Kyrgyz SSR	9.5	0.04
			Kazakh SSR	89	0.89
Mid-stream	Main stem	Kayrakum: TS-4.0; AS-2.6 Farkhad TS-0.15	Uzbek SSR	629.7	7.19
			Tajik SSR	87.6	1.03
			Kazakh SSR	117	1.34
	Small Tributaries (0.3)	–	Uzbek SSR	33.6	0.3
			Tajik SSR	30.5	0.23
Artur	Arys (1.2)	–	Kazakh SSR	200	–
Downstream	Main stem	Chardara:TS-5.7; AS-4.7	Kazakh SSR	374	–

However, Table 1 also reveals that some irrigation districts are transboundary. Hence, within irrigation districts there is a second layer of potential water insecurity for transboundary parties. Within the Ferghana Valley irrigation district some areas experienced more potential water insecurity than others. These potential insecurities are not related to the shift of Toktogul reservoir operation but more due to smaller transboundary infrastructure. Examples of these smaller infrastructures are the Big Namangan Canal, which is mainly supplying farmers in Uzbekistan. The diversion structure for the canal is located in Kyrgyzstan. In addition, within the Ferghana Valley and within smaller tributary basins downstream Uzbek areas are potentially water insecure, such as in the Isfara tributary [15]. Irrigated areas within Tajikistan are potentially water insecure if they depend on transboundary infrastructure, for example Tajikistan is at the tail-end of the Big Ferghana and North Ferghana Canals [15]. Although Kyrgyzstan is mainly upstream from the Ferghana valley, areas in Kyrgyzstan receive water through pump stations located in Uzbek territory or diversion from transboundary main canals, such as the South Ferghana Canal [14].

Within the mid-stream irrigation district, the Dustlik canal is transboundary and shared between Uzbekistan at the head-end and Kazakhstan at the tail-end [79,80]. Within the Dustlik canal Uzbekistan irrigates 98 thousand ha and Kazakhstan 125 thousand ha. Wegerich [86], comparing data from the years 1990 and 1991 with the years 2004 and 2005, shows that after independence Kazakhstan received less water and later during the cultivation period. Hence, for Kazakhstan along the Dustlik canal not the operation of Kayrakum reservoir appears to be the main factor, but the withdrawals along the Dustlik canal within Uzbekistan.

The Chakir irrigation district is independent of the operation of the Toktogul reservoir and is mainly based on water of tributaries, with only small diversion through lift from the Syr Darya to Uzbekistan (Dalverzin canal's command area) [87]. Within the irrigation district, Uzbekistan transfers water through three canals (Zakh, Khanym, and Big Keles) from the Chirchik river to the Keles massif irrigation system in Kazakhstan irrigating about 66 thousand ha [88], the same canals return water to Uzbekistan to irrigate about 17 thousand ha in Tashkent and Kibray districts [89]. Dukhovny *et al.* [88] report variation of water supplied to Kazakhstan. In the period 1995–2003 water supply varied between 347 and 595 km³/year. It is not evident whether the mentioned 89 thousand ha [84] in Kazakhstan are all within the Keles massif and that therefore a major reduction of water supply occurred after independence. The major reduction could either be caused by a reduced total flow, or again as in the case of the Dustlik canal by withdrawals along the three main canals within Uzbekistan.

Overall, the irrigation district which was most water insecure after independence was dependent on one transboundary source (the main stem of the Syr Darya) and one transboundary infrastructure (reservoir). Similarly, areas within irrigation districts which were most water insecure to water supply shortages were dependent on transboundary infrastructure (main canals, pump stations) and had access to one transboundary water source (small transboundary tributary (STT) or main stem of the Syr Darya). The irrigation district which is the most water insecure is Mid-stream, and the areas being most water insecure within irrigation districts are some Uzbek and Tajik areas within the Ferghana Valley, Kazakh areas within the Mid-stream, and Kazakh as well as minor Uzbek areas within the Chakir irrigation districts.

4. Past and Current Water Security Approaches: Capturing Winter Flow and Alternative Sources

4.1. Past Water Security Approaches–Example the Ferghana Valley Irrigation District

Looking particularly at the Ferghana Valley, in the past, different strategies have been used to facilitate adaptation to seasonal fluctuations. Early on, Soviet Engineers started linking the main tributaries with smaller tributaries of the Syr Darya through main canals [90–92]. Later on, water security was increased through the construction of small reservoirs for capturing winter flows of tributaries [16,17]. Finally, with the increase of irrigated areas in small tributaries within upstream Kyrgyzstan, pump stations were constructed to lift water from main canals towards small tributaries [15,93]. Hence, within the Ferghana Valley, a meshed system was constructed which allowed switching from the main water source to an alternative water source and winter flow on tributaries as well as main stem was captured. Therefore, water security within the Ferghana Valley was achieved through tapping from alternative water sources through additional infrastructure (duplication) and storing winter flow.

4.2. Current Water Security Approaches

After independence from the Soviet Union, whenever possible, all states tried and are still trying to find solutions for becoming independent from each other. These solutions are based on capturing current unused winter flow or through duplication of access infrastructure for the exploration of alternative water sources. Dukhovny [94] highlights that "Uzbekistan is striving for almost full satisfaction of its demand for additional water through releases from the Andijan reservoir and, partially, through construction of in-stream reservoirs".

Within the Ferghana Valley irrigation district, Uzbekistan constructed the Rezaksay reservoir (0.3 km^3) between the North Ferghana and the Big Namangan Canals and the Markaziy (Central) reservoir (0.35 km^3) along the Big Andijan Canal to store unused winter flow of Toktogul. Different riparian states are also trying to attempt water security through duplication of access infrastructure and shifting to alternative sources. Uzbekistan plans additional pump stations from the North Ferghana to the Big Namangan Canal to compensate for the inoperative diversion structure for the Big Namangan Canal in Kyrgyzstan. In addition, Uzbekistan plans new pump stations for utilizing ground water in the Ferghana province, here the aim is to compensate for less water received along smaller tributaries (key informant from BVO Syr Darya, January 2014). Still within the Ferghana Valley irrigation district, Tajikistan first started to negotiate water allocation issues regarding Big Ferghana Canal with Uzbekistan through issue linkages with the Kayrakum reservoir. More recently, Tajikistan started to divert water from the Isfara tributary directly into the tail-end part of the Big Ferghana Canal which is within its territory. Hence, although water allocations between Tajikistan and Uzbekistan were negotiated, Tajikistan was still not able to receive its water share on the Big Ferghana Canal. As consequence of the recent Tajik strategy, Uzbekistan's irrigated areas in the downstream Isfara tributary became more water insecure [15].

In the late 1990s, within the Midstream irrigation district, Uzbekistan anticipated to use the flood spills towards Arnasai Lake for irrigation [82,95]. Hence, Uzbekistan anticipated to making use of an existing "reservoir" to facilitate alternative water supply or duplication to existing canal infrastructure. More recently, Uzbekistan started to construct the Sardova reservoir (1.0 km^3) along the South Golodnyesteppe canal (key informant from BVO Syr Darya, January 2014). The reservoir will enable Uzbekistan to secure winter flow below the Kayrakum reservoir. Within the Downstream irrigation district, Kazakhstan built the Koksarai reservoir (3.0 km^3) below Chardara, and has planned two additional reservoirs for flood protection and to store unused winter flow (Figure 3). Regarding alternative water supply and duplication, Kazakhstan made use of its existing Chardara reservoir and constructed pump-stations towards the Dustlik canal to reduce its dependence on transboundary infrastructure.

So far, no alternative sources or duplication seem to be anticipated for the Kazakh part of the Chakir irrigation district. However, Dukhovny *et al.* [88] mentions that Kazakhstan is planning the expansion of its irrigated area to 98 thousand ha, with a total withdrawal of 1140 Mm^3/year. To secure the additional water needs Kazakhstan might consider a similar approach as taken in the Mid-stream and Downstream irrigation districts and might attempt to reduce dependence on transboundary infrastructure. The potential source, could be the Chirchik directly, since, 0.75 km^3/year of its flow generation is within Kazakhstan [58].

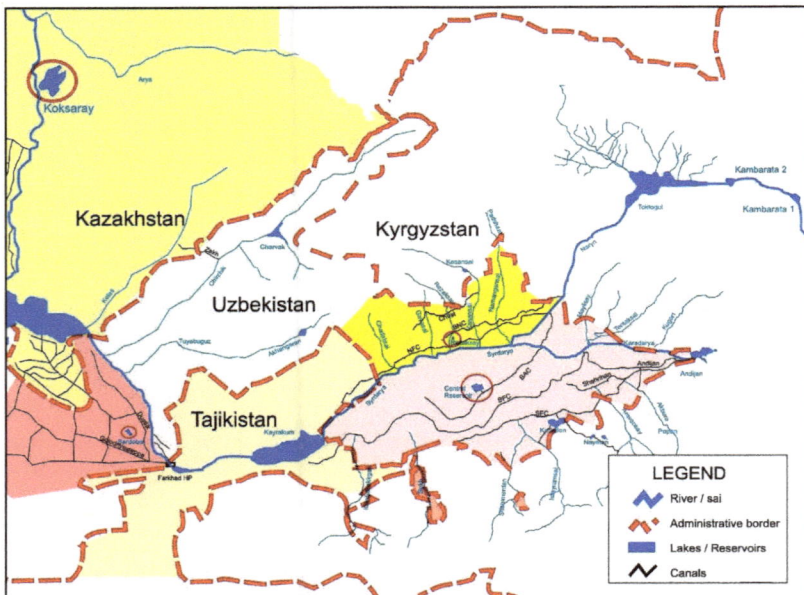

Figure 3. New reservoirs in the Syr Darya. (Source: Based on information compiled through GIS maps, as well as key informer BVO Syr Darya).

4.3. Short Discussion: Downstream Countries Increasing their Indirect Water Security

The two downstream riparian states, Uzbekistan and Kazakhstan, which were negatively affected by the operation of Toktogul within irrigation districts which depended to a larger extent on the main stem (Ferghana Valley, Midstream and Downstream) developed similar strategies regarding water security. Both are creating water storage within their territories for not having to depend on the Toktogul or Kayrakum reservoirs operation. With the creation of additional storage, the basin's downstream riparian states avoid having to negotiate summer operations and therefore paying for electricity from Toktogul. Within the Downstream irrigation district the reservoir has multiple functions; storing winter flows for flood mitigation and securing irrigation needs during the summer. However, water to the Koksarai reservoir has to be pumped. Overall, the creation of capacity to store winter flow not only reduces the dependence on other riparian states, but also decreased the bargaining power of upstream states. Given that off-season flows were already allocated within the closed Syr Darya basin, the creation of national storage, particularly in the midstream country Uzbekistan, might off-set the existing but not anymore operationalized riparian states allocations to downstream states. In addition, the additional storage within the midstream and downstream countries (Uzbekistan and Kazakhstan) might put into question the water delivery to the Northern Aral Sea in Kazakhstan.

The creation of access to alternative water resources such as groundwater resources within the Uzbek part or the diversion of the transboundary tributary (Isfara) within the Tajik part of the Ferghana Valley irrigation district highlights the dividing up of the transboundary irrigation district according to national boundaries. Similarly, the creation of a pump station along the Chardara reservoir for supplying the Kazakh part of the Dustlik canal highlights the merging of parts of the former Midstream and Downstream irrigation districts along national boundaries. Hence, the identified water security solutions rely on national water security solutions, rather than on transboundary solutions. The implication is that although pumping costs such as on the Chardara might be economically higher, and the diversion of the STT might be more unstable due to seasonality, these solutions might provide more stability and reliability compared to the past transboundary water supply solutions.

Overall, these new water security solutions for water shortages are building on past security approaches practiced within Soviet Central Asia. However, these new solutions focus primary on national water security. While these are technical solutions to water shortages, it is questionable whether the water bureaucracy can safeguard the availability of water resources for their users.

5. Surveillance and Guards–Irrigation Bureaucracy (Example Ferghana Province)

While the previous section focused on poly-centers within the basin and on more passive security measures, here the focus turns to the meso-level, the Ferghana Province, and direct security measures, such as water metering devices, surveillance and guards. As mentioned in the introduction, here a historic approach is taken by looking at long term trends [96] (1978 to 2010) of the water supply control side of the irrigation department of Ferghana Province. Therefore it is first necessary to highlight the changes on the demand side during the Soviet period and after independence.

5.1. Geographic Background to Ferghana Province

Ferghana Province is located within the Uzbek part of the Ferghana Valley. The province occupies 6800 km^2 and consists of fifteen districts, four major cities and has a total population of about three million. The province borders Kyrgyzstan to the south-east, Tajikistan to its western side and two Uzbek provinces Andijan and Namangan to the east and north respectively. The province has access to different water sources, the Syr Darya, the Big Ferghana Canal (BFC)-diverting water from the Naryn (controlled by Toktogul reservoir in upstream Kyrgyzstan), Karadarya (controlled by Andijan Reservoir operated by Uzbekistan), the South Ferghana Canal (SFC) (taking water directly from the Andijan Reservoir) and the Big Andijan Canal (BAC) (also diverting water from the Naryn), as well as five Smaller Transboundary Tributaries (STTs): Kuvasai, Isfayramsai, Shakhirmadansai, Sokh and Isfara (from east to west), which all, with the exception of the Kuvasai, intersect with either the SFC or BFC [14]. On all main canals and small tributaries Ferghana Province is at the tail-end. With independence, the water situation for Ferghana Province was aggravated on the main canals BFC and BAC as well as some STTs (Figure 4).

Before independence, within Ferghana Province the irrigated area increased from 285,000 ha in 1969 up to 368,300 ha in 1988. After independence in 1991, the irrigated area first declined, but stabilized at about 361,000 ha from 2006 onwards. According to Bucknall *et al.* [97] about one third of the irrigated area in the province, 115,000 ha, is supplied via pumps and pump stations (lift). Recently Wegerich *et al.* [93] showed that 151,000 ha are supplied via pump stations and of these 69,000 ha have a lift of over 50 meters. About one third of these pump stations can be classified as transboundary pump stations, which were constructed to mitigate upstream expansion in transboundary tributaries [93].

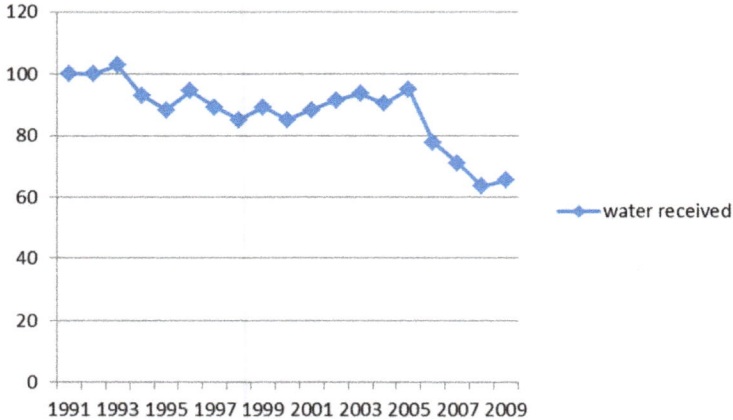

Figure 4. Total water received and utilized in Ferghana Province from 1991 to 2009 (deviation from 1991), (Source: compiled from data of the irrigation department of Ferghana Province).

5.2. Demand Side Changes-Farming Units and Crops

During the Soviet period, agricultural production was organized in crop specialized state owned large scale collective farms, varying in sizes between 2000 and 8000 ha. Within Ferghana Province were a total of 120 collective farms in 1975, which changed to 162 by 1991. During the 1990s, collective farms were transformed into semi-cooperative farms, with an average size of 2000 to 3000 ha. Within Ferghana Province a total of 164 semi-cooperative farms were registered. Although already in 1992 the law on peasant farms [98] (peasant/dehkan farms) was issued, privatization did not kick off until 2001, based on a new law concerning farms in 1998 [99]. Within the province, the number of private farms rose from about 3000 in 2000 to below 26,000 in 2007. In 2009, a Presidential Decree [100] on farm optimization was issued, which led to decrease in the number of private farms. Already prior to the Presidential Decree [100], the number of private farms dropped again in Ferghana Province. The total number was below 12,000 in 2010 (Figure 5).

Usually, when reference is made to the Uzbek SSR and its agricultural production during the Soviet period, cotton monoculture and alfalfa are mentioned. It is also argued that because of state planning the Uzbek SSR increased its irrigated area for further expanding its cotton monoculture [91,101]. The data of Ferghana Province shows that from 1978 the increases of irrigated area did not lead to an increase of the area under cotton cultivation, the area under cotton even decreased. According to Anderson [102] after the cotton scandal in the 1980s and to soften the social conditions the Uzbek SSR Leader (Nishonov) asked permission to reduce the cotton quota for Uzbekistan. Consequently, the area under cotton decreased further in the Ferghana Province from 196,000 ha (56 percent) in 1987 to 164,000 (46 percent) in 1990.

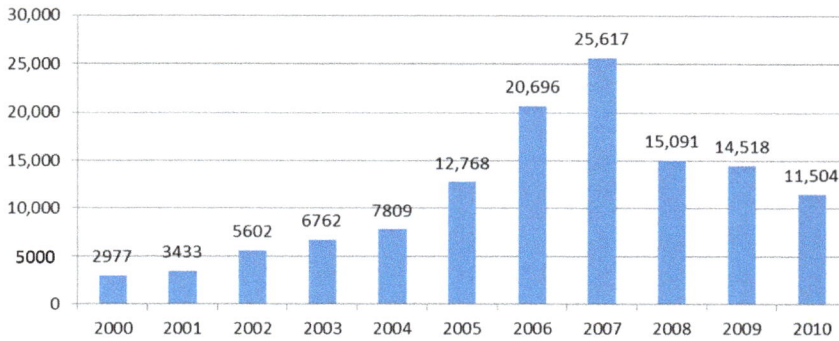

Figure 5. Dynamics of private farms in Ferghana Province 2000–2010 (total numbers).

5.3. Current Water Security Approaches Focusing on the Demand Side

After independence, Uzbekistan shifted to a policy of food self-sufficiency and therefore expanded the area under wheat cultivation. Although usually emphasized as food security policy, one could argue that the food security policy was in fact a water security policy. Within Ferghana Province winter wheat was grown from 1995 onwards. In the period from 1995 to 2010, the area allocated to cotton decreased from 36% to about 30% and the area allocated to winter wheat increased from 0% in 1994 to 31% in 2010 (Figure 6). Within Ferghana Province, winter wheat has mainly replaced alfalfa. Although winter wheat would imply less water demand during the summer season, farmers utilize the period between harvest and sowing to grow a second crop [103,104]. Recent studies have shown that the ratio of second crops after winter wheat is between 60% and 80% within Ferghana province [105]. Given the large ratio of second crops the potential for water savings is reduced (Figure 7).

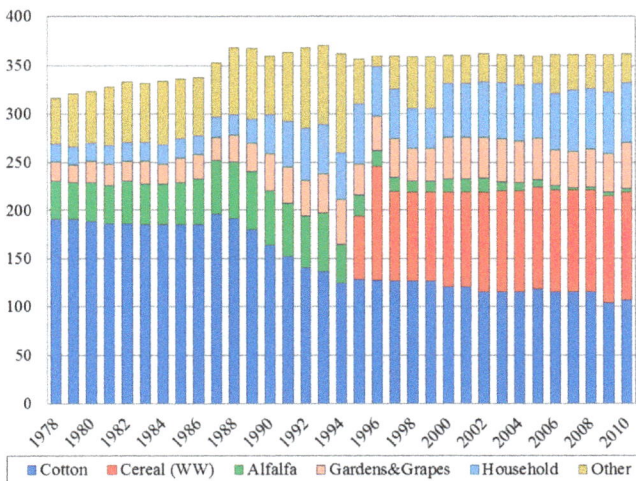

Figure 6. Changes of cropping patters 1978 to 2010 (1,000 ha).

Being concerned about the fragmentation of former collective farms and based on international recommendations, the Cabinet of Ministers of the Republic of Uzbekistan approved the Procedure for organizing Water User Associations (WUAs) in 2002 [106]. Within Ferghana Province, the Integrated

Water Resources Management project, funded by the Swiss Agency for Development and Cooperation (SDC), established WUAs along the SFC and Shakhirmadansai STT. By 2011, Ferghana Province had 119 WUAs. WUAs have been mainly established on the territory of the former semi-cooperative farms (with the exception in the donor funded project). Because of the difference in numbers, it is not evident whether the process of creating WUAs was completed within the province. WUAs are newly created organizations and therefore it is questionable whether they can plan and allocate water according to requests of farmers and available water resources supplied by the irrigation department.

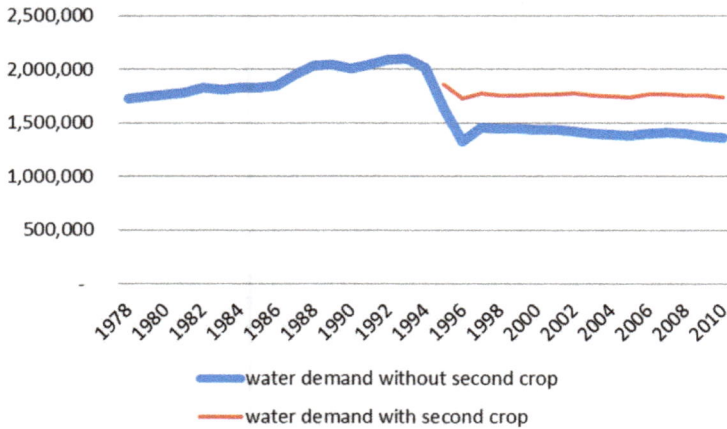

Figure 7. Irrigation norms and trends of water demand during summer season 1978 to 2010 (m^3).

5.4. Current Water Security Approaches Focusing on the Supply Side

There were few projects that focused on the water supply organizations (irrigation departments). A SDC funded project, focusing on main canal management, brought governance issues forward and therefore established a "union of canal water users" along the SFC (2002 to 2012), with less emphasis on infrastructure or finance which was demanded by the irrigation department [33]. An additional SDC project focused on main canal automation along the SFC (2005 to 2010). Towards the end of the main canal automation project key problems were raised regarding sustainability of operation and maintenance as well as capacity of irrigation departments' operating staff [107]. A Deutsche Gesellschaft für Internationale Zusammenarbeit (GIZ) project (2009 to 2011) focused on GIS capacity building of water supply organizations and on creating transparency of water flow information [108]. However, also here, long term sustainability was voiced regarding staff issues. Given the early start of these different projects, one has to note that when the first SDC project started in 2002, there were no international publications on the irrigation departments in Central Asia, except some donor reports.

5.5. Supply Side Changes—The Irrigation Bureaucracy

According to Dukhovny and de Schuetter [101] "the beginning of the 1980s saw the first signs that governments were paying attention to the problems of managing the large river basins (Amu Darya and Syr Darya) in Central Asia". During the 1980–1985 period "more than 70 rubles (US$ 45 in 1980, which converts to US$ 137 buying power in 2014) per irrigated hectare were annually allocated to water management organizations. Accordingly fixed assets at the inter-farm level increased by 36%, the number of service staff at the inter-farm level increased by 20%, the number of inter-farm irrigation networks equipped with water-measuring structures increased by 93% and water-distributing structures increased by 94%" [101]. After independence, the situation changed. Thurman [109] highlighted limiting factors of the irrigation departments for controlling water supply to the users, "very low salaries, small operational budgets, and very little equipment". Wegerich [110,111] looking at staffing and logistics of the irrigation department of Khorezm province argued that the past procedure of controlling off-takes from main canals was not anymore possible. Other studies have highlighted that the 1997 merger between the Ministry of Agriculture and the Ministry of Water Resources, led to a downgrading of the Water Ministry as merely dependent department [70,111–115].

5.6. Case Study the Ferghana Province Irrigation Department

At the time of basin closure in the beginning of the 1980s, the Ferghana Province Irrigation Department controlled water supply through hydroposts (measuring infrastructure within main canal and at off-take level) and flexible guarding (motor bikes and staff) (Figure 8). The long-term trend up to 1997 shows that the number of staff of the irrigation department was set on expansion. This is similar to the trend with the number of hydroposts, showing a rapid increase from the mid-1980s to 1990, and a slowing increase up to 1997. Flexible guarding through motor bikes expanded rapidly in 1983 and 1984, and stabilized at a lower level in 1987. However, after independence in 1991, motor bikes were not mentioned anywhere within the annual reports of the irrigation department.

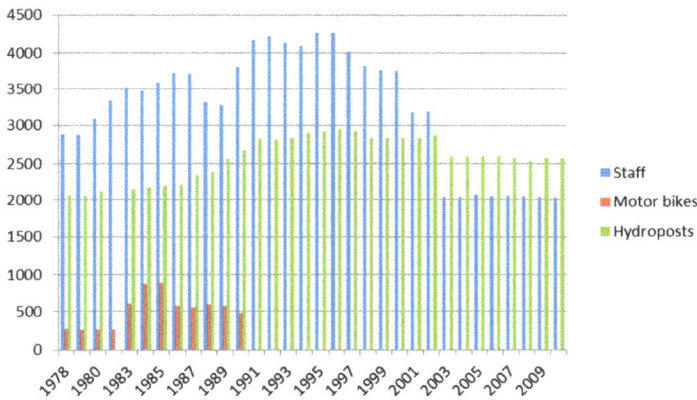

Figure 8. Dynamics of supply side control-surveillance and guards in Ferghana Province (total numbers).

While the most staff increases of the irrigation department can be attributed to the construction of pump stations during the 1980s [93], the rapid fall of staff numbers in 1988 to 89 and 1992 to 93 appear to be due to wider social, nationalistic and economic changes [116–118]. In 1997 the Ministry of Water Resources merged with the Ministry of Agriculture 1997 marks the turning point for trigging a downward trend regarding staffing as well as hydroposts. In 2003, a main canal dispatch center was created in Ferghana Valley, which as consequence had a reduction of the number of staff as well as the reported number of hydroposts under the Ferghana Province Irrigation Department.

As mentioned before, there has been a major increase of operation and maintenance from 1980s onwards [101]. Until 1985, expenditure on operation and maintenance appears to have been nearly stable with rapid increases during the period from 1986 to 1990 (Figure 9). Hence, it appears that the Soviet Union put high emphasis on water supply security and control of water supply. During the economic crisis which followed independence, the operation and maintenance expenditure decreased rapidly, and regained the level of 1986 only by 1996. However, with the merger between the two Ministries operation and maintenance as well as rehabilitation expenditure declined to insignificance. Although from 1996 onwards the Uzbek Gross Domestic Product (GDP) started to increase again [119], this increase has not triggered a reinvestment in the Ferghana Province Irrigation Department.

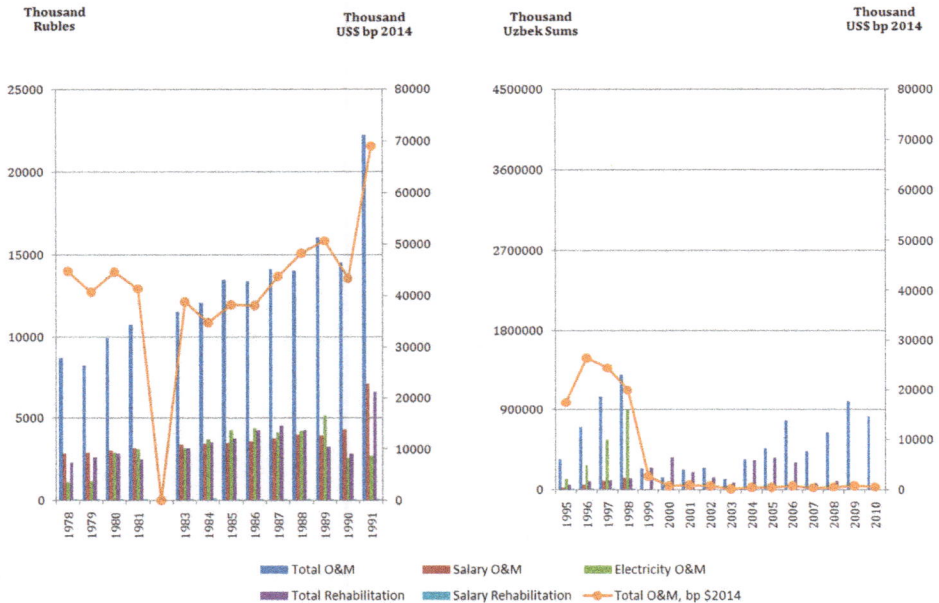

Figure 9. Expenditure for Operation and Maintenance and Rehabilitation (in 1,000 Soviet Union Rubles for the period 1978–1991 and 1,000 Uzbek Sums for the period 1995–2010, the secondary axis is in 1,000 USD buying power as of 2014).

The disappearance of motor bikes and the increasing total number of staff and hydroposts could suggest that there was a shift from flexible to static (staff being posted at the hydroposts directly) water supply control after independence up to 1997. However, the declined salary level combined with an overall rising staff number suggest that full-time employment within the irrigation department might not have guaranteed livelihood security. Kandiyoti mentions that often employees were only formally employed but in fact did not receive salaries [120]. Therefore, it is questionable whether after independence surveillance and guarding continued. Similarly, the decreasing expenditure on maintenance after independence suggests that the number of functioning hydroposts has declined, but that dysfunctional hydroposts are still recorded.

5.7. Short Discussion: Losing Direct Water Security

During the time of the Soviet Union, rising water demands (due to expansion as well as change in cropping patterns) triggered an adaptation process of the irrigation department. During this period there was an increase of water supply control noticeable with the increasing numbers of staff, hydroposts and motorbikes. More funding was allocated to operation and maintenance as well as rehabilitation of water supply infrastructure. After independence water demand continued to increase; however this was not anymore matched with increases in water supply or an increased water supply security. Therefore, there appears to be an apparent mismatch between official figures of water supplied and evidence of water utilized.

An adaptation process for the new situation came only in 1995 with the introduction of winter wheat and therefore an assumed reduction of overall water demand. The adaptation of crops cannot only be interpreted as a response to a decrease of water supply, but possibly also as a political attempt to avoid strengthening the organizational capacity of the irrigation department, through increases of finances and logistics. In this respect, the merger between the two ministries, which followed the introduction of winter wheat, could be interpreted not only in terms of budget savings, but also that the solution for water supply insecurity was seen through controlling agriculture rather than water supply. However, given that about 60% to 80% of winter wheat area is utilized for second crops, this attempt did not reduce the need for water supply security solutions but aggravated the situation further, since the second crop diversity demands more irregular irrigation compared to mono cropping of cotton. The merger triggered further budget and staff cuts within the irrigation department and therefore the water supply security was lost completely. With the implementation of land reforms, the need for water supply security would have increased significantly. However, there is no evidence that water supply security has been strengthened.

Looking at past projects (SDC and GIZ) it is evident that the first focus on governance through "union of canal water users" was misplaced, since water supply security was not possible. Looking at the total operation and maintenance budget of the irrigation department, it is also evident that canal automation would have added an additional burden, which was unlikely to be sustainable. Looking at the salary of irrigation department staff, it is also evident that staff after having gained additional knowledge would look for other job opportunities. Therefore, unfortunately, these projects appear to have been too ambitious and possibly too premature, assuming that the irrigation department would have an existing capacity of supply security. It appears that these projects focused on a piecemeal approach, omitting key questions of capacity, and mainly not considering that irrigation departments are government organizations, and the potential of donor influence on these established bureaucracies could be limited if it is single focused and project based.

Given the deteriorating position of funding and therefore capacity of the irrigation department to control and the negative aspects for its staff, regarding salary, it is evident that the diversity and power inequities of WUAs along main canals might increase. Given the lack of capacity to control, it is likely that the foremost rising inequity will be based on the location along main canals. In addition, Platonov *et al.* [105] have highlighted that within Ferghana Valley the inequity depends also on the off-take infrastructure. The implication is, that irrigated area which rely on more costly infrastructure (such as pump stations), which are operated and maintained by the irrigation department, are less likely to be able to produce profitable second crops. In addition, there is already evidence, that the lack of control on the main canal level has negative effects and has increased power disparities within WUAs. Mukhamedova and Wegerich [45] highlight how water scarcities are inequitably distributed within WUAs and affecting mainly the most vulnerable part of the communities, kitchen gardens in villages.

Although this second section only presented the case of the Uzbek Ferghana Province, it is assumed that Ferghana Province is not only representative regarding its budget limitations for Uzbekistan, but also for other riparian states (Kazakhstan, Kyrgyzstan and Tajikistan). However, Uzbekistan and Tajikistan compared to Kazakhstan and Kyrgyzstan are late implementers of land reforms as well as lifting slowly the restrictions on agricultural production. Therefore, in Uzbekistan, agricultural water demand and the subsequent need for increasing water supply has likely expanded more slowly compared to its neighboring countries.

6. Discussion Linking Irrigation District to Meso-Level Water Security

The first section demonstrated that after independence in 1991, water supply security was high on the agenda with building resilience through additional or alternative water supply sources (from existing national reservoirs, groundwater sources, or small transboundary tributaries) as well as duplication of or less reliance on critical infrastructure (through the construction of smaller reservoirs within the country, pump-stations or diversion canals for small transboundary rivers). The second section showed for the Uzbek Ferghana Province that less attention was paid to surveillance and guards in the irrigation schemes, and that consequently the irrigation department lost its capacity to secure water supply to the water users. Hence, water supply security appears to focus on technical solutions (new infrastructure), while old and deteriorating infrastructure as well as operational sustainability of the water departments are neglected. Possibly, the focus on capturing more water and setting up additional supply lines might postpone the strengthening of existing organizations. Consequently, the chosen approach might secure more water resources, but possibly will not lead to more equitable water distribution within irrigation systems. Furthermore, there appears to be little reflection of the government on past infrastructure strategies and subsequent consequences for the irrigation departments. The past strategy on duplication (switching sources through the construction of pump stations) proved to be very costly and financially unsustainable for the Ferghana Province irrigation department [93]. Although these strategies could imply more water security through avoidance of transboundary dependence, the short and long term financial sustainability is questionable. Although these strategies increase national water security they also increase the energy demand for supporting irrigated agriculture. Therefore these strategies move the water-energy nexus from the basin to the national or provincial level.

The different water security approaches after independence were: (1) introduction of winter wheat in 1995 (adjusting demand); (2) creation of WUAs in 2002 (direct security approach at the local level) and (3) infrastructure projects starting by 2010 (indirect security approach at the meso-level). None of these initiatives directly addressed the main water supply organizations. Hence, there was no direct security approach at the meso-level. It is likely that this was a conscious decision. Possibly for Uzbekistan the merger between the Ministry of Water Resources and the Ministry of Agriculture decreased the focus on water supply organizations. A direct security approach would have increased the potential power of the former Water Resources Ministry. However, since globally public irrigation management is viewed negatively and as having failed [26], it is likely that also from the donor side strengthening the irrigation departments was off the agenda. At least, looking at Kyrgyzstan, the donor community attempted to circumvent irrigation departments by establishing top-down "bottom-up" WUA federations for replacing irrigation departments (as in the case of World Bank projects) [121]. Nevertheless, it is not evident, how WUA federations would be able to take over managerial responsibility or would have the capacities to cover operation and maintenance costs, especially looking at the high past expenditures of the Ferghana Province Irrigation Department.

7. Conclusions

So far, neither polycentric water management nor water supply organizations (irrigation departments) have received broad international attention within the Syr Darya Basin. Instead, so far the main focus has been on the creation of WUAs at the local level and basin management at the international level. The implication is that there is a widening gap regarding promoted and actual water security approaches as well as a missing link, which in the past have been the province irrigation departments. Hence, by looking at meso-level water security ("irrigation districts" as well as irrigation departments) the paper has attempted to close an important gap of the current water security focus in the Syr Darya Basin.

Irrigation departments are negatively viewed as public sector organizations, and are perceived as having failed to adapt to wider socio-political or environmental changes [26]. However, the long term data on the Ferghana Province highlighted that the irrigation department was capable of adapting by increasing staffing and mobility as well as creating resilience through new infrastructure during the Soviet period. Only after independence and due to the economic crisis, the administrative changes (merger between the two Ministries) and possibly the exposure to the global "neo-liberal" donor community focusing on water user governance, led to the decline of the irrigation department's capacity. There is great potential for the current discourse on water security and the water, energy and food nexus to refocus attention to the challenges of existing water supply organizations. However, at least based on suggestions from some of the global literature [4,6], the essential element of water supply security, have been either taken for granted or overlooked. Similar, although Sadoff *et al.* [28] puts key emphasis on water supply organizations, the mentioned "institutional tripod" (water users, states and markets) might imply an emphasis on governance, without strengthening the capacity of the water supply organizations first. The findings of the case study demonstrated the loss of technical and organizational water control and therefore the loss of the capacity of the irrigation department. The implication of not strengthening the water supply organizations are already evident; increased inequality of water distribution along the main canals, which negatively affects water distribution within WUAs. Given the already identified weaknesses of governance due to the diversity and power inequities of users, these inequities could further increase.

As indicated by our analysis the weakest link for water security is the public administrations, it is therefore essential to finally engage with the water bureaucracy. This calls for a comprehensive analysis regarding past and current internal as well as external challenges for the water bureaucracy for enabling its revitalization on key water security challenges. This call for revitalizing the water bureaucracy challenges the neo-liberal paradigm. Given that the bureaucracy is a public administration, a revitalization will not be possible in a piecemeal approach of donor sponsored "projects" (like the SDC and GIZ project mentioned in the case study) but instead calls for a long-term approach for reinvestment and modernization and therefore strengthening the public administration.

Acknowledgments: During the article preparation, Kai Wegerich was based at the International Water Management Institute (IWMI). The data analyzed in this article were gathered during IWMI's Irrigation Bureaucracy project in Central Asia funded by the Consultative Group on International Agricultural Research (CGIAR-wide) Research Program on Water, Land and Ecosystems (WLE).

Author Contributions: Kai Wegerich developed the initial and final versions of the framework, analyzed the data, organized the systematic discussion and led the drafting process of the manuscript. He incorporated the study incorporating the contributions from co-authors as well as from reviewers at TU Berlin and IWMI. Daniel Van Rooijen provided substantial comments on the initial framework section as well as the initial version of the manuscript. Ilkhom Soliev and Nozilakhon Mukhamedova contributed by providing their expertise and insights on the study area as well as on transboundary arrangements as well as local level land and water reforms.

Conflicts of Interest: The authors declare no conflict of interest.

References

1. Wolf, A. Conflict and cooperation along international waterways. *Water Policy* **1998**, *1*, 251–265. [CrossRef]
2. Deininger, R.A.; Meier, P.G. Sabotage of public water supply systems. In *Security of Public Water Supplies*; Deininger, R.A., Literathy, P., Bartram, J., Eds.; Springer: Berlin, Germany, 2000; pp. 241–248.
3. Gleick, P.H. Water and terrorism. *Water Policy* **2006**, *8*, 481–503. [CrossRef]
4. United Nations University. *Water Security and the Global Water Agenda, a UN-Water Analytical Brief*; United Nations University: Hamilton, PA, USA, 2013.
5. Grey, D.; Sadoff, C. Sink or swim? Water security for growth and development. *Water Policy* **2007**, *9*, 545–571. [CrossRef]
6. Hoff, H. Understanding the Nexus. In Proceedings of the Bonn Conference: The Water, Energy and Food Security Nexus, Stockholm, Sweden, 16–18 November 2011.
7. Faures, J.M.; Svendson, M.; Turral, H. *Water for Food, Water for Life: A Comprehensive Assessment of Water Management in Agriculture*; Earthscan: London, UK, 2007; pp. 353–394.
8. Tortajada, C.; Biswas, A.K. Editorial: Infrastructure and development. *Int. J. Water Resour. Dev.* **2014**, *30*, 3–7. [CrossRef]
9. Lankford, B.; Hepworth, N. The cathedral and the bazaar: Monocentric and polycentric river basin management. *Water Altern.* **2010**, *3*, 82–101.
10. Lautze, J.; Wegerich, K.; Kazbekov, J.; Yakubov, M. International river basin organizations: Variation, options and insights. *Water Int.* **2013**, *38*, 30–42. [CrossRef]
11. Wescoat, J.L.; Halvorson, S.J.; Mustafa, D. Water management in the Indus basin of Pakistan: A halfcentury perspective. *Int. J. Water Resour. Dev.* **2000**, *16*, 391–406. [CrossRef]
12. Stucki, V.; Sojamo, S. Nouns and numbers of the water-energy-security nexus in central Asia. *Int. J. Water Resour. Dev.* **2012**, *28*, 399–418. [CrossRef]
13. Granit, J.; Jägerskog, A.; Lindström, A.; Björklund, G.; Bullock, A.; Löfgren, R.; de Gooijer, G.; Pettigrew, S. Regional options for addressing the water, energy and food nexus in central Asia and the Aral sea basin. *Int. J. Water Resour. Dev.* **2012**, *28*, 419–432. [CrossRef]

14. Wegerich, K.; Kazbekov, J.; Kabilov, F.; Mukhamedova, N. Meso-level cooperation on transboundary tributaries and infrastructure in the Ferghana Valley. *Int. J. Water Resour. Dev.* **2012**, *28*, 525–543. [CrossRef]
15. Pak, M.; Wegerich, K.; Kazbekov, J. Re-examining conflict and cooperation in central Asia: A case study from the Isfara river, Ferghana Valley. *Int. J. Water Resour. Dev.* **2014**, *30*, 230–245. [CrossRef]
16. Pak, M.; Wegerich, K. Competition and benefit sharing in the Ferghana Valley: Soviet negotiations on transboundary small reservoir construction. *Cent. Asian Aff.* **2014**, *1*, 225–246. [CrossRef]
17. Soliev, I.; Wegerich, K.; Kazbekov, J. The costs of benefit sharing: Historical and institutional analysis of shared water development in the case of the Ferghana Valley in the Syr Darya basin. *Water* **2015**, *7*, 2728–2752. [CrossRef]
18. Ertsen, M.W. Colonial irrigation: Myths of emptiness. *Landsc. Res.* **2006**, *31*, 147–167. [CrossRef]
19. Ertsen, M.W. The development of irrigation design schools or how history structures human action. *Irrig. Drain.* **2007**, *56*, 1–19. [CrossRef]
20. Vermillion, D.L. Impacts of Irrigation Management Transfer: A Review of the Evidence. Available online: http://www.iwmi.cgiar.org/Publications/IWMI_Research_Reports/PDF/pub011/REPORT11.PDF (accessed on 7 February 2007).
21. Water Sector Policy Review and Strategy Formulation: A General Framework. Available online: http://www.fao.org/docrep/v7890e/v7890e00.htm (accessed on 9 February 2007).
22. Irrigation Management Transfer: Worldwide Efforts and Results. Available online: http://www.fao.org/3/a-a1520e.pdf (accessed on 21 April 2014).
23. Irrigation Reform in Asia: A review of 108 Cases of Irrigation Management Transfer. Available online: https://ideas.repec.org/p/iwt/rerpts/h042851.html (accessed on 20 April 2014).
24. Wegerich, K.; Warner, J.; Tortajada, C. Water sector governance: A return ticket to anarchy. *Int. J. Water Gov.* **2014**, *2*, 7–20. [CrossRef]
25. Rap, E.; Wester, P. The practices and politics of making policy: Irrigation management transfer in Mexico. *Water Altern.* **2013**, *6*, 506–531.
26. Surhardiman, D.; Giordano, M. Is there an alternative for irrigation reform? *World Dev.* **2014**, *57*, 91–100. [CrossRef]
27. Molle, F.; Mollinga, P.P.; Wester, P. Hydraulic bureaucracies and the hydraulic mission: Flows of water, flows of power. *Water Altern.* **2009**, *2*, 328–349.
28. Spoor, M.; Krutov, A. XI. The power of water in a divided central Asia. *Perspect. Glob. Dev. Technol.* **2003**, *2*, 593–614. [CrossRef]
29. Abdullaev, I.; de Fraiture, C.; Giordano, M.; Yakubov, M.; Rasulov, A. Agricultural water use and trade in Uzbekistan: Situation and potential impacts of market liberalization. *Int. J. Water Resour. Dev.* **2009**, *25*, 47–63. [CrossRef]
30. Hirsch, D. Water User Associations in Uzbekistan: Theory and Practice. Ph.D. Thesis, University of Bonn, Cuvillier Verlag, Göttingen, Germany, June 2006.
31. Abdullaev, I.; Kazbekov, J.; Manthritilake, H.; Jumaboev, K. Water user groups in central Asia: Emerging form of collective action in irrigation water management. *Water Resour. Manag.* **2009**, *24*, 1029–1043. [CrossRef]
32. Abdullaev, I.; Kazbekov, J.; Manthritilake, H.; Jumaboev, K. Participatory water management at the main canal: A case from South Ferghana canal in Uzbekistan. *Agric. Water Manag.* **2009**, *96*, 317–329. [CrossRef]
33. Abdullaev, I.; Kazbekov, J.; Jumaboev, K.; Manthritilake, H. Adoption of integrated water resources management principles and its impacts: Lessons from Ferghana Valley. *Water Int.* **2009**, *34*, 230–241. [CrossRef]
34. Nexus Nirvana or Nexus Nullity? A Dynamic Approach to Security and Sustainability in the Water-Energy-Food Nexus. Available online: http://steps-centre.org/wp-content/uploads/Water-and-the-Nexus.pdf (accessed on 30 April 2015).
35. Critical Infrastructure Partnership Advisory Council Water Sector Strategic Planning Working Group. Available online: https://www.gov.uk/government/uploads/system/uploads/attachment_data/file/170644/28307_Cm_8583_v0_20.pdf (accessed on 29 April 2015).
36. Contest: The United Kingdom's Strategy for Countering Terrorism Annual Report. Available online: https://www.gov.uk/government/uploads/system/uploads/attachment_data/file/170644/28307_Cm_8583_v0_20.pdf (accessed on 29 April 2015).

37. Lautze, J.; Manthrithilake, H. Water Security. In *Key Concepts in Water Resource Management*; Lautze, J., Ed.; Routledge: Oxford, UK, 2014.

38. Sadoff, C.W.; Hall, J.W.; Grey, D.; Aerts, J.C.J.H.; Ait-Kadi, M.; Brown, C.; Cox, A.; Dadson, S.; Garrick, D.; Kelman, J.; *et al. Securing Water, Sustaining Growth: Report of the GWP/OECD Task Force on Water Security and Sustainable Growth*; University of Oxford: Oxford, UK, 2015.

39. Saldías, C.; Boelens, R.; Wegerich, K.; Speelman, S. Losing the watershed focus: A look at complex community-managed irrigation systems in Bolivia. *Water Int.* **2012**, *37*, 744–759. [CrossRef]

40. Dill, B.; Crow, B. The colonial roots of inequality: Access to water in urban East Africa. *Water Int.* **2014**, *39*, 187–200. [CrossRef]

41. Boelens, R.; Seemann, M. Forced engagements: Water security and local rights formalization in Yanque, Colca Valley, Peru. *Hum. Organ.* **2014**, *73*, 1–11. [CrossRef]

42. Boelens, R. *The Shotgun Marriage: Water Security, Cultural Politics, and Forced Engagements between Official and Local Rights Frameworks*; Routledge: Oxford, UK, 2013.

43. Tortajada, C. Water management for a megacity: Mexico City metropolitan area. *AMBIO J. Hum. Environ.* **2003**, *32*, 124–129. [CrossRef]

44. Molle, F.; Berkoff, J. *Cities Versus Agriculture: Revisiting Intersectoral Water Transfers, Potential Gains and Conflicts*; International Water Management Institute: Colombo, Sri Lanka, 2006.

45. Mukhamedova, N.; Wegerich, K. Integration of villages into WUAS—The rising challenge for local Water management in Uzbekistan. *Int. J. Water Gov.* **2014**, *2*, 153–179. [CrossRef]

46. Earle, A. The role of cities as drivers of international transboundary water management processes. In *Water Security: Principles, Perspectives and Practices*; Routledge: Oxford, UK, 2013.

47. Allan, T.; Keulertz, M.; Woertz, E. The water-food-energy nexus: An introduction to nexus concepts and some conceptual and operational problems. *Int. J. Water Resour. Dev.* **2015**, *31*, 301–311. [CrossRef]

48. Keulertz, M.; Woertz, E. Financial challenges of the nexus: Pathways for investment in water, energy and agriculture in the Arab world. *Int. J. Water Resour. Dev.* **2015**, *31*, 312–325. [CrossRef]

49. Boelens, R.; Doornbos, B. The battlefield of water rights. rule making amidst conflicting normative frameworks in the ecuadorian highlands. *Hum. Organ.* **2001**, *60*, 343–355. [CrossRef]

50. Swyngedouw, E. Dispossessing H_2O: The contested terrain of water privatization. *Capital. Nat. Soc.* **2005**, *16*, 81–98. [CrossRef]

51. Wegerich, K. The Afghan water law: "A legal solution foreign to reality"? *Water Int.* **2010**, *35*, 298–312. [CrossRef]

52. Dimitrov, R.S. Water, conflict and security: A conceptual minefield. *Soc. Nat. Resour.* **2002**, *15*, 677–691. [CrossRef]

53. Zeitoun, M. The web of sustainable water security. In *Water Security: Principles, Perspectives and Practices*; Routledge: Oxford, UK, 2014.

54. Gleick, P.H. The changing water paradigm: A look at twenty-first century water resource development. *Water Int.* **2000**, *25*, 127–138. [CrossRef]

55. Lankford, B. Infrastructure hydromentalities: Water sharing, water control and water (in) security in Lankford. In *Water Security: Principles, Perspectives and Practices*; Routledge: Oxford, UK, 2014; pp. 256–272.

56. Molden, D.J.; Vaidya, R.A.; Shrestha, A.B.; Rasul, G.; Shrestha, M.S. Water infrastructure for the Hindu Kush Himalayas. *Int. J Water Resour. Dev.* **2014**, *30*, 60–77. [CrossRef]

57. Biswas-Tortajada, A. The Gujarat state-wide water supply grid: A step towards water security. *Int. J. Water Resour. Dev.* **2014**, *30*, 78–90. [CrossRef]

58. Bottrall, A.F. Comparative Study of the Management and Organization of Irrigation Projects. Available online: http://documents.worldbank.org/curated/en/1981/05/438966/comparative-study-management-organization-irrigation-projects (accessed on 25 April 2015).

59. Chambers, R. *Managing Canal Irrigation*; Oxford & IBH Publishing: New Delhi, India, 1988.

60. Uphoff, N.T. *Managing Irrigation: Analyzing and Improving the Performance of Bureaucracies*; Sage publication: Thousand Oaks, CA, USA, 1991.

61. Uphoff, N.T. *Improving International Irrigation Management with Farmer Participation: Getting the Process Right*; Westview Press: Boulder, CO, USA, 1986.

62. Mollinga, P.P. *On the Water Front*; Orient Longman: New Delhi, India, 2003.

63. Ali, I. *The Punjab under Imperialism 1885–1947*; Princeton University Press: Princeton, NJ, USA, 1988.

64. Bolding, A.; Mollinga, P.P.; van Straaten, K. Modules for modernization: Colonial irrigation in India and the technological dimension of agrarian change. *J. Dev. Stud.* **1995**, *31*, 805–844. [CrossRef]
65. Ul-Haq, A. *Case Study of the Punjab Irrigation Department*; International Water Management Institute: Colombo, Sri Lanka, 1998.
66. Wade, R. Irrigation reform in the conditions of populist anarchy: An Indian case. *J. Dev. Econ.* **1984**, *14*, 285–303. [CrossRef]
67. Deininger, R.A.; Literathy, P.; Bartram, J. *Security of Public Water Supplies*; Springer: Berlin, Germany, 2000.
68. Davis, J.; Whittington, D. Challenges for water sector reform in transition economies. *Water Policy* **2004**, *6*, 381–395.
69. Managing Singapore's Water. Available online: http://www.businesstimes.com.sg/opinion/managing-singapores-water (accessed on 24 April 2015).
70. Wegerich, K. Shifting to hydrological/hydrographic boundaries: A comparative assessment of national policy implementation in the Zerafshan and Ferghana Valleys. *Int. J. Water Resour. Dev.* **2015**, *31*, 88–105. [CrossRef]
71. Bakker, K. Archipelagos and networks: Urbanization and water privatization in the South. *Geogr. J.* **2003**, *169*, 328–341. [CrossRef]
72. Shah, T. *Taming the Anarchy: Groundwater Governance in South Asia*; Routledge: Oxford, UK, 2010.
73. Molle, F.; Wester, P.; Hirsch, P. River basin closure: Processes, implications and responses. *Agric. Water Manag.* **2010**, *97*, 569–577. [CrossRef]
74. Saiko, T.A.; Zonn, I.S. Irrigation expansion and dynamics of desertification in the Circum-Aral region of Central Asia. *Appl. Geogr.* **2000**, *20*, 349–367. [CrossRef]
75. Zonn, I.S.; Glantz, M.; Kosarev, A.N.; Kostianoy, A.G. *The Aral Sea Encyclopedia*; Springer: Berlin, Germany, 2009.
76. Kasymova, V. National Constraining Factors to the Agreement on Water and Energy Use in the Syr Darya Basin. Available online: http://rmportal.net/library/content/tools/environmental-policy-and-institutional-strengthening-epiq-iqc/epiq-environmental-policy-and-institutional-strengthening-cd-vol-1/epiq-cd-1-tech-area-dissemination-of-policy-knowledge-environmental-communication/national-constraining-factors-to-the-agreement-on-water-and-energy-use-in-the-syr-darya-basin-the-tajik-republic/view (accessed on 30 January 2015).
77. Weinthal, E. Sins of omission: Constructing negotiating sets in the Aral Sea basin. *J. Environ. Dev.* **2001**, *10*, 50–79.
78. The World Bank. *Water Energy Nexus in Central Asia: Improving Regional Cooperation in the Syr Darya Basin, Europe and Central Asia Region*; The World Bank: Washington, DC, USA, 2004.
79. Ryabtsev, A.D. Threats to Water Security in the Republic of Kazakhstan in the Transboundary Context and Possible ways to Eliminate them. 2008. Available online: http://www.icwc-aral.uz/workshop_march08/pdf/ryabtsev_en.pdf (accessed on 7 February 2009).
80. Libert, B.; Orolbaev, E.; Steklov, Y. Water and energy crisis in central Asia. *China Eurasia Forum Q.* **2008**, *6*, 9–20.
81. Antipova, E.; Zyryanov, A.; McKinney, D.; Savitsky, A. Optimization of Syr Darya water and energy uses. *Water Int.* **2002**, *27*, 504–516. [CrossRef]
82. McKinney, D.; Kenshimov, A.K. Optimization of the Use of Water and Energy Resources in the Syr Darya Basin under Current Conditions. Available online: http://www.caee.utexas.edu/prof/mckinney/papers/aral/00-06-W/00-06-W_eng/Vol-1/Frontle.htm (accessed on 24 April 2015).
83. Wegerich, K.; Kazbekov, J.; Lautze, J.; Platonov, A.; Yakubov, M. From monocentric ideal to polycentric pragmatism in the Syr Darya: Searching for second best approaches. *Int. J. Sustain. Soc.* **2012**, *4*, 113–130. [CrossRef]
84. Pak, M. International River Basin Management in the Face of Change: Syr Darya Basin Case Study. Ph.D. Thesis, Oregon State University, Corvallis, OR, USA, November 2014.
85. Strengthening Cooperation for Rational and Efficient Use of Water and Energy Resources in Central Asia. Available online: http://www.unece.org/fileadmin/DAM/env/water/damsafety/effuse_en.pdf (accessed on 10 May 2010).

86. Wegerich, K. Have your cake and eat it too: Agenda setting in central Asian transboundary rivers. In *Water, Environmental Security and Sustainable Rural Development: Conflict and Cooperation in Central Eurasia*; Arsel, M., Spoor, M., Eds.; Routledge: Oxford, UK, 2010.

87. Dukhovny, V.A.; Sorokin, A.G.; Tuchin, A.I.; Sorokin, D.A.; Temlyantseva, Y. Concept of Integrated Model for Chirchik-Ahangaran-Keles Sub-Basin. Available online: http://cawater-info.net/rivertwin/documents/pdf/concept6_e.pdf (accessed on 30 March 2015).

88. Dukhovny, V.A.; Sorokin, A.G.; Tuchin, A.I.; Rysbekov, U.H.; Stulina, G.V.; Nerosin, S.A.; Rusiev, I.B.; Sorokin, D.A.; Katz, A.; Shahov, V.; Solodky, G. A regional model for integrated water management in twinned river basins. Available online: http://s3.amazonaws.com/zanran_storage/www.rivertwin.de/ContentPages/2493962769.pdf (accessed on 30 March 2015).

89. Rysbekov, Y.K. Rivertwin Central Asia: The basic Results of Work Package 7. Available online: http://www.cawater-info.net/rivertwin/documents/pdf/rysbekov2_presentation_e.pdf (accessed on 30 March 2015).

90. Benjaminovich, Z.M.; Tersitskiy, D.K. *Irrigation of Uzbekistan II*; Fan Publishing House: Tashkent, Uzbekistan, 1975. (In Russian)

91. Weinthal, E. *State Making and Environmental Cooperation*; The MIT Press: Cambridge, MA, USA, 2002.

92. Wegerich, K.; Kazbekov, J.; Mukhamedova, N.; Musayev, S. Is it possible to shift to hydrological boundaries? The ferghana valley meshed system. *Int. J. Water Resour. Dev.* **2012**, *28*, 545–564. [CrossRef]

93. Wegerich, K.; Soliev, I.; Akramova, I. Estimation of the long term costs to cope with upstream irrigation expansion in the transboundary setting of Ferghana Province. *Int. J. Water Gov.* **2015**. submitted for publication.

94. Dukhovny, V.A. Water and globalization: Case study of Central Asia. *Irrig. Drain.* **2007**, *56*, 489–507. [CrossRef]

95. Central Asia: Water and conflict. Available online: http://www.crisisgroup.org/en/regions/asia/central-asia/034-central-asia-water-and-conflict.aspx (accessed on 9 December 2014).

96. Ferghana Province Irrigation Department. *Annual Reports 1978–2010*; Ferghana Province Irrigation Department: Ferghana, Uzbekistan, 2011.

97. Bucknall, J.; Klytchnikova, I.; Lampietti, J.; Lundell, M.; Scatasta, M.; Thurman, M. Irrigation in Central Asia: Social, Economic and Environmental Considerations. Available online: http://siteresources.worldbank.org/ECAEXT/Resources/publications/Irrigation-in-Central-Asia/Irrigation_in_Central_Asia-Full_Document-English.pdf (accessed on 5 October 2014).

98. Law of the Republic of Uzbekistan. In *On Dehkan Farms*; No. 654-XII; The Senate of the Republic of Uzbekistan: Tashkent, Uzbekistan, 1992. (In Uzbekistan)

99. Law of the Republic of Uzbekistan. In *On Farms*; No. 599-I; The Senate of the Republic of Uzbekistan: Tashkent, Uzbekistan, 1998. (In Uzbekistan)

100. Orders of Cabinet of Ministers and Presidential Decree. In *On Measures to Further Optimize the Size of Land Plots of Farms*; No. F-3287 and F-3212; The Senate of the Republic of Uzbekistan: Tashkent, Uzbekistan, 2009. (In Uzbekistan)

101. Dukhovny, V.A.; de Schuetter, J.L.G. *Water in Central Asia: Past, Present, Future*; CRC Press: Boca Raton, MA, USA, 2011.

102. Anderson, J. *The International Politics of Central Asia*; Manchester University Press: Manchester, UK, 1997.

103. Pomfret, R. Agrarian reform in Uzbekistan: Why has the Chinese model failed to deliver? *Econ. Dev. Cult. Chang.* **2002**, *48*, 269–282. [CrossRef]

104. Djanibekov, N.; Rudenko, I.; Lamers, J.P.A.; Bobojonov, I. Pros and cons of cotton production in Uzbekistan, case study of the program: "Food policy for development countries: The role of government in the global food system. 2010. Available online: http://chatt.hdsb.ca/~blairj/course_files/FOV1-001625C2/cotton%20case%20study.pdf (accessed on 14 January 2014).

105. Platonov, A.; Wegerich, K.; Kazbekov, J.; Kabilov, F. Beyond the state order? Second crop production in the Ferghana Valley, Uzbekistan. *Int. J. Water Gov.* **2014**, *2*, 83–104. [CrossRef]

106. Resolution of the Cabinet of Ministers Republic of Uzbekistan. In *On Measures for the Reorganization of Agricultural Enterprises into Farms*; No. 8; Cabinet of Ministers Republic of Uzbekistan: Tashkent, Uzbekistan, 2002. (In Uzbekistan)

107. Swiss Agency for Development and Cooperation (SDC). *Regional Water Sector Programme Central Asia—Water Management Backstopping (Irrigation)*; Swiss Agency for Development and Cooperation: Bern, Switzerland, 2008.

108. Abdullaev, I.; Rakhmatullaev, S.; Platonov, A.; Sorokin, D. Improving water governance in central Asia through application of data management tools. *Int. J. Environ. Stud.* **2012**, *69*, 151–168. [CrossRef]

109. Thurman, M. *Irrigation and Poverty in Central Asia: A Field Assessment*; World Bank Group: Washington, DC, USA, 2001.

110. Wegerich, K. Organizational problems of water distribution in Khorezm, Uzbekistan. *Water Int.* **2004**, *29*, 130–137. [CrossRef]

111. Wegerich, K. What happens in a merger? Experiences of the state department for water resources in Khorezm, Uzbekistan. *Phys. Chem. Earth* **2005**, *30*, 455–462. [CrossRef]

112. Yalcin, R.; Mollinga, P.P. Institutional Transformation in Uzbekistan's Agricultural and Water Resources Administration: The Creation of a New Bureaucracy. Available online: http://www.econstor.eu/dspace/handle/10419/88374 (accessed on 11 June 2008).

113. Djalalov, A.A.; Mirzaev, N.N.; Horst, M.G.; Stulina, G.V.; Abdurazakov, Z.B.; Ergashev, O.; Khoshimov, A.; Rasulov, F.; Alesandrova, N.G. *Comprehensive Hydrographic Study of the Ferghana Valley: Vision of the Integrated Water Resources Management based on the IWRM-FV Project Experience in Uzbekistan*; Scientific Information Centre of the Interstate Commission on Water Coordination: Tashkent, Uzbekistan, 2011.

114. Dukhovny, V.A.; Sokolov, V.I.; Galustyan, A.; Djalalov, A.A.; Mirzaev, N.N.; Horst, M.G.; Stulina, G.V.; Muminov, S.; Ergashev, I.; Kholikov, A.; *et al. Report on Comprehensive Hydrographic Study of the Ferghana Valley*; Scientific Information Centre of the Interstate Commission on Water Coordination: Tashkent, Uzbekistan, 2012.

115. Suhardiman, D.; Giordano, M.; Rap, E.; Wegerich, K. Bureaucratic reform in irrigation: A review of four case studies. *Water Altern.* **2014**, *7*, 442–463.

116. Bichsel, C. *Conflict Transformation in Central Asia: Irrigation Disputes in the Ferghana Valley*; Routledge: Oxford, UK, 2009.

117. Inagamov, S.R. The Uzbek model, the Sunday Indian. Available online: http://www.thesundayindian.com/en/story/the-uzbek-model/7/26310/ (accessed on 5 July 2014).

118. Peyrouse, S. *The Russian Minority in Central Asia: Migration, Politics, and Language*; Woodrow Wilson International Center for Scholars: Washington, DC, USA, 2008.

119. Taube, G.; Zettelmeyer, J. Output Decline and Recovery in Uzbekistan: Past Perfarmance and Future Prospects. Available online: https://www.imf.org/external/pubs/cat/longres.aspx?sk=2745.0 (accessed on 14 August 2014).

120. Kandiyoti, D. Poverty in transition: An ethnographic critique of household surveys in post-soviet central Asia. *Dev. Chang.* **1999**, *30*, 499–524. [CrossRef]

121. WUA Workshop-Country Summary: Kyrgyz Republic. Available online: http://go.worldbank.org/W02GK2LE70 (accessed on 2 May 2015).

water

MDPI

Article

Sharing Benefits in Transboundary Rivers: An TExperimental Case Study of Central Asian Water-Energy-Agriculture Nexus

Shokhrukh-Mirzo Jalilov *, Olli Varis and Marko Keskinen

Water & Development Research Group, Aalto University, Aalto 00076, Finland; olli.varis@aalto.fi (O.V.); marko.keskinen@aalto.fi (M.K.)
* Correspondence: shokhrukh.jalilov@aalto.fi; Tel.: +358-46-6171001

Academic Editor: Miklas Scholz
Received: 17 June 2015; Accepted: 27 August 2015; Published: 2 September 2015

Abstract: Cooperation in transboundary river basins is challenged by the riparian countries' differing needs for water use. This is the case especially in Amu Darya Basin in Central Asia, where upstream Tajikistan is building the Rogun Hydropower Plant (RHP) to increase its energy security, while the downstream countries oppose the plant due to the feared negative impacts to their irrigated agriculture. Several experimental scenarios illustrate how the concept of benefit sharing could be used as a framework to investigate these water-energy-agriculture linkages in a transboundary context. Using a hydro-economic model, we investigate the economic benefits of various scenarios emphasizing agricultural and/or energy production, thus benefiting the riparian countries uniquely. Subsequently, we discuss how benefit-sharing arrangements with different forms of compensations could be used as a mechanism to facilitate transboundary cooperation. Our results indicate that several scenarios have a potential to increase the total energy-agriculture benefits in the basin. Yet, agreeing on the actual benefit-sharing mechanism between the countries poses special challenges as each may require countries to give up some of their anticipated maximum potential benefits. The presented scenarios provide a potential starting point for debates over benefit-sharing arrangements across countries needing to address the water-energy-agriculture nexus.

Keywords: benefit-sharing; water-energy-agriculture nexus; water allocation; hydro-economic modeling; Rogun; Amu Darya

---

1. Introduction

The principle of benefit-sharing has been proposed as a strategy to facilitate the cooperation on transboundary river basins to address a wide range of water uses such as agriculture and energy production [1–4]. The main idea behind benefit-sharing is not to share physical quantities of water, but rather to share the benefits arising from water development and use [2]. Benefit-sharing thus builds on the assumption that the beneficiaries—in transboundary river basins, the riparian countries—are more interested in the economic value and benefits that water and its development creates, rather than dividing up a fixed quantity of water itself. In ideal benefit-sharing situations, the riparian countries view the benefits of water development and use as a positive-sum game associated with benefits optimization, rather than the zero-sum game associated with simple water sharing [5]. Such benefits are often economic, but can also include, for example, social, political or environmental benefits, and various mixes between these [6].

Sadoff and Grey [1] identified four types of cooperative benefits: benefits to the river, benefits from the river, reduced costs because of the river and benefits beyond the river. This study focuses on the second type of benefits (from the river), where cooperative management of shared rivers can yield major benefits in terms of increased food and energy production. Benefit-sharing may also include some form of redistribution of benefits or compensation for lost benefits. While, the actual form of compensation is highly situation specific, it can involve for example monetary transfers, granting of rights to use water, financing of investments, or the provision of non-related goods and services [2].

This article presents an experimental case study looking at the possibility for benefit sharing in the Amu Darya River Basin (ADRB). The focus of the study is on the economic benefits that the riparian countries could receive from the planned Rogun Dam and its operation on the Vakhsh River, tributary to Amu Darya. The Rogun Hydropower Plant (RHP) is a controversial project, and there have been active debates (mostly on websites such as www.eurasianet.org, www.fergananews.com and centrasia.ru) about its benefits, revolving around Tajikistan's aim to enhance its energy security and Uzbekistan's concerns about the negative impacts on its agriculture.

Despite the on-going debates, there has been little scientific research investigating the actual level or distribution of economic benefits that various operation schemes of the Rogun Reservoir enables for riparian countries. In addition, existing studies e.g., [7–9] look mainly at basin-wide benefits, thus assuming that the riparian countries would be willing to negotiate and agree on the most optimal operation scheme. Yet, this is not obvious in the study context, as there is currently no sign of willingness among the riparian countries to reach mutual agreement on the operation of the Rogun Reservoir [10,11].

As a result, this study takes the view of quantifying the economic benefits for operation schemes emphasising each riparian country's priorities and then comparing those with operation scheme optimising basin-wide benefits. After this, we use the concept of benefit-sharing to investigate different possibilities that could serve as a basis for the operation of the Rogun Dam and Hydropower Plant in a way that brings about a Pareto improvement (economic welfare gain) satisfying all riparian countries. We hope that such approach will provide a meaningful contribution to current policy debates around the Rogun Dam, initiating discussion, dialogue, and debate about innovative measures to investigate at the benefits derived across the water-energy-agriculture nexus in the Amu Darya River Basin.

2. Material and Methods

2.1. Case Study Basin

The Amu Darya River is the largest river in Central Asia in terms of its length of 2540 km [12], as well as its average annual supply of some 65 cubic kms [13]. The mainstream is supplied by the confluence of two main tributaries, the Vakhsh and Pyandj Rivers, and the river terminates in the Aral Sea (Figure 1). The river basin is shared by five riparian countries of Afghanistan, Kyrgyzstan, Turkmenistan, Tajikistan, and Uzbekistan. The Basin's drainage area includes about 309,000 km² [14] and is home to approximately 55 million people [15]. On its route from the headwaters to the Aral Sea, the River borders Afghanistan and Tajikistan as well as Afghanistan and Uzbekistan. For this reason, there have been considerable policy debates for years on approaches to manage water for the competing users and uses in the Basin.

The basin presents a unique case of transboundary water management, as before 1991 the area was part of the former Soviet Union and the majority of the river basin (except for Afghanistan) was managed by a single, central government under a common political system. At that time, the water use in the basin emphasized large-scale agricultural production (primarily cotton and wheat) and reservoirs were built in the upstream Tajikistan to provide water storage for the downstream irrigation demands see also [14,16–18]. As such, the situation before 1991 thus resembled a kind of benefit-sharing scheme, where the Republics of the Soviet Union shared and optimized benefits from the river based on barter relations.

Figure 1. Map of Central Asia with the location of the Rogun Dam.

After the dissolution of the Soviet Union in 1991, the Amu Darya became a transboundary river flowing through five independent states. Kyrgyzstan supplies 2% of total flow in the basin [19] and according to the "Agreement on Cooperation in Joint Management, Use and Protection of Interstate Sources of Water Resources" signed in February 1992 entitled to use only 0.6% of annual flow. Due to such small shares this country has not been considered in the study. The water management practices changed fundamentally, as the newly-independent countries focused on maximizing their individual benefits that the river and its water resources could provide, often at the expense of their neighbors.

To resolve the problem with its growing energy deficit, in 2008 Tajikistan announced its intention to resume the construction of 335 m high Rogun Dam on the Vakhsh River, a tributary to the Amu Darya River [20]. The construction of the Dam was already started during the Soviet era time in 1976, but halted in 1991 with the breakup of the Soviet Union. The purpose of the Rogun Dam is to supply the upstream Tajikistan with hydropower, but downstream countries oppose the dam due to possible negative impacts that the changed flow regime is expected to cause to downstream agriculture. Hence, while Tajikistan is in a hydrologic position to decide unilaterally on the operation of the Rogun Dam, it could also make friends by operating the dam to facilitate transboundary cooperation by considering the benefits of other riparian countries as well.

2.2. The Modeling Framework

An integrated river basin model was developed to estimate the benefits resulting from the various possible operation schemes (scenarios) of the Rogun Dam. Such operation schemes investigated three individual (country-specific) and two cooperative schemes, and focused on two of the most important water uses, *i.e.*, hydropower generation and agricultural production in the Amu Darya River Basin. In addition, environmental water need is included into the model through a constraint that ensures a 10% minimum flow at the confluence of Vakhsh and Pyanj must inflow to the Aral Sea at any given time period.

The model is an extension of the previous Amu Darya model [9], in which the main tributary flows are allocated to aggregate downstream water demands for beneficial use. The previous study provided empirical evidence for ways to achieve basin-wide Pareto efficiency by improving outcomes over shared transboundary waters, outcomes in which all riparian countries could be made better off by sharing the benefits of water development and allocation. In this study, the model framework is updated to include new operation schemes looking at country-specific benefits: under such schemes, the model maximizes the economic benefits for each country with no consideration of the total basin-wide benefits.

The model considers the economic importance of irrigated agriculture in the four basin countries, (Afghanistan, Tajikistan, Turkmenistan, Uzbekistan) in addition to the potential for hydropower production in upstream Tajikistan. The model allocates water for energy and agricultural production over a 10-year period with a monthly time step. The model considers two cropping seasons, with the first (early planting) crop season starting in March and lasting until August and the second, *i.e.*, so-called mid-term crop starting in May (late planting) and lasting until early autumn. Altogether three key crops are included: cotton as a strategic cash crop for Uzbekistan and Turkmenistan and wheat and potato as crops ensuring food security for all the riparian countries.

The integrated hydro-economic river basin model consists of hydrologic, agronomic, and economic elements, with an emphasis on the economic element. It is nonlinear, and programmed using the GAMS language [21]. The basin scale integrated model maximizes discounted net present value (NPV) across all water uses (in this case agriculture and energy), and all time periods subject to hydrologic and institutional constraints. NPV is the sum of present values of future benefits and costs. Discounted net present value is equal to the sum of agricultural and energy benefits. The model allocates water among the basin's water uses, locations, and time periods to maximize net present value, subject to the described constraints.

The river basin framework is developed as a node link network, which is a representation of the spatial objects in the river basin. Nodes represent river flows, reservoir, and demand objects, and links represent the linkages between these objects (Figure 2). Runoff from headwaters in the river basin constitutes inflows to these nodes. Balance between flows is calculated for each node at each time period, and flow move is calculated on the spatial connection in the river basin geometry (Appendix A).

2.3. Data Used in the Model

Figure 2 shows schematic of the Amu Darya and information on average annual water supply by source as well as the design data for Rogun Dam and Reservoir. The average annual water runoff in the Amu Darya River equals 65.58 km^3. The two major tributaries of the Amu Darya, the Pyanj and Vakhsh rivers, constitute 49% and 30% of the main river's total flow, respectively.

The economic benefits of hydropower and irrigated agriculture are derived from the use of water for energy and crop production. Crop water use data were used for existing cropping patterns by country and crop, and combined with agricultural production details including crop prices, production costs, and crop yields (Table 1). Net agricultural income per hectare and total land in production by country, crop, and season were identified. Agricultural income per land unit for each crop was defined as crop price multiplied by yield minus costs of production.

Discounted net farm income was summed over crops, time periods, and countries, subject to water supply and sustainability constraints described subsequently. Consistent with neoclassical price theory, reduced water quantities supplied to agricultural users decrease crop production and raise crop prices. Prices are based on previously published work that estimated price elasticities of demand and a linear demand price response at historically-observed prices and production levels in the entire region. This means that the model considers all countries as a unified and linked market and allocates water to the country where cost of production is lower and profit is higher.

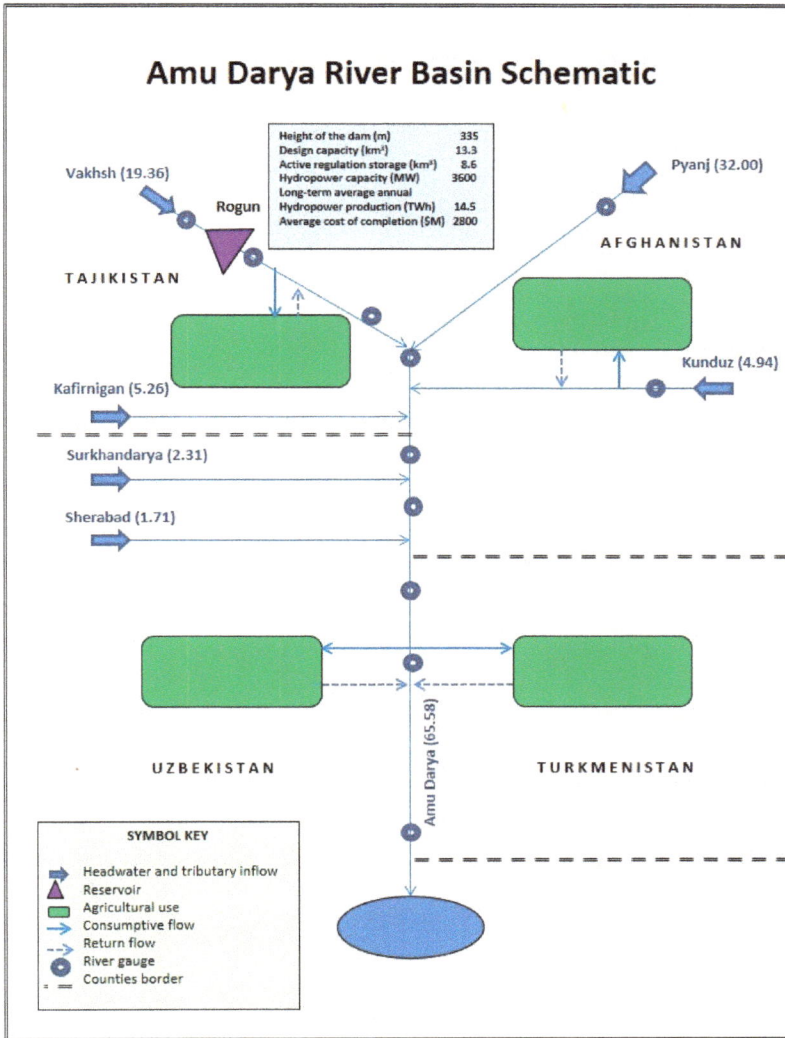

Figure 2. The schematic of the ADRB, including the design characteristic of the Rogun Hydropower Plant [9] and average annual water flow (billion cubic meters/year) of the Amu Darya tributaries used in the model [22].

Since the energy price is currently regulated by the Tajik government, the benefits from the energy production (hydropower) were derived by multiplying energy production by a constant energy price. Energy production is unknown in advance and part of the optimized model results, but as a matter of principle, it depends on the head of falling water, water discharge, gravitational acceleration, and a coefficient representing the technical efficiency of turbines by which falling water is converted into power. The model of energy production closely reproduced the maximum energy capacity of the Rogun Reservoir that is used in most of the recently-published planning documents [24].

Table 1. Agricultural data by country and crop. The total land area in production indicates the sum for all three crops. Source: [23].

Country	Crop	Yield (tons/ha)	Cost ($US/ha)	Water Requirements (m³/ha per year)	Total Land Area in Production within Amu Darya Basin (million ha)
	cotton	1.8	444	12	
Tajikistan	wheat	1.5	168	8	0.5
	vegetable	12	500	12	
	cotton	1.8	444	12	–
Afghanistan	wheat	1.6	165	8	0.4
	potato	12	503	12	–
	cotton	2.3	390	14	–
Uzbekistan	wheat	1.5	283	6	2.3
	vegetable	11	702	11	–
	cotton	2.2	392	14	–
Turkmenistan	wheat	1.5	283	6	1.1
	vegetable	11	702	11	–

The analysis requires the Rogun Reservoir to be filled to its maximum capacity by the last period (end of the 10th year). By establishing this constraint on the terminal period's reservoir level, sustainable water supplies and use and operation modes are protected. In addition, the model was calibrated by fitting model predictions suitably close to observed historical values of crop production and crop land pattern in each basin country in the baseline scenario.

2.4. Scenarios Examined

One baseline and five alternative model scenarios were considered as options to examine the economic benefits from the RHP for the riparian countries: (1) Baseline without the Rogun Dam; (2) Upstream Energy Priority (Tajikistan); (3) Uzbekistan Priority; (4) Turkmenistan Priority; (5) Downstream Agriculture Priority (Uzbekistan and Turkmenistan); and (6) No Priority ("optimal") (Figure 3). Afghanistan was not included in these operation schemes as it is not a major water user in the basin, yet its water use was included in the actual model results (Table 2).

Table 2. Total discounted economic benefits over 10 years for five scenarios, by country and the basin (million US$, NPV 5%).

Country	Scenario	Agricultural Benefits	Energy Benefits	Total benefits	
				$US	% Change from the Baseline
	Baseline without Rogun Dam	2268	–	2268	100
	Upstream Energy Priority	2210	3679	5889	160
Tajikistan	Uzbekistan Priority	1676	3981	5657	149
	Turkmenistan Priority	1703	3970	5673	150
	Downstream Priority	1310	3488	4798	112
	Optimal-No Priority	3084	3485	6568	190
	Baseline without Rogun Dam	192	–	192	100
	Upstream Energy Priority	339	–	339	76
Afghanistan	Uzbekistan Priority	188	–	188	−2
	Turkmenistan Priority	192	–	192	0
	Downstream Priority	184	–	184	−4
	Optimal-No Priority	462	–	462	140

Table 2. *Cont.*

Country	Scenario	Agricultural Benefits	Energy Benefits	Total benefits	
				$US	% Change from the Baseline
Uzbekistan	Baseline without Rogun Dam	26,588	–	26,588	100
	Upstream Energy Priority	29,579	–	29,579	11
	Uzbekistan Priority	37,895	–	37,895	43
	Turkmenistan Priority	9097	–	9097	−66
	Downstream Priority	35,668	–	35,668	34
	Optimal-No Priority	30,155	–	30,155	13
Turkmenistan	Baseline without Rogun Dam	2063	–	2063	100
	Upstream Energy Priority	6446	–	6446	212
	Uzbekistan Priority	1856	–	1856	−10
	Turkmenistan Priority	29,425	–	29,425	1326
	Downstream Priority	4864	–	4864	136
	Optimal-No Priority	8701	–	8701	322
Total over countries	Baseline without Rogun Dam	31,110	0	31,110	100
	Upstream Energy Priority	38,575	3679	42,254	36
	Uzbekistan Priority	41,616	3981	45,597	47
	Turkmenistan Priority	40,417	3970	44,387	43
	Downstream Priority	42,027	3488	45,515	46
	Optimal-No Priority	42,402	3485	45,887	47

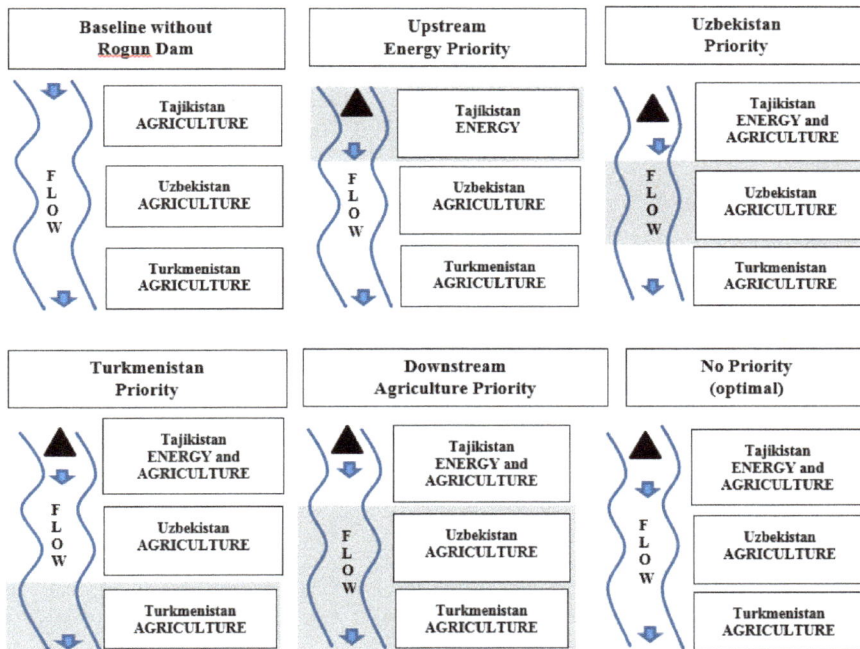

Figure 3. The scenarios modeled in the framework. Shaded area indicates the priority under each scenario, while black triangle indicates the Rogun Dam.

The baseline was included in the model by constraining the irrigation in the basin to closely reproduce historical agricultural land under production in each riparian country without the Rogun

Dam. This created the baseline within which the five scenarios (with the Rogun Dam) can be compared. Below, the constraints included in the scenarios are presented.

2.4.1. Upstream Energy Priority (Tajikistan)

The first scenario ensures Tajikistan's energy benefits by first optimizing its stated electricity demand and then taking into account the agricultural water use in Tajikistan, as well as other riparian countries. This scheme was modeled so that the Rogun Hydropower Plant must produce at least 70% of total energy requirements of Tajikistan for each month of a year. The Tajik Aluminum Enterprise (TALCO), which is the only major industrial plant in Tajikistan, consumes about 40% of the country's electricity [25]. Characteristic to the seasonal pattern of electricity consumption in Tajikistan are two consumption peaks; the bigger in winter due to heating and the lower in summer due to agricultural water pumping [25].

2.4.2. Uzbekistan Priority

The second scenario maximizes the agricultural benefits for downstream Uzbekistan with the Rogun Dam. The scheme was modeled so that the agricultural benefits of Uzbekistan with the Rogun Dam must be equal or higher than agricultural benefits without the dam. In addition, other riparian countries' agricultural land area was constrained so that it must be equal or less than their agricultural land without the Rogun Dam.

2.4.3. Turkmenistan Priority

The third scenario maximizes the agricultural benefits for downstream Turkmenistan with the Rogun Dam. The scheme was modeled in a similar manner than for Uzbekistan, meaning that the agricultural benefits of Turkmenistan with the Rogun Dam must be equal or higher than agricultural benefits without the dam. In addition, other riparian countries' agricultural land area was constrained so that it must be equal or less than their agricultural land without the Rogun Dam.

2.4.4. Downstream Agriculture Priority (Uzbekistan and Turkmenistan)

This scenario first maximizes Uzbekistan's and Turkmenistan's agricultural benefits and thereafter the water uses in Tajikistan and Afghanistan. The Rogun Dam thus operates in so-called irrigation mode, providing water during the vegetation period downstream and storing water in winter. This scheme was modeled so that the agricultural benefits for both Uzbekistan and Turkmenistan with the Rogun Dam must be equal or higher than the agricultural benefits without it. In addition, a constraint was introduced requiring agricultural land area in Tajikistan and Afghanistan with the Rogun Dam to be equal or less than the area without the dam. Such a constraint ensures that there is no shift between crop areas in upstream countries, and allow greater flexibility in choosing which crop to grow for downstream countries.

2.4.5. No Priority (optimising basin-wide benefits)

The last scenario is a so-called optimal scenario where all model constraints except hydrological are removed: the model thus seeks to maximize the basin-wide benefits, instead of the optimal benefits for individual riparian countries. This operation scheme is thus expected to create the largest total economic benefits (agriculture and energy) for the whole basin.

3. Results

3.1. Economic Benefits from Different Modeled Scenarios

The key results for the economic benefits of the five scenarios are summarized in Table 2, showing the economic benefits summed over 10 years for each country separately as well as total basin-wide benefits. Figure 4 complements the table by summarizing the model results across the years (averaged over the modeled period *i.e.*, 10 years) in terms of the reservoir monthly water discharge and corresponding energy production.

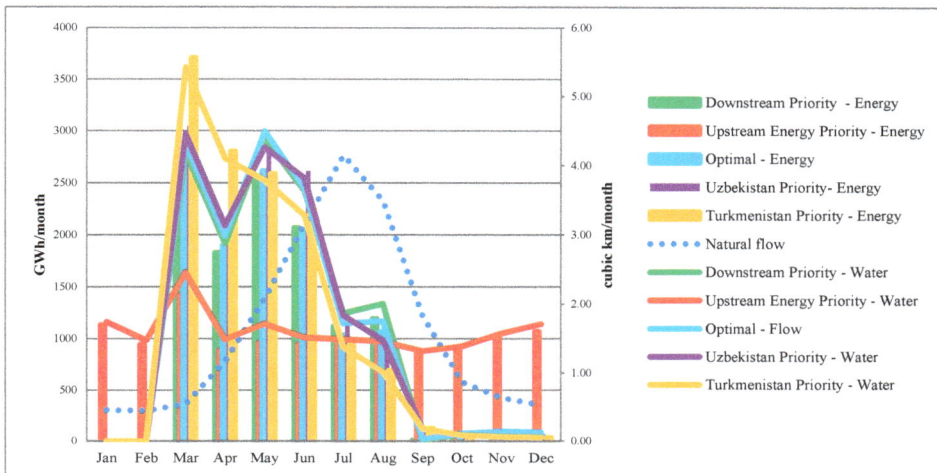

Figure 4. The model results showing monthly water discharge (on right *y*-axis) and corresponding energy production (on left *y*-axis) of the Rogun Hydropower Plant under five scenarios and the baseline (natural flow), averaged over 10 years. As can be seen, all five scenarios would significantly alter the natural flow, with "Upstream Energy Priority" leveling natural flow significantly.

The first key finding emerging from Table 2 is that the total basin-wide economic benefits (for all countries combined) for all five considered scenarios are considerably higher than the baseline, with the increase ranging from 36% (under "Upstream Energy Priority") to 47% (under "Uzbekistan Priority" and "No Priority" scenarios). In other words, the Rogun Dam has, clearly, a major potential to make the entire Amu Darya River Basin better off.

Another key finding is that the scenario with no country priority creates only relatively little additional total benefits when compared to other scenarios. Looking more closely to the results, an interesting finding emerges: "No Priority" brings clearly less benefits to Uzbekistan than to Afghanistan, Tajikistan, and Turkmenistan (13% against 140%, 190%, 322% correspondingly). In addition, "No Priority" creates more total benefits for upstream Tajikistan than scenario maximizing Tajikistan's energy production. Yet, looking only at total benefits is not enough, but the results must also be looked at in terms of how water is being allocated during the year, as this has major impact on the energy and agricultural production (Figure 4). The implications of this are discussed below with the chapter presenting more detailed results for each riparian country. Results regarding changes in land patterns of various scenarios is presented in Appendix B.

3.1.1. Tajikistan

The modeling results indicate that Tajikistan is clearly better off under any scenarios than currently, *i.e.*, under Baseline, with increased total benefits ranging from 112% ("Downstream Agricultural Priority") to 190% ("No Priority"). Remarkable is that the total economic benefits are highest under "No Priority" (due to major increase in agricultural benefits), and not under the scenario prioritizing Tajikistan's energy use. Similarly remarkable is that Tajikistan's energy benefits are actually highest under two scenarios prioritizing the water use of Uzbekistan and Turkmenistan. Yet, in reality such schemes would not be optimal for Tajikistan due to non-optimal timing of the energy production during the year (Figure 4). As can be seen from the figure, only "Upstream Energy Priority" gives Tajikistan opportunity to produce energy throughout the year according to its likely electricity demand. However, besides significant energy surplus, "Upstream Energy Priority" results in a small 3% decrease to baseline agricultural benefits, compared to a significant increase in the most downstream Turkmenistan (212%) and small 11% increase in Uzbekistan.

When considering also the timing of the hydropower production, the most preferable scenario for Tajikistan would be the one prioritizing its energy use, *i.e.*, "Upstream Energy Priority." While other scenarios also increase its total benefits (with the highest total benefits under "No Priority"), they are less optimal due to clearly less optimal timing of energy production, as well as due to reduced agricultural benefits (except for "No Priority").

3.1.2. Uzbekistan

Uzbekistan's total benefits increase in all modeled scenarios except "Turkmenistan Priority." This scenario means a 66% decrease of agricultural benefits in Uzbekistan, as water is redirected to Turkmenistan's agricultural sector. Among all considered scenarios the most preferable for Uzbekistan would be scenarios of prioritizing its own benefits and "Downstream Agricultural Priority", as they bring 43% and 34% increases, correspondingly. While Uzbekistan would experience an increase in benefits in scenarios of "Upstream Energy Priority" and "No Priority" (11% and 13%, correspondingly), it seems that the increase is very small to convince this country to make a switch and agree for upstream water infrastructure development.

Thus, the most preferable scenarios for Uzbekistan are "Uzbekistan Priority" and "Downstream Agricultural Priority." The difference between these scenarios relates to total economic benefits as well as to the possible compensations: if the country would choose "Uzbekistan Priority", it had to compensate looses for both the upstream Tajikistan and downstream Turkmenistan (and possibly Afghanistan), which could be too expensive, while in "Downstream Agricultural Priority" Uzbekistan has to make payments to upstream countries only as Turkmenistan is also better off in this scenario.

3.1.3. Turkmenistan

According to the model results, Turkmenistan experiences increase in agricultural benefits in all scenarios except "Uzbekistan Priority" that gives a 10% decrease in total benefits. Agricultural benefits would increase by a remarkable 1326% under "Turkmenistan Priority", under "Downstream Agricultural Priority" by 136%, under "Upstream Energy Priority" by 212%, and under "No Priority" by 322%. Such remarkable increases are possible, as Turkmenistan has relatively small agricultural area (particularly when compared to Uzbekistan). As a result, the new water regime enabled by the RHP lead in the modeled scenarios to major increases in Turkmenistan's cropping area, from 0.78 million hectares to up to 1.71 million hectares under "Turkmenistan priority" (Appendix B). As result, Turkmenistan would be fine with any scenarios except "Uzbekistan Priority", although it would obviously prefer the scenario of "Turkmenistan Priority".

3.1.4. Afghanistan

The model results show that Afghanistan would benefit from "No Priority" (+140%) and "Upstream Energy Priority" (76%) and would negligibly lose in "Uzbekistan Priority" (−2%) and in "Downstream Priority" (−4%); "Turkmenistan Priority" doesn't change the status quo of this country. Thus, Afghanistan would be better off under the scenarios emphasizing total basin-wide benefits and Tajikistan's energy production, although it would not experience remarkable economic losses under any scenario.

3.2. Opportunities for Benefit-Sharing Based on Total Economic Benefits

How do the modeling results presented above link to the concept of benefit-sharing? This chapter aims to answer this question by discussing alternative ways that our modeling results indicate for benefit-sharing in the Amu Darya River Basin. The starting point for such a discussion is the conclusion that the planned Rogun Dam has a potential to increase the total economic benefits to all the riparian countries. The follow-up question is thus: what kind of scenarios and related compensation mechanisms are most realistic to provide economic benefits for all riparian countries?

To initiate such discussion, let us first look at how preferable the different operation schemes are for different countries based on the total economic benefits (Table 3). As can be seen from Tables 2 and 3, Tajikistan has, in a way, the most optimal situation, as all scenarios i.e. operation schemes are clearly more beneficial for it in terms of total economic benefits. The situation is different for Uzbekistan, Turkmenistan and Afghanistan that all have some operation schemes that are at least slightly detrimental when compared to the current situation, *i.e.*, the baseline. Yet, a closer look at Tajikistan's situation reveals that, due to less optimal timing of energy production (Figure 4), it would not favor any other schemes except "Upstream Energy Priority" unless its energy demand is compensated in some other way.

As a result, the key message from Table 3 is that there is no single operation scheme to simultaneously satisfy the needs of all riparian countries, indicating the need for some kind of benefit-sharing and related compensation mechanisms.

Table 3. Preferable scenarios for the riparian countries in terms of total economic benefits as derived from the model (green—desirable, yellow—acceptable, red—detrimental). *: scenario not seen desirable because of the less optimal timing of energy production.

Scenario/Country	Tajikistan	Uzbekistan	Turkmenistan	Afghanistan
Upstream Energy Priority	green	yellow	green	green
Uzbekistan Priority	yellow	green	red	yellow
Turkmenistan Priority	yellow	red	green	yellow
Downstream Agricultural Priority	yellow	green	green	green
No Priority-Optimal	yellow *	green	green	green

When looking at the different operation schemes and their potential for benefit-sharing (Tables 2 and 3, Figure 4), we can first exclude "Uzbekistan Priority" and "Turkmenistan Priority", as both these scenarios result in clearly detrimental impact (red color) to some other country.

Consequently, three scenarios that have the biggest potential for benefit-sharing and monetary compensation are: "Upstream Energy Priority", "Downstream Agricultural Priority", and "No Priority." These are discussed in more detail below.

3.2.1. Upstream Energy Priority

This scenario provides the most obvious starting point for benefit-sharing negotiations, as this scenario corresponds most closely with the probable operation of the Rogun Hydropower Plant. Interestingly, the modeled economic benefits for all four riparian countries are, under this scenario, greater than with the baseline, with Tajikistan and Afghanistan benefiting the most.

Yet, while the total economic benefits for Tajikistan and Uzbekistan are greater than currently (Baseline), the additional benefits for both countries are considerably less than with their country priority or, in the case of Uzbekistan, also with the Downstream Agricultural Priority. In addition, this scenario would result in remarkable changes in the downstream countries' land use and agricultural practices, with the timing for both cotton and wheat changed (Appendix B). As a result, the actual benefit-sharing negotiations under this scenario would most probably focus on the downstream agriculture, and the actual possibilities of Uzbekistan and Turkmenistan to initiate such a radical change in their cropping practices.

3.2.2. Downstream Agricultural Priority

This scenario provides another, very different way to look at the possibilities for benefit-sharing. Under this operation scheme all countries except Afghanistan (with remarkably small decline by 4%) have considerable growth in total benefits: Tajikistan by 112%, Uzbekistan by 34%, and Turkmenistan by 136%. While total economic benefits are considerable even for Tajikistan, this scenario is challenging for Tajikistan due to the less optimal timing of Rogun's energy production. Consequently, one solution for this is to broaden the discussion from mere water use to broader aspects of energy production, and to link the solution to regional energy trade.

In such a setting the downstream countries of Uzbekistan and Turkmenistan need to agree to buy surplus energy produced by the Rogun Dam in summer months (as water is released for agricultural needs) and also agree to provide to Tajikistan additional energy (electricity, coal, oil, natural gas) in the winter period (when Tajikistan cannot operate Rogun, but needs to store its water for irrigation). Such a benefit-sharing scheme is actually not something new: it worked well in the former Soviet Union period (see e.g., [26]). Today, such arrangements just need political will and trust among riparian countries.

Table 4 helps to illustrate what this kind of compensation mechanism could look like in practical terms. The table presents the average energy production (GWh/month) for "Downstream Agricultural Priority" and "Upstream Energy Priority", with latter reflecting Tajikistan's ideal energy use over a year. It is clearly seen from Table 4 that if countries would pursue the "Downstream Agricultural Priority" scenario, Tajikistan would have a deficit of energy in the winter period (September–February) and a surplus of energy in the summer period (March–August). Therefore, the country will need compensation to satisfy its energy demands in the winter which, if summed up, presents 5805 GWh/winter of electricity. Depending on an agreement, the compensation could be in the form of barter supplies (like in the former Soviet Union period—coal, natural gas or oil to burn in the country's power plants) or monetary payments so that Tajikistan could buy necessary energy from other countries. According to our estimation the monetary value could be as much as USD 230 million (at price USD0.04/kWh).

In addition, the countries would need to agree what will happen to Tajikistan's summer electricity surplus (5252 GWh per summer): Tajikistan could either sell it through regional energy markets or then Uzbekistan and Turkmenistan could agree to buy it with an agreed rate. With price USD0.04/kWh, the cost of this summer electricity surplus would be USD 210 million. Overall, the monetary transactions of the downstream countries could be close to half of billion US dollars per year, with part of it going for buying Tajikistan's summer electricity surplus and part of it compensating Tajikistan for its winter energy deficit.

Table 4. Rogun reservoir energy production in two scenarios by months, averaged over future years (GW hours/month).

Month	Downstream Priority	Upstream Energy Priority	Difference between Scenarios
January	0	1134	−1134
February	0	945	−945
March	2670	1421	1250
April	1831	888	943
May	2536	962	1573
June	2070	945	1125
July	1120	976	144
August	1195	977	217
September	9	887	−878
October	54	920	−866
November	68	1030	−961
December	50	1071	−1021

3.2.3. No Priority (Optimising Basin-Wide Benefits)

The scenario of "No Priority", meaning maximization of basin-wide benefits, could also be considered as a possible case for further benefits-sharing elaboration. Similarly to the "Upstream Energy Priority" scenario, also this scenario brings increase in total economic benefits for all basin countries. When compared with "Downstream Agricultural Priority" scenario, however, the total economic benefits of Uzbekistan are in this scenario almost three times less and the benefits of Turkmenistan three times more. Such a change would most likely have implications on how the downstream countries' compensation to Tajikistan is shared.

Similarly to the "Downstream Agricultural Priority" scenario, the main shortcoming of this scenario is that Tajikistan's seasonal electricity production does not follow its energy needs. As a result, also in this scenario the possible benefit sharing scheme would include compensations from downstream countries to Tajikistan to ensure its energy needs. Table 5 presents the average energy production (GWh/month) for "No Priority" and "Upstream Energy Priority", with latter reflecting Tajikistan's ideal energy use over a year. Under the "No Priority", Turkmenistan and Uzbekistan would need to compensate Tajikistan a total of 5789 GWh of electricity per winter with an estimated price of USD 232 million (at price USD0.04/kWh). In addition, Tajikistan would also produce a total of 5229 GWh of "surplus electricity" during the summer (electricity it does not fully need), and the downstream countries could agree to buy this energy with an estimated total cost of USD 209 million. Downstream countries would have to spend almost half of billion US dollars a year for Tajikistan (part of it buying electricity, part of it as direct compensations) in order to make sure more optimal water allocation for their irrigated agriculture.

It seems that this scenario is almost the same as the previous case, but the difference exists—in this case, benefits of Uzbekistan have been reduced by almost three times and benefits of Turkmenistan have been increased by three times, compared with the "Downstream Agricultural Priority" scenario. This could mean a change in the compensation mechanism from each country to Tajikistan and, also, between them.

Table 5. Rogun reservoir energy production in two scenarios by months, averaged over future years (GW hours/month).

Month	No Priority	Upstream Energy Priority	Difference between Scenarios
January	0	1134	−1134
February	0	945	−945
March	2747	1421	1326
April	1900	888	1012
May	2611	962	1649
June	2043	945	1097
July	1031	976	55
August	1067	977	89
September	15	887	−872
October	60	920	−861
November	72	1030	−958
December	51	1071	−1020

4. Discussion

4.1. Methodological Implications: Beyond the Economic Benefits

This study has presented an experimental case study of modeling the economic benefits related to a water infrastructure development in a transboundary river, namely the Rogun Dam in Amu Darya River Basin. For us, the largest limitation of the study is its strong focus on economic benefits alone. Yet, the benefits-sharing scheme can also include several other benefits such as environmental, social, and political, which are difficult but still possible to assign monetary value. In the specific context of the Amu Darya River Basin, the political relations between the riparian countries are currently very complex, and the geopolitical relations and power asymmetries are the greatest hindrance for any cooperative agreement.

Both Uzbekistan and Turkmenistan oppose the Rogun Hydropower Plant as it would increase Tajikistan's control over Amu Darya's flows, and in this way provide a way to impact downstream countries. Similarly, Tajikistan would most likely not be willing to agree with the "Downstream Agricultural Priority" scenario, as the compensations included in such a scenario (whether through energy bartering or monetary payments) would increase its dependency on both Uzbekistan and Turkmenistan. Despite these limitations, we believe that highlighting the common economic gains that can be achieved from the operation of Rogun also motivate the countries to try to find the political will to look at the possible benefits of transboundary cooperation.

It should also be noted that the benefits are in the model are calculated only for agricultural and energy sectors. The study does not take into consideration other economic benefits, such as those related to domestic, industrial, recreational, and navigational use. It must also be noted that benefit-sharing in its broadest sense considers all actions that change the allocation of costs and benefits associated with transboundary cooperation. Yet, the study does not, for example, take into account who would bear the cost of the Rogun Dam, which is currently financed by the government of Tajikistan and is estimated to be remarkable for this poor country (around USD 3 billion [9]).

Finally, modeling always involves inaccuracies and simplifications, arising from the limited information available. The experimental case presented here consists of a number of assumptions and simplifications that need to be taken into account when considering the results. For example, the data concerning crops, prices, water use, and return flows was sparse, while the study excluded entirely groundwater use for irrigation due to lack of reliable data. In addition, the results are based on the assumption that there exist institutions and markets that efficiently use and allocate land, water, crop, and energy as described in the scenarios. Importantly, it is also assumed that the riparian countries are able to change their agricultural cropping patterns in ways outlined in Appendix B. Although it is unfeasible to change in a short run due to regional and/or national infrastructure, soil conditions, labor and farming methods, financial support from the local governments, etc., certainly various options to do this in a long run are available.

There were several important aspects that were excluded from the model. First, the impact of climate change to the Amu Darya River Basin and its flows was not included. A comprehensive analysis would be needed to account for range of risks associated with future climate variability. Second, the optimization framework was modeled to ensure energy requirements according to the published *i.e.*, historical monthly electricity demand in Tajikistan [27]. As a result, if electricity demand changes, also the actual operation scheme of the RHP could change: this would obviously impact also downstream flows. Third, the model did not consider the impacts from drought periods that are also likely to change the Rogun's operation scheme. Fourth, the model neglects the benefits that the establishment of common trade area such as common markets of agricultural commodities and energy would result in the region.

Given the experimental nature of the presented study there is a big room for improvements and further extension of the modeled framework. Even within the economic benefits, such as energy and agricultural ones considered in the current study, there are possibilities to include new dimensions and policy options, such as the establishment of a common trade, area such as common markets of agricultural commodities and energy, which are currently absent. Surely, these new policies would give a new look at ways to solve the existing problems and, hopefully, would push the basin countries to cooperate and negotiate new agreements on the ways to the sustainable management of water resources.

4.2. Reflections on Benefit-Sharing

The Amu Darya River Basin is considered as one of the complex food-water-energy nexus situations in transboundary river basins [28]. Today's complicated conditions are mostly related to the past situation when water, energy and agriculture of the region was managed as one integrated scheme from a central body in Moscow, which ruled "water quotas and energy barter deals to capitalize on an abundance of water resources in the upstream territories and a wealth of fossil-fuel resources in the downstream territories, respectively." [28]. As the riparian countries gained independence and the barter relations disappeared. The situation changed profoundly as several riparian countries are currently involved in a conflict over the use of scarce water, energy and food resources of the region. Desire of upstream countries to build new dams added a spark to the already conflict-prone regional affairs. In the context where each country aims for a kind of resource capture in relation to water and is not willing to negotiate cooperative water resource management, this study looks at how the idea of benefit-sharing would help the region to proceed from the current political deadlock. We believe that the concept could be useful in bringing riparian countries with opposite interests to the possible cooperation scheme.

The benefit-sharing process consists of three main phases: motivation, design, and implementation [29]. In the case of the Amu Darya River Basin, however, none of these phases from riparian countries really exists today. Given the complex political situation, our study could form a basis for a "pre-motivation" for benefit sharing. The starting point for such a phase would be the notion that the Rogun Hydropower Plant has a potential to make all riparian countries better off, and that a common benefit sharing scheme and related compensations mechanisms are actually economically feasible. The region can also search for lessons learnt from other river basins where dams have created common benefits, potential examples including the dams on the Senegal River, the Canadian dams on the Columbia River, the Lesotho Highlands Water Project on the River Orange-Senqu, the Kariba Dam on the River Zambezi, the Itaipu Dam on the Rio Parana, *etc.*

The possible "pre-motivation" for the benefit-sharing could be offered by the "Upstream Energy Priority", "Downstream Agricultural Priority", and "No Priority" scenarios. "Upstream Energy Priority" scenario presents the most natural starting point, as it would ensure Tajikistan's energy production and at the same time increase the agricultural benefits to downstream countries (although only through quite remarkable changes in land use and cropping practices). Yet, it would also provide a Tajikistan a means to control part of the Amu Darya's flow, and in this way to influence the critical agricultural sectors downstream. On the other hand, the energy benefits gained by Tajikistan could even lead the country to provide some support for the downstream countries of Uzbekistan and Turkmenistan to restructure their agriculture to better cope with the changed water flows.

Another potential options for pre-motivation are presented by the "Downstream Agricultural Priority" and the "No Priority" scenarios. Both scenarios could be viewed as a return to the benefit-sharing logic used during the Soviet period, where water is stored in upstream during the winter and released then during summer to ensure downstream irrigation. As a return, downstream states provide upstream states with energy resources (natural gas, oil, coal) during winter. While both scenarios are structurally similar, they differ in terms of the distribution of the economic benefits between downstream countries as well as in upstream Tajikistan (where the "No Priority" would lead to remarkable increase in agricultural benefits). This would also have an impact on the actual compensations mechanisms *i.e.*, *who* and *how much* should be compensated. To help such discussion, we also compared the energy production under these scenarios with "Upstream Energy Priority" and gave also monetary estimates for their seasonal differences.

Naturally, each country wishes to maximize its own individual benefits from the river. Without any basin agreement, Tajikistan is likely to build and operate Rogun to maximize its energy benefits. Tajikistan is thus not likely to agree on any other operation scheme unless other countries offer Tajikistan some kind of compensation. However, the situation is not so simple, as in reality Tajikistan has major challenges to go ahead with Rogun without the consent of other riparian countries. Unlike upstream countries in some other river basins (for example Turkey in Tigris-Euphrates or China in Mekong River), Tajikistan is geopolitically or economically not in a strong position to pursue its own interests without agreement with its more powerful neighboring countries.

While it is at this stage difficult to foresee even the scale of future negotiations and size of monetary compensation, it seems possible that Uzbekistan could require some benefits from upstream Tajikistan and possibly also from downstream Turkmenistan, as otherwise it could use water in own agriculture without letting it flow to Turkmenistan. All basin countries would anyway need to develop some kind of agreement on the ways to utilise the changed water flows that the RHP will cause. Obviously the basis for such an agreement could stem from the Convention on the Protection and Use of Transboundary Watercourses and International Lakes of the UN System (often called the UNECE Convention or the Helsinki Convention), as the downstream countries Uzbekistan and Turkmenistan already have ratified the convention at the time of writing of this paper [30].

5. Conclusions

With increasing resource scarcities, the water, food and energy nexus is growing in importance. Addressing conflict among these closely interrelated domains is considered increasingly important for equity and sustainability in transboundary river basins. The issue complex but it requires careful consideration by reviewing existing best practices, identifying new approaches, and implementing state-of-the-art principles and processes. This article argues that despite the basin-scale and sectoral challenges in the Amu Darya River Basin, the concept of benefit sharing could be utilized to find a way that enhances the economic benefits and, ultimately, water, energy and food security for all riparian countries. Due to its history as "one basin' under the Soviet regime, the basin provides an interesting case on how the water use has developed. After the collapse of the Soviet Union, newly independent Central Asian nations prioritized the own food, water and energy security as well as agricultural trade to the detriment of basin-wide integrated management.

Amu Darya River Basin is shared by four riparian countries with different interests in utilizing the River's water for agricultural and energy production. Water, food and energy security issue in the region has been extensively under consideration in recent years. The conflicting situation between upstream countries willing to enhance energy security and downstream countries willing to continue with status quo because of fearing for their food security could worsen current geopolitical situation. Well-known and increasingly popular concepts of basin-wide benefits optimization may not work in such conflict prone conditions without a process of negotiation with sound motivation and commitment. Even at the national level, management of water, energy and agriculture remains almost exclusively sector based. The water, energy and agriculture nexus can thus be used as a mechanism for both conflict resolution and development. However, current situation presents clear conflict between upstream energy development and downstream agricultural development, energy versus food security in relevant countries. Current political situation leaves almost no room for possible maneuver.

This study provides an experimental case study on how the concept of benefit-sharing could be used to motivate the riparian countries to enhance their cooperation in the development of their shared water resources. The used hydro-economic model used provides specific estimates on the total economic benefits that different operation schemes (scenarios) of the planned Rogun Hydropower Plant could bring for each riparian country. While the model can suggest an optimal scenario for the entire basin (with no country priorities), it may not be optimal for a particular basin country. Yet, we recognised three different scenarios—"Upstream Energy Priority", "Downstream Energy Priority", and "No Priority"—that could all be used as a starting point for further negotiations about benefit-sharing and related compensation mechanisms. The scenarios help each riparian country to see the potential for benefits—and their limits—associated with cooperating in a regional, basin-wide agreement across water-energy-agriculture nexus.

Acknowledgments: The study was done under the Academy of Finland funded NexusAsia project (269901). The authors wish like to thank the colleagues at Aalto University's Water & Development Research Group for their support. Special thanks go to Ms Miina Porkka for providing map of Central Asia (Figure 1).

Author Contributions: Shokhrukh-Mirzo Jalilov developed the hydro-economic model used in the analysis, producing and interpreting its results, and writing majority of the article. Marko Keskinen and Olli Varis contributed in defining the scope of the article and wrote parts of the Introduction, Discussion and Conclusions sections.

Conflicts of Interest: The authors declare no conflict of interest.

Appendix A

The model and its documentation were originally developed for application to the Amu Darya Basin in the Central Asia. However, it is adaptable to the hydrology, water infrastructure development, land use patterns, economics, and institutions of any basin. The essential principle of the hydrology component of model is mass balance—for surface flow, water diversions and water depletions for use in irrigated agriculture. Important variables tracked include water storage capacity, crop mix, land in production, and farm income. The model structure is defined below using the GAMS notation, described by the vendor at gams.com.

A.1. Sets

Sets are the dimensions over which the storage scaling model is defined. A similar structure could be used for reservoir capacity analysis anywhere new water storage is planned. The following sets and set elements are used:

i	Flows	/inflow, river, divert, use, return, release/
u	Stocks	/reservoir/
t	Months	/January–December/
y	Years	/1–10/
j	crop	/cotton, wheat, vegetables/
k	crop season	/first, second/
n	water supply scenario	/base, dry/
p	Policy	/Baseline, Energy, UZ, TK, Downstream, OPTIM/
s	Region countries	/Afghanistan, Tajikistan, Turkmenistan, Uzbekistan/

A.2. Data

Some of the following parameters (data) terms end in p to distinguish parameters (unknown terms) from unknown variables. Parameters are:

Bu (divert, use)	defines consumptive use as a percent of diversion
Br (divert, return)	defines surface return flow as a percent of diversion
BLv (rel, u)	links reservoir releases to downstream flows
source (inflow ,y, t, n, p)	annual basin inflows at headwaters in scenario (cubic km per month)
yield_p (use, j, k)	crop yield (tons per hectare)
cost_p (use, j, k)	crop cost of production (USD per ton)
price_elast (j)	price elasticity of demand
P_p (j)	observed crop price (USD per ton)
Bu_p (i, j, t)	crop water demand per hectare (divert + use + return) per month
Capacity (res, p)	reservoir maximum capacity by stages
Z0 (res)	initial reservoir level at stock node
h0_p (res, y, t, n, p)	dam's maximum height in stages
ID_ru (return, use)	identity matrix connects return nodes to use nodes
ID_du (divert, use)	identify matrix connects divert nodes to use nodes
Landrhs_p (use)	Irrigated land area by countries (million hectares)
hydro_price (res, t)	price of hydropower (constant USD per kWh)

A.3. Variables (Unknowns)

Some of unknown variable ends in v, to distinguish variables from known data. The model solves for the optimal value of each of these variables, for which the goal is to maximize total basin net economic benefits.

A.3.1. Positive Variables

Z (res, y, t, n, p)	reservoir water stocks
reservoirs_h_v (res, y, t, n, p)	reservoirs height in each month
supply_v (inflow, y, t, n, p)	supplies
hectares_v (use, j, k, y, t, n, p)	area under each crop in each country in time
land_v (use, y, t, n, p)	land in production
production_v (use, y, t, n, p)	crop produced in each country
T_production_v (j, k, y, t, n, p)	crop produced
energy_prod_v (res, y, t, n, p)	energy production
energy_ben_v (res, y, t, n, p)	energy production benefits in Rogun in each month

A.3.2. Free variables

X (i, y, t, n, p)	water flows (inflow, river, divert, use, return, release)
Con_surp_v (j, y, t, n, p)	consumer surplus
ag_ben_v (use, y, t, n, p)	net income over crops by node and period
tot_agben_v (use, n, p)	net agricultural benefits by country
con_surpl_v (j, y, t, n, p)	consumer surplus
Totben_v (n, p)	total benefits

A.4. Equations

A.4.1. Hydrology

The essential principle of the hydrology is mass balance, both for surface flow interactions, reservoir levels. The hydrology uses mass balance principles to account for headwater flows, river flows, reservoir levels, water from surface applied to various uses, and the impact of surface flows on current and future reservoir storage levels.

A.4.1.1. Headwater Runoff

$$X \text{ (inflow, y, t, n, p)} = \text{source (inflow, y, t, n, p)} \tag{A1}$$

A.4.1.2. River Flow

$$\begin{aligned} X \text{ (river, y, t, n, p)} = \text{sum (inflow, Bv (inflow, river)} \times X \text{ (inflow, y, t, n, p))} + \\ \text{sum (riverp, Bv (riverp, river)} \times X \text{ (riverp, y, t, n, p))} + \\ \text{sum (divert, Bv (divert, river)} \times X \text{ (divert, y, t, n, p))} + \\ \text{sum (return, Bv (return, river)} \times X \text{ (return, y, t, n, p))} + \\ \text{sum (rel, Bv (rel, river)} \times X \text{ (rel, y, t, n, p))} \end{aligned} \tag{A2}$$

A.4.1.3. Water Diverted

$$\begin{aligned} X \text{ (divert, y, t, n, p)} = \text{sum ((j,k), Bu\_p (divert, j, t)} \times \text{sum (use, ID\_du (divert, use)} \\ \times \text{hectares\_v (use, j, k, y, t, n, p)} \end{aligned} \tag{A3}$$

A.4.1.4. Gross Surface Returns to River

$$X \text{ (return, y, t, n, p)} = \text{sum ((j,k), Bu\_p (return, j, t)} \times \text{sum (use, ID\_ru (return, use)} \\ \times \text{ hectares\_v (use, j, k, y, t, n, p)} \quad \text{(A4)}$$

A.4.1.5. Water Consumed

Any water use node's consumptive use, it is an empirically-determined proportion of total water applied. For irrigation, consumptive is the quantity of water lost through plant evapotranspiration (ET) to any future use in the system. For agricultural nodes, water use is measured as:

$$X \text{ (use, y, t, n, p)} = \text{sum ((j,k), Bu\_p (use, j, t)} \times \text{hectares\_v (use, j, k, y, t, n, p)} \quad \text{(A5)}$$

For hydropower generation use, consumptive use is the quantity of water flowing through turbines. However, that water quantity could be reused for irrigation if it fits right time. That water use generates energy, which cannot be negative. It is measured as:

$$Z \text{ (res, y, t, n, p)} = Z0 \text{ (res)} - \text{sum (rel, BLV (rel, res)} \times X \text{ (rel, y, t, n, p))} \quad \text{(A6)}$$

Energy production is total water flow to generate energy in month of year, in scenario and policy. Remaining coefficients are: g-gravitational constant (g = 9.8 N/kg); E, Efficiency, which can vary from 0 to 1, and transformation coefficients.

$$\text{Energy\_prod\_e (res, y, t, n, p)} = X \text{ (y,t,n,p)} \times 9.8 \times 0.75 \times 24 \times (365/12) \quad \text{(A7)}$$

A.4.1.6. Reservoir Storage

Reservoir contents are:

$$Z \text{ (res, y, t, n, p)} = \text{sum ((y2, t2), Z (res, y2, t2, n, p)} - \text{sum (rel, BLv (rel, res)} \times X \text{ (rel, y, t, n, p)} \quad \text{(A8)}$$

Electric power comes from building a Reservoir on a river that has a large drop in elevation. There are few hydroelectric plants in flat places. The dam stores water behind it in the reservoir, and a higher storage volume of water in the reservoir means that the water falls a greater distance and reaches a greater velocity when passing through the turbines. The turbine converts the energy of flowing water into mechanical energy. The hydroelectric generator converts this mechanical energy into electricity. The hydraulic head for the reservoir's dam in the month of year, scenario and policy was empirically estimated to fit conditions for the Rogun reservoir:

$$\text{Reservoir\_h (res, y, t, n, p)} = 3698.10229 - 3451.26986 \times (1/(z \text{ (res, y, t, n, p)} + 0.001) \times 0.01)) \quad \text{(A9)}$$

The idea of these coefficients is: on the basis of known limited data on Rogun reservoir's water volume which depends on height of dam to find nonlinear relation between storage volume and head of the dam, so we could exactly predict what would be head having certain water volume. Contents of the reservoir in the initial period (0), Z0:

$$Z0 = 0 \quad \text{(A10)}$$

The upper bound on the reservoir's contents is defined as:

$$Zmax = C \quad \text{(A11)}$$

This equation guarantees that the reservoir's level never exceeds its capacity. Policies that would change a reservoir's capacity, such as dredging or adding to a dam's height, are simulated by altering the value of C.

A.5. Land Use

Land use patterns affect the demand for water. For irrigated agriculture, total land in production is expressed as:

$$\text{Sum ((j,k), hectares\_v (use, j , k, y, t, n, p) = land\_v (use, y, t, n, p)} \tag{A12}$$

This states that irrigated land in production by node, crop, season, and time, summed over crops and seasons cannot exceed available land (RHS) by node, time period for any given scenario and policy. In most dry rural regions of the world, like the Amu Darya Basin, water is often more limiting than land. While we used the maximum current capacity in irrigated land for countries of the Basin as the upper limit on available land, more area will likely become available if greater long term water supplies can be secured and if institutions adjust to permit the extra water to be used by agriculture.

The baseline policy analysis is constrained to replicate historical irrigated land by country and crop. For the two alternative policies, those constraints are removed by allowing water tradeoffs to occur, either within a single or among irrigated areas. Either policy permits existing water to be reallocated to higher economic valued water uses where the economics would support such a reallocation.

A.6. Economics

Economic benefits are produced by water depletions at use nodes for irrigated agriculture and by the water flowing through turbines to generate energy at reservoir nodes. For agricultural uses, the willingness to pay is measured by the contribution of water to net farm income which equals crop price multiplied by yield minus cost of production plus any unpriced consumer surplus. Consumer surplus is an unpriced value, equal to the amount by which power buyers' economic welfare exceeds the actual price charged. It is measured as the area beneath the demand function and above actual price charged:

$$\text{Con\_surp\_v (j, y, t, n, p)} = 0.5 \times \text{(b0\_p (j)} - \text{Crop\_price\_v (j, y, t, n, p)]} \times \text{sum} \\ \text{(k, T\_production\_v (j, k, y, t, n, p))} \tag{A13}$$

For energy benefits, total revenue is measured as the price of electricity multiplied by the quantity produced. In the current implementation of the model, that electricity price is set at recent observed levels in the basin. Reduced prices from additional hydropower will raise consumer surplus, while increased prices will reduce consumer surplus. For regions of the basin that currently have little access to power, increases in consumer surplus are economically and politically very important to achieve:

$$\text{Energy\_ben\_v (res, y, t, n, p) = energy\_prod\_v (res, y, t, n, p)} \times \text{hydro\_price (res, t)} \tag{A14}$$

Agricultural benefits are measured as:

$$\text{Ag\_ben\_v (use, y, t, n, p) = sum ((j,k), (P\_p (j)} \times \text{yield\_p (use, j,t)} - \text{cost\_p (use, j, t))} \tag{A15}$$

Price of particular crop is a negatively sloping demand function, which means that one price is set for each crop for all riparian countries, so any crop could be grown in the most favorable conditions. What this means is that:

$$\text{Crop\_price\_v (j, y, t, n, p) = b0\_p(j) + b1\_p(j)} \times \\ \text{sum(k, T\_production\_v (j, k, y, t, n, p));} \tag{A16}$$

The empirically estimated coefficients b0_p and b1_p are linearized demand function based on estimated price elasticities.

To measure total crop production following equation is used:

$$T\_production\_v = sum \; (use, \; production\_v \; (use, \; j, \; k, \; y, \; t, \; n, \; p)) \tag{A17}$$

A.7. Discounted Net Present Value

Finally, the basin scale integrated model maximizes discounted net present value across all water uses, water environments, and time periods subject to hydrologic and institutional constraints:

$$DNPV = sum \; ((u, \; t, \; y) \times Ag\_ben\_v \; (use, \; y, \; t, \; n, \; p) \; / \; (1+r_u)^t + \\ Energy\_ben\_v \; (res, \; y, \; t, \; n, \; p)/(1 + r_u)^t \tag{A18}$$

This says that the net present value of total water-based benefits for all nodes in Amu Darya Basin sums income over countries and time-periods, which discounts future incomes more heavily when there is a higher discount rate. The current model implementation uses a 5% discount rate. The model allocates water among the basin's water uses, locations, and time periods to maximize *DNPV*, subject to stated constraints.

Appendix B

Table B1 shows results of land area under crop production by country, scenario, crop, and cropping season. As can be seen most scenarios give chance to the basin countries to increase their irrigated land compared to the baseline scenario. Upstream Tajikistan and Afghanistan don't have any areas under cotton, because the model optimizes its production to downstream countries where cost of production is smaller and yield is higher; also, upstream mountainous and cool climate areas could be not very suitable for cotton production. Among all countries only Afghanistan shows no reduction of land under production in any scenarios. As this country doesn't have large irrigated land in the basin and doesn't withdraw much water directly from Amu Darya on compare to other riparians, thus below analysis will be about other three basin countries.

Table B1. Land under production by country, crop, season, and scenario, averaged over future years (millions of hectares/season).

Country	Scenario	Cotton		Wheat		Vegetables		Total Land
		Planted in March	Planted in May	Planted in March	Planted in May	Planted in March	Planted in May	
Tajikistan	Baseline-No dam	0.00	0.00	0.22	0.00	0.03	0.06	0.30
	Upstream Energy Priority	0.00	0.00	0.18	0.00	0.07	0.02	0.27
	Uzbekistan Priority	0.00	0.00	0.22	0.00	0.03	0.03	0.28
	Turkmenistan Priority	0.00	0.00	0.22	0.00	0.03	0.03	0.28
	Downstream Priority	0.00	0.00	0.22	0.00	0.03	0.02	0.26
	No Priority-Optimal	0.00	0.00	0.53	0.00	0.09	0.01	0.64
Afghanistan	Baseline-No dam	0.00	0.00	0.08	0.00	0.00	0.00	0.08
	Upstream Energy Priority	0.00	0.00	0.12	0.00	0.01	0.00	0.13
	Uzbekistan Priority	0.00	0.00	0.08	0.00	0.00	0.00	0.08
	Turkmenistan Priority	0.00	0.00	0.08	0.00	0.00	0.00	0.08
	Downstream Priority	0.00	0.00	0.08	0.00	0.00	0.00	0.08
	No Priority-Optimal	0.00	0.00	0.22	0.01	0.01	0.00	0.23

Table B1. *Cont.*

Country	Scenario	Cotton		Wheat		Vegetables		Total Land
		Planted in March	Planted in May	Planted in March	Planted in May	Planted in March	Planted in May	
Uzbekistan	Baseline-No dam	0.04	0.74	1.08	0.02	0.00	0.00	1.88
	Upstream Energy Priority	0.50	0.33	0.62	0.45	0.00	0.00	1.90
	Uzbekistan Priority	0.99	0.11	0.08	1.33	0.05	0.00	2.57
	Turkmenistan Priority	0.04	0.19	1.08	0.02	0.00	0.00	1.33
	Downstream Priority	0.95	0.11	0.17	1.40	0.00	0.01	2.62
	No Priority-Optimal	0.80	0.08	0.31	1.29	0.00	0.00	2.49
Turkmenistan	Baseline-No dam	0.32	0.00	0.31	0.15	0.00	0.00	0.78
	Upstream Energy Priority	0.45	0.01	0.17	0.31	0.00	0.00	0.94
	Uzbekistan Priority	0.01	0.00	0.31	0.45	0.00	0.00	0.77
	Turkmenistan Priority	0.90	0.00	0.01	0.76	0.04	0.00	1.71
	Downstream Priority	0.04	0.00	0.39	0.35	0.06	0.00	0.84
	No Priority-Optimal	0.45	0.02	0.08	0.40	0.00	0.00	0.95

Tajikistan could experience a reduction of irrigated land under production in all scenarios except "No Priority" scenario. This happens due to fact that water is redirected to produce energy ("Upstream Energy Priority") or to irrigate more profitable downstream cotton and wheat areas in "Uzbekistan Priority", "Turkmenistan Priority", and "Downstream Priority" scenarios. Noticeably, "No Priority" scenario allows Tajikistan more than double irrigated land and increase land under wheat production which means increase food security for the country.

Uzbekistan could have reduction of cropping area in the scenario of Turkmenistan agriculture prioritization but would have considerable land increase in all scenarios except "Upstream Energy Priority." This means if the reservoir under construction would be managed in a way to pursue agricultural interests of Uzbekistan or both Uzbekistan and Turkmenistan or basin-wide benefits it would potentially be source for economic development in this country, particularly in its Amu Darya region.

Finally, Turkmenistan could have minor reduction of irrigated area in scenario "Uzbekistan Priority" but could have increase in all other scenarios. Thus, the Rogun reservoir would bring agricultural development to this country as well.

As far as crop differentiation by countries, Uzbekistan is the major cotton producer in the region with almost a million hectares of land occupied by this strategic cash crop. Uzbekistan is followed by Turkmenistan in cotton production in the region. Wheat is the major food crop for all countries in the Amu Darya basin and performs an important role in regional food security. Similarly to cotton, Uzbekistan has the most land under wheat cultivation, followed by Turkmenistan. However, since wheat is essential for food security, it also makes up an important part of the crop mix in both Tajikistan and Afghanistan.

References

1. Sadoff, W.C.; Grey, D. Cooperation on International Rivers: A Continuum for Securing and Sharing Benefits. *Water Int.* **2005**, *30*, 420–427. [CrossRef]
2. Sadoff, C.W.; Grey, D. Beyond the river: The benefits of cooperation on international rivers. *Water Policy* **2002**, *4*, 389–403. [CrossRef]
3. Klaphake, A. *Kooperation an Internationalen Flüssen aus Ökonomischer Perspektive: Das Konzept des Benefit Sharing*; Deutsches Institut für Entwicklungspolitik: Bonn, Germany, 2005; Available online: http://www.die-gdi.de/uploads/media/6-2005.pdf (accessed on 28 August 2015).

4. Phillips, D.J.H.; Daoudy, M.; Öjendal, J.; Turton, A.R.; McCaffrey, S. *Trans-boundary Water Cooperation as a Tool for Conflict Prevention and for Broader Benefit-Sharing*; Expert Group on Development Issues, Department for International Development Cooperation; Ministry of Foreign Affairs: Stockholm, Sweden, 2006; Available online: http://www.eldis.org/go/home&id=22735&type=Document#.Vd_-DdKl_mM (accessed on 28 August 2015).

5. Qaddumi, H. *Practical Approaches to Transboundary Water Benefit Sharing*; Overseas Development Institute: London, UK, 2008.

6. Lee, S. Benefit sharing in the Mekong River basin. *Water Int.* **2014**, *40*, 139–152. [CrossRef]

7. Bekchanov, M.; Bhaduri, A.; Ringler, C. *Is Rogun a Silver Bullet for Water Scarcity in Central Asia?* Center for International Development and Environmental Research: Giessen, Germany, 2013; Available online: http://ageconsearch.umn.edu/bitstream/159075/2/Bekchanovetal2013aRogunimpactfinal.pdf (accessed on 31 May 2014).

8. Jalilov, S.; Amer, S.; Ward, F. Reducing conflict in development and allocation of transboundary Rivers. *Eurasian Geogr. Econ.* **2013**, *54*, 78–109.

9. Jalilov, S.; Amer, S.; Ward, F. Water, food, and energy security: An elusive search for balance in central Asia. *Water Resour. Manag.* **2013**, *27*, 3959–3979. [CrossRef]

10. Kim, Y.; Indeo, F. The new great game in central Asia post 2014: The US "New Silk Road" strategy and Sino-Russian rivalry. *Commun. Post Commun. Stud.* **2013**, *46*, 275–286. [CrossRef]

11. Arbour, L. Next year's wars. *Foreign Policy*. 22 February 2011. Available online: foreignpolicy.com/2011/12/27/next-years-wars-2/ (accessed on 28 August 2015).

12. Wegerich, K. Coping with disintegration of a River-Basin management system: Multidimensional issues in central Asia. *Water Policy* **2004**, *6*, 335–344.

13. Spoor, M.; Krutov, A. The power of water in a divided central Asia. *Perspect. Glob. Dev. Technol.* **2003**, *2*, 593–614. [CrossRef]

14. Wegerich, K. Hydro-hegemony in the Amu Darya Basin. *Water Policy* **2008**, *10*, 71–88. [CrossRef]

15. Central Intelligence Agency (CIA). The World Factbook. Uzbekistan. Available online: https://www.cia.gov/library/publications/resources/the-world-factbook/geos/uz.html (accessed on 28 August 2015).

16. Glantz, M. Water, climate, and development issues in the Amu Darya Basin. *Migr. Adapt. Strateg. Glob. Chang.* **2005**, *1*, 23–50. [CrossRef]

17. Schlüter, M.; Herrfahrdt-Pähle, E. Exploring resilience and transformability of a river basin in the face of socioeconomic and ecological crisis: An example from the Amu Darya River Basin, central Asia. *Ecol. Soc.* **2011**, *16*, 32–45.

18. Schlüter, M.; Savitsky, A.; McKinney, D.; Lieth, H. Optimizing long-term water allocation in the Amu Darya River delta: A water management model for ecological impact assessment. *Environ. Model. Softw.* **2005**, *20*, 529–545. [CrossRef]

19. CA Water Info. Scientific-Information Center of the Interstate Commission for Water Coordination in Central Asia. 2015. Available online: http://www.cawater-info.net/index_e.htm (accessed on 28 August 2010).

20. Schmidt, R. Onwards and Upwards. Water Power Magazine. 2008. Available online: http://www.waterpowermagazine.com/features/featureonwards-and-upwards/ (accessed on 28 August 2015).

21. Brooke, A.; Kendrick, D.; Meeraus, A.; Raman, R. *GAMS Language Guide*; GAMS Development Corporation: Washington, DC, USA, 2006.

22. UNECE. Our Waters: Joining Hands across Borders. First Assessment of Transboundary Rivers, Lakes and Groundwaters. 2007. Available online: www.unece.org/env/water/blanks/assessment/assessmentweb_full.pdf (accessed on 28 August 2015).

23. World Bank. Irrigation in Central Asia. Social, Economic and Environmental Considerations. 2003. Available online: http://siteresources.worldbank.org/ECAEXT/Resources/publications/Irrigation-in-Central-Asia/Irrigation_in_Central_Asia-Full_Document-English.pdf (accessed on 28 August 2015).

24. The Rogunskaya Hydro Power Station, Performance Characteristics. Dushanbe. Tajikistan Open Joint Stock Holding Company "Barki Tojik". 2009. Available online: http://www.tjus.org/Copy%20of%20Rogynskaya%20GES%20_English%20version1.pdf (accessed on 25 September 2012).

25. FACT SHEET. TALCO Energy Audit: Improved Efficiency Could Help Solve Winter Electricity Shortages. December 2012; in Financial Assessment of Barki Tojik, World Bank, October 2013. Available online: http://www.worldbank.org/content/dam/Worldbank/document/tj-talco-energy-audit-fact-sheet.pdf (accessed on 28 August 2015).

26. Soliev, I.; Wegerich, K.; Kazbekov, J. The Cost of Benefit Sharing, the Case of the Ferghana Valley in the Syr Darya Basin. *Water* **2015**, *7*, 2728–2752. [CrossRef]

27. The International Bank for Reconstruction and Development; The World Bank. Tajikistan's Winter Energy Crisis: Electricity Supply and Demand Alternatives. November 2012. Available online: http://siteresources.worldbank.org/ECAEXT/Resources/TAJ_winter_energy_27112012_Eng.pdf (accessed on 30 May 2014).

28. Ozment, S.; DiFrancesco, K.; Gartner, T. *The Role of Natural Infrastructure in the Water, Energy and Food Nexus, Nexus Dialogue Synthesis Papers*; IUCN: Gland, Switzerland, 2015; Available online: http://www.iwa-network.org/downloads/1438744856-Natural%20Infrastrucure%20in%20the%20Nexus_Final%20Dialogue%20Synthesis%20Paper%202015.pdf (accessed on 28 August 2015).

29. Sadoff, C.; Greiber, T.; Smith, M.; Bergkamp, G. *Share-Managing Water across Boundaries*; IUCN: Gland, Switzerland, 2008.

30. United Nations Treaty Collection. *Chapter XXVII Environment, 5. Convention on the Protection and Use of Transboundary Watercourses and International Lakes*; United Nations Office of Legal Affairs: New York, NY, USA, 2015; Available online: https://treaties.un.org/Pages/ViewDetails.aspx?src=TREATY&mtdsg_no=XXVII-5&chapter=27&lang=en (accessed on 27 August 2015).

water

MDPI

Article

Water-Energy-Food Nexus in a Transboundary River Basin: The Case of Tonle Sap Lake, Mekong River Basin

Marko Keskinen [1],*, Paradis Someth [2], Aura Salmivaara [1] and Matti Kummu [1]

[1] Water & Development Research Group, Aalto University, P.O. Box 15200, Aalto 00076, Finland;
aura.salmivaara@aalto.fi (A.S.); matti.kummu@aalto.fi (M.K.)

[2] Department of Rural Engineering, Institute of Technology of Cambodia, P.O. Box 86, Russian Federation Boulevard, Phnom Penh 12152, Cambodia; someth@itc.edu.kh

* Correspondence: marko.keskinen@aalto.fi; Tel.: +358-50-3824626

Academic Editor: Miklas Scholz
Received: 1 June 2015; Accepted: 31 August 2015; Published: 12 October 2015

Abstract: The water-energy-food nexus is promoted as a new approach for research and policy-making. But what does the nexus mean in practice and what kinds of benefits does it bring? In this article we share our experiences with using a nexus approach in Cambodia's Tonle Sap Lake area. We conclude that water, energy and food security are very closely linked, both in the Tonle Sap and in the transboundary Mekong River Basin generally. The current drive for large-scale hydropower threatens water and food security at both local and national scales. Hence, the nexus provides a relevant starting point for promoting sustainable development in the Mekong. We also identify and discuss two parallel dimensions for the nexus, with one focusing on research and analysis and the other on integrated planning and cross-sectoral collaboration. In our study, the nexus approach was particularly useful in facilitating collaboration and stakeholder engagement. This was because the nexus approach clearly defines the main themes included in the process, and at the same time widens the discussion from mere water resource management into the broader aspects of water, energy and food security.

Keywords: transboundary water-energy-food nexus; climate change; science–policy–stakeholder interaction; cross-sectoral collaboration; integrated planning; Integrated Water Resources Management (IWRM); transboundary rivers

1. Introduction

The water-energy-food nexus ("the nexus") and its different variations (including, e.g., security [1], climate change [2] or ecosystems [3]) are discussed by a variety of actors [4–17]. "Nexus" literally means "a means of connection between things in a series" [18], and the water-energy-food nexus focuses on the linkages between the three nexus themes. The main rationale for those promoting the water-energy-food nexus is that as the different nexus themes are so closely related, they should be looked at simultaneously to encourage win-win situations, avoid negative impacts and, ultimately, enhance sustainability. As Hoff (2011) [1] puts it: "*The nexus focus is on system efficiency, rather than on the productivity of isolated sectors*".

Such objectives are by no means completely new, but share similarities with the aims of integrated approaches such as Integrated Water Resources Management (IWRM) [19,20]. They are, however, perhaps more important than ever given the growing scarcity of land, water and related resources caused by population growth and increasing demand for food and energy. When discussing the nexus, it is useful to recognise that it has different dimensions: we discuss what we consider to be the two key dimensions of the nexus. The first dimension promotes the nexus as an approach for research and analysis by, for example, quantifying the linkages between the nexus sectors [1,6,21–23]. The second dimension presents the nexus more as a policy-making tool that is seen to facilitate cross-sectoral collaboration and integrated planning and policy-making [3,4,9,24]. The key in both is to focus on the linkages between the nexus themes, rather than on separate themes and sectors alone.

This article looks at the water-energy-food nexus in Cambodia's Tonle Sap Lake area. Tonle Sap Lake is closely connected to the transboundary Mekong River, and the annual floods of the Mekong are the main driving force for the Tonle Sap flood pulse. The flood pulse extends the lake to the vast floodplains, and brings fertile suspended solids and fish larvae to the lake-floodplain system. For the same reason, the hydrological and environmental changes happening in the Mekong River Basin have direct impacts to the Tonle Sap system [25–27]. As a result, while the Tonle Sap system forms an important source of local and national food security, its food production capacity is threatened by one of the most intensive hydropower development plans in the world [28]. The area thus presents a highly topical example of a transboundary water-energy-food nexus context, but the area and its future development has not yet been assessed using such an approach.

This article aims to fill this research gap by presenting the key findings from a research project focusing on the Tonle Sap Lake [29]. The project was active in 2010–2013 and had the water-energy-food nexus as its starting point [30], seeking to increase the understanding of the Tonle Sap system in terms of water, energy and food security as well as climate change. While the project built on the earlier research by the authors in the area [31–35], the nexus approach was expected to provide new insights as well as to facilitate science-policy-stakeholder interaction. Both were seen to be important given the significance of the Tonle Sap system, the changes it is experiencing and the management challenges of the area [36]. The analysis made use of new information and novel research methods, including detailed, cumulative assessment of the impacts of hydropower development and climate change on the Tonle Sap as well as the first-ever trend analysis of key demographic and socio-economic indicators using the Population Census for years 1998 and 2008. Special attention was paid to the policy relevance of the research results, and the project was done in close collaboration with the governmental Tonle Sap Authority (TSA).

In this article, we discuss the key findings from our research project in both practical terms (the Tonle Sap system and its future, with emphasis on the nexus) and in methodological terms (the role that nexus had in our study; also when compared to our earlier experiences using IWRM). The aim of the article is thus two-fold: To increase understanding of the Tonle Sap system and to contribute to the on-going discussion about the nexus as an approach.

2. Materials and Methods

2.1. Defining the Linkages between the Nexus Themes

The nexus approach selected for this study put the emphasis on the interconnections between water, energy and food security. Due to the unique character of the Tonle Sap system, the roles that water, energy and food have in our study as well as the ways they are linked were relatively different from each other. Energy security was considered largely through hydropower development, with such development taking place in the upstream parts of the Mekong River Basin and in Laos in particular [37]. Together, energy and climate change were seen mainly as external transboundary drivers that have impacts on the Tonle Sap system. The local-scale analysis in the Tonle Sap area focused then on the possible changes in water resources and food security as well as on the recent trends in demography and livelihoods. The research on these themes was carried out under two research components focusing on: (1) hydrology and water resources and (2) livelihoods and food security.

The key linkages between the nexus themes are expressed as a flow chart in Figure 1, with water acting as the connecting factor between energy and food security as well as climate change across the three geographical scales. The approach was selected in July 2010, before most of the key water-energy-food nexus publications, and it therefore differs from depictions of the common nexus approach, where the connections between water, energy, and food security are usually described in equal terms and visualised with a triangle [1,5,9].

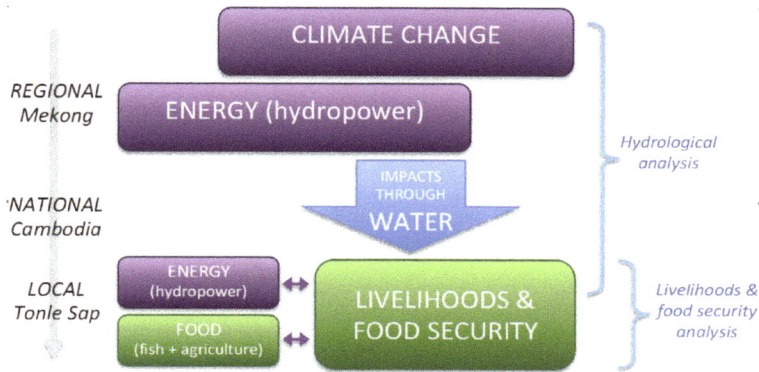

Figure 1. The key connections between the nexus themes of water, energy and food security as well as climate change, together with three geographical scales (left) and two research components (right).

2.2. Research Component on Hydrology and Water Resources

The first research component created the link from energy (*i.e.*, Mekong hydropower development) and climate to water resources by analysing the impacts of regional level changes to the Tonle Sap's flow regime (Figure 2) [38]. The impacts from the Mekong hydropower development and climate change were analysed both separately and together, using the ten-year period from 2032 to 2042.

The analysis under this research component was done with the help of two mathematical models. A distributed hydrological model called VMod [39] was established for the whole Mekong River Basin. The VMod hydrological model is a dynamic model constructed from square grid cells, and it was used to simulate the cumulative impacts of hydropower development and climate change on the Mekong flow regime. A 5 km × 5 km resolution raster dataset was constructed for the VMod model using SRTM 90 m elevations [40], Global Land Cover 2000 [41], and the Food and Agriculture Organisation of the United Nations (FAO) soil map of the world [42]. The main meteorological and discharge data were acquired from the Mekong River Commission (MRC) database [43].

After the model grid and the data were set up, the VMod model was calibrated against measured discharge for the calibration period (1982–1991). The validity of the model calibration was checked by computing the validation period (1993–1999) using the calibrated parameters, and comparing the fit from the validation period to calibration period results in selected river stations. The modelled daily discharge over the validation period agreed very well with the measured ones: Nash-Sutcliffe efficiency (NSE) varied from 0.78 to 0.94, depending on the station [38]. The relationship is shown visually in Figure S1 in the supplementary material.

The VMod model was then linked to a more detailed EIA 3D floodplain model [38] that simulated the impacts of basin-wide cumulative changes on the Tonle Sap flood pulse. The EIA 3D model is a three-dimensional baroclinic multilayer model that numerically solves the simplified Navier-Stokes equations using the implicit finite difference method [44]. Incoming flow at Kratie and from the Tonle Sap Catchment based on simulations from [38] and water levels at the South China Sea near the Mekong Delta based on [45] were used as upstream and downstream boundary conditions, respectively. The EIA 3D floodplain model was calibrated against measured water levels. As shown in Figures S2–S3 of the supplementary material, the model is able to regenerate the measured flood pulse rather well.

(a) Mekong Basin   **(b)** Tonle Sap Catchment

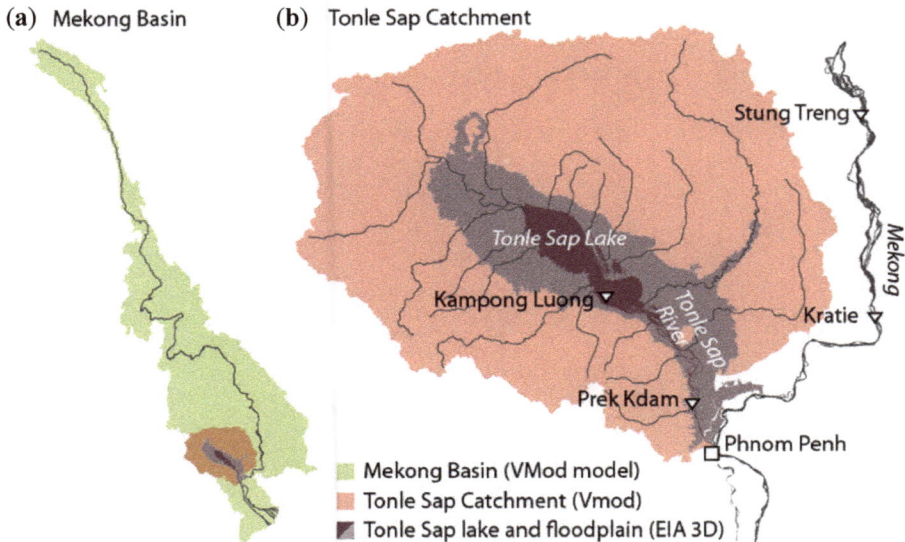

Figure 2. The three geographical scales of hydrological analysis and the related drivers of change: (a) Mekong River Basin, (b) Tonle Sap Lake catchment as well as the lake and its floodplain. The model used at each scale is indicated in parentheses.

The regional climate change scenarios for the hydrological model were obtained by downscaling five different Global Circulation Models (GCMs) for the Mekong Basin. For the selected GCMs, two different emission scenarios, *i.e.*, B1 (550 ppm stabilisation) and A1b (720 ppm stabilisation), were used. The MRC hydropower database was used to account for future development of large-scale hydropower in the Mekong Basin. To define the operation of hydropower reservoirs, a linear programming optimisation was used with the monthly outflows for each reservoir estimated separately, starting upstream.

2.3. Research Component on Livelihoods and Food Security

The second research component looked at local water and food security through analysis of the demographic and socio-economic situation and its trends in the Tonle Sap area. The analysis built on extensive spatial analysis of the key socio-economic databases, *i.e.*, Population Census 1998 and 2008 [46], and it was complemented by the CSIRO Tonle Sap Household Survey covering 1000 households in the area [47]. Due to Cambodia's tumultuous history [48], the Population Censuses of 1998 and 2008 are the first demographic datasets since 1960s, providing a major opportunity to look at the key demographic and socio-economic trends in the country. There have been very few analyses done on such trends anywhere in Cambodia, e.g., [49], and none in the specific context of the Tonle Sap Lake.

The Census data includes vast number of indicators, and for our analysis we selected demographic and socio-economic indicators that are particularly relevant for livelihoods, well-being and overall development. Given the close connection between the lake and the livelihoods of the area, many indicators have also a strong link to water and food security. The analysis looked both the current status (as of 2008) and the recent trends (from 1998 to 2008) of these key indicators. The change from 1998 to 2008 was looked in two ways: in terms of quantities (e.g., number of people having fishing as their main source of livelihood) and in terms of proportion (e.g., percentage of economically active population engaged in fishing).

The selected indicators were grouped and analysed spatially, with the Tonle Sap area divided into three spatial zones: Zone 1 (Lower Floodplain), Zone 2 (Upper Floodplain), and Zone 3 (Urban) (see also [32]). Such a spatial approach was used to establish a connection with the hydrological analysis and the lake itself. To provide a link with policy-making, the spatial zones were further divided into 18 sub-zones based on the administrative boundaries of the six Tonle Sap provinces. The statistical meaningfulness of the spatial zones was tested with k-means clustering of the Population Census data; the test confirmed that the Census data can be organised with spatially interpretable linkages to water and main livelihood sources [46].

The analysis of local water and food security was complemented by mapping of local water infrastructure in the Tonle Sap area. This mapping exercise focused on food security and energy security, and included major irrigation structures and hydropower development in the Tonle Sap Basin collected from a variety of information sources. While such infrastructure is a major component of local water, energy and food security, this was first time that location-specific information on such infrastructure was publicly presented in a single map [50].

2.4. Scenario Formulation: Four Alternative Futures for the Tonle Sap

The two research components were supported by a scenario process that was used to describe and discuss the connections between water, energy, food and climate change. The scenarios thus created a kind of synthesising nexus framework for the study, connecting two research components that looked only at selected nexus themes. The aim of the scenario process was to create plausible alternative futures for the Tonle Sap in year 2040, taking into account both external and internal drivers affecting the area.

The scenario process included three phases and was built on a set of stakeholder workshops with the participation of a number of experts on different nexus themes. The process was initiated by the creation of alternative scenario narratives for the Tonle Sap during the first two stakeholder workshops [51,52]. These scenario narratives helped to focus the actual research, with initial research results being discussed in the following two stakeholder workshops. The research findings were synthesised and connected to policy-making through the creation of four alternative scenarios for the Tonle Sap in 2040 that were elaborated during the last workshop.

This also means that our final scenarios can be seen to be close to data- and analysis-driven forecasts that build on our modelling estimates and socio-economic trend analyses. While there are also other, less rigid ways to create scenarios [53–55], the main characteristics of our scenarios are in line with the general scenario description by the Intergovernmental Panel on Climate Change (IPCC): "*A scenario is a coherent, internally consistent and plausible description of a possible future state of the world. It is not a forecast; rather, each scenario is one alternative image of how the future can unfold*" [56].

3. Key Results: Water-, Energy-, Food- and Climate-Related Drivers in the Tonle Sap

This section presents the key results from the two research components (Figure 1) as well as from the scenario formulation process. The results are based on Keskinen *et al.* [29], with emphasis on their implications for the water-energy-food nexus. For more detailed description of the results, see [29,38,46].

3.1. Hydrological Analysis of Energy- and Climate-Related Drivers

The hydrological research component analysed how energy- and climate-related drivers impact the Tonle Sap system's hydrology and, consequently, food security. The modelling results clearly indicate that the flood pulse of the Tonle Sap is likely to change in the future, with the planned hydropower dam development in the Mekong River Basin causing dramatic changes to the flood dynamics and hydrology. The operation of planned hydropower dams will flatten the hydrograph by causing higher dry season water levels and lower flood peaks (Figure 3).

The analysis shows that climate change will cause changes to the rainfall and temperature in the area, impacting the runoff and water levels in the Mekong mainstream and in the Tonle Sap system. However, our analysis indicates that the exact impact of climate change remains unclear, mainly due to differences in the different GCMs applied to the Mekong Basin. Remarkably, even the direction of the flow changes caused by climate change differs depending on the emission scenario and GCM used. Consequently, it is impossible to even say whether climate change will increase or decrease the flood season water level in the Tonle Sap.

Within the timeframe used in the model, *i.e.*, by the year 2042, climate change alone does not have a considerable impact on the dry season water level in the lake. The climate change impact on the flooded area, which is an important element for ecosystem productivity, is thus uncertain, with the estimates for the future floodplain area varying from 92% (8832 km^2) to 109% (10,464 km^2) of the current average floodplain area of 9600 km^2.

In contrast, the cumulative impacts of hydropower operation and climate change have a clear impact on dry season water levels, which are estimated to be 0.5–0.9 m above the current levels. This would mean that the permanent lake area would increase 18%–31%, submerging important habitats, such as flooded forest (Figure 3). The cumulative impacts are also significant for the flood season water levels, with our modelling estimates indicating lower flood peaks, although with large uncertainty due to the differences in GCMs used in the study. Overall, the floodplain area is expected to significantly reduce due to cumulative impacts from climate change and hydropower reservoirs, with the minimum area being around 75% of the current floodplain area.

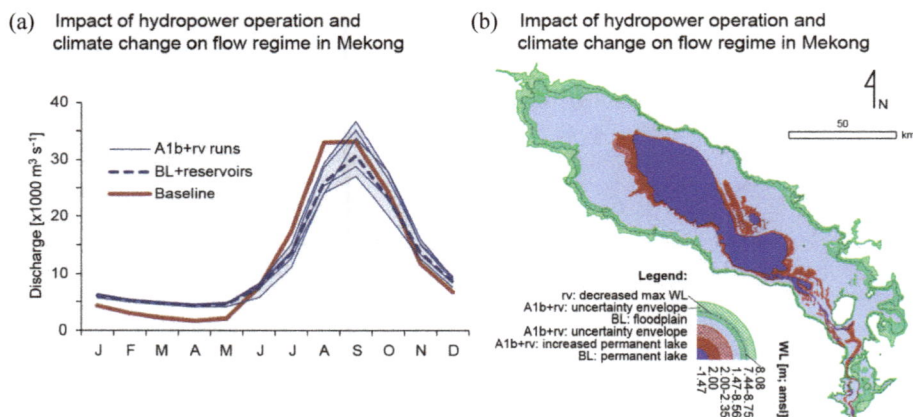

Figure 3. Estimated future changes to discharge and water levels. (**a**) Impact of reservoir operation (rv) and climate change (A1b scenario) on flow regime in the Mekong mainstream in Kratie, Cambodia (see location in Figure 2); figure modified from [38]; (**b**) Map of the estimated future changes (2032–2042) caused by the cumulative impacts of climate change (A1b) and reservoir operation (rv) in the permanent lake and flooded areas compared to the baseline (BL; 1982–1992).

3.2. Livelihood & Food Security Analysis

The livelihood and food security analysis looked at the current demographic and socio-economic situation and trends in the Tonle Sap area and related them to food security by focusing on agriculture and fishing as main livelihood sources. The three zones (Lower floodplain, Upper floodplain and Urban areas) created differ greatly in terms of their population, area as well as relationships to the lake and its flood pulse. While Zone 2 clearly has the biggest population, Zone 1 is the largest area-wise and is most directly connected to the lake and its flood pulse due to its location.

Altogether, the Tonle Sap study area has some 1,707,000 people (according to Population Census 2008) living in 1555 villages and settlements, with 1244 of those being rural and 311 urban. The very clear dominance of youth in the total population is demographically remarkable: the two biggest age groups are between 15–19 years and 20–24 years. These young people are thus just entering the work force, searching for meaningful work opportunities.

People living in and around the Tonle Sap Lake have adapted to the enormous annual variation of the lake's water level; many even live in floating houses on the lake itself. People's livelihoods are closely connected to the lake and the natural resources it provides and supports. From the nexus point of view, the strong connection between livelihoods and food security is remarkable: Agriculture clearly dominates (61% of total work force has it as main livelihood), with fishing coming third (4.5%) after trade (11.5%) (Table 1). Over 65% of the Tonle Sap's work force thus has either agriculture or fishing as their main livelihood source, and both sectors form significant secondary livelihood sources as well. In addition, agriculture and fishing are the most important contributors to national food security, with rice and fish being the staple food of Cambodia.

Our findings indicate that the total work force in the agricultural and fishing sector increased dramatically between 1998 and 2008 due to rapid population growth. Although the proportion of the workforce engaged in the agricultural sector decreased slightly from 66% in 1998 to 61% in 2008, the number of people workforce in the sector increased by 130,000 people. A similar finding was evident in the fishing sector, where the absolute number of people with fishing as their main economic activity increased by 10,700 people, although the proportion of people involved in the sector decreased from 4.7% in 1998 to 4.5% in 2008 (Figure 4). Hence, over 140,000 new people established agriculture and fishing as their main economic activity between 1998 and 2008. Such an increase obviously puts pressure on existing resource use but may also enhance food security if it leads to increased and more diverse food production.

At the same time, it is important to note that there are remarkable differences between the livelihood structures in different zones and sub-zones of the Tonle Sap (Table 1). While fishing is the most important livelihood activity in the areas closest to the lake, urban areas in particular have much more diverse livelihood structures. In addition, there are considerable differences across the six Tonle Sap provinces. The comparison between Population Census 2000 and 2008 also show that the general livelihood structure is slowly diversifying, with an increasing number of people gradually transferring from traditional, agriculture-based livelihoods to other sources of livelihood. This is particularly visible in the urban areas, with the provincial capitals and Siem Reap in particular also attracting migrants from the rural areas.

Table 1. Involvement in key livelihood sectors by 3 zones and 18 sub-zones. Data source: [57].

Zones	Agriculture (Include Forestry)		Fishing		Wholesale		Construction		Transport, Storage and Communication		Manu-Facturing		Total for Six Sectors	
	%	Number of People	%	Number of People	%	Number of People	%	Number of People	%	Number of People	%	Number of People	%	Number of People
Zone 1: Lower Floodplain	**28.3**	**12,589**	**60.2**	**26,801**	**6.7**	**2998**	**0.2**	**103**	**1.3**	**568**	**0.9**	**408**	**97.6**	**43,467**
Banteay Meanchey	97.4	1,190	0.4	5	0.3	4	0.0	0	0.0	0	0.2	2	98.3	1,201
Battambang	2.4	145	85.6	5,253	7.7	472	0.4	24	0.8	51	0.6	34	97.4	5,979
Kampong Chhnang	43.7	5,212	43.2	5,161	7.3	873	0.4	43	2.2	267	0.8	90	97.6	11,646
Kampong Thom	17.3	887	76.8	3,935	3.8	195	0.0	0	0.0	2	0.3	15	98.3	5,034
Pursat	21.4	1,756	67.5	5,533	7.1	579	0.1	5	1.1	92	1.1	90	98.3	8,055
Siem Riep	28.5	3,399	58.1	6,914	7.3	875	0.3	31	1.3	156	1.5	177	97.0	11,552
Zone 2: Upper Floodplain	**80.9**	**433,777**	**1.3**	**6898**	**5.6**	**30,253**	**2.5**	**13,246**	**1.5**	**8095**	**2.1**	**11,253**	**93.9**	**503,522**
Banteay Meanchey	81.9	58,794	0.5	374	4.6	3,323	2.0	1,449	1.7	1,212	2.7	1,912	93.4	67,064
Battambang	76.6	90,193	1.5	1,789	7.2	8,422	3.4	4,043	2.2	2,594	2.4	2,845	93.3	109,886
Kampong Chhnang	81.3	35,404	2.2	956	4.9	2,114	1.1	464	1.1	478	3.1	1,341	93.6	40,757
Kampong Thom	85.7	59,329	1.2	834	4.5	3,104	1.0	660	1.3	896	1.7	1,159	95.3	65,982
Pursat	84.8	79,451	1.4	1,278	5.0	4,646	1.0	951	1.1	1,053	1.1	1,049	94.4	88,428
Siem Riep	79.0	110,606	1.2	1,667	6.2	8,644	4.1	5,679	1.3	1,862	2.1	2,947	93.9	131,405
Zone 3: Urban	**27.6**	**74,411**	**1.7**	**4549**	**24.0**	**64,677**	**6.8**	**18,275**	**7.8**	**21,043**	**6.3**	**16,917**	**74.1**	**199,872**
Banteay Meanchey	32.9	12,667	0.6	213	27.0	10,391	6.9	2,658	7.6	2,910	6.6	2,540	81.6	31,379
Battambang	30.3	24,180	0.7	530	24.1	19,256	6.5	5,217	7.5	6,008	7.6	6,046	76.6	61,237
Kampong Chhnang	18.3	3,644	12.0	2,381	27.7	5,502	4.0	804	7.9	1,562	9.5	1,897	79.5	15,790
Kampong Thom	53.6	13,701	0.7	177	15.7	4,014	2.6	661	4.5	1,156	3.7	944	80.8	20,653
Pursat	38.4	7,547	0.3	57	20.3	3,995	4.8	940	5.5	1,084	4.6	897	73.9	14,520
Siem Riep	53.2	126,677	4.1	9,772	13.0	31,038	5.8	13,705	4.3	10,341	3.2	7,726	83.7	199,259
Total for Tonle Sap	**61.3**	**520,777**	**4.5**	**38,248**	**11.5**	**97,928**	**3.7**	**31,624**	**3.5**	**29,706**	**3.4**	**28,578**	**87.8**	**746,861**

Note: Column with % indicates the proportion of livelihood from the economic active population in the zone/sub-zone.

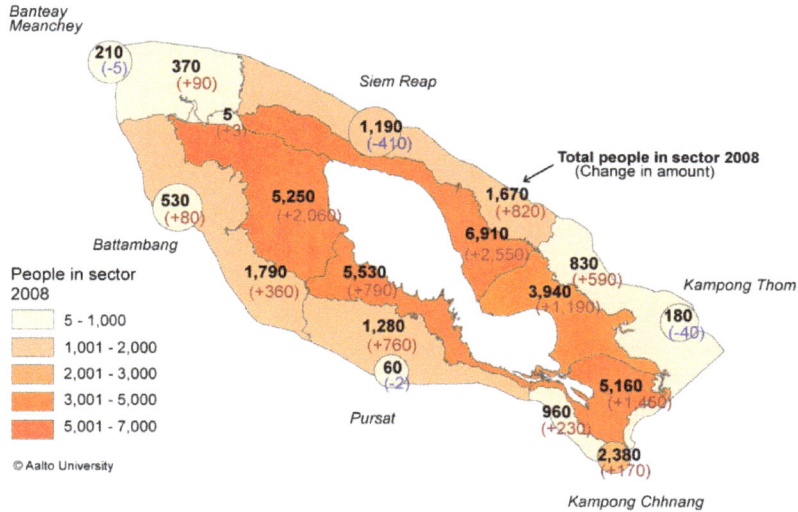

Figure 4. The map of Tonle Sap area and its zoning, with six sub-zones of Lower Floodplain closest to the lake, Upper Floodplain further and Urban areas indicated as six circles around provincial capitals. The numbers indicate the number of people with fishing as their main source of livelihood. Bold figures indicate the total number in 2008, while the figures in parentheses show the change from year 1998 based on comparison between Population Census 1998 and 2008. Blue numbers indicate decline and red increase in the number of people. Reproduced with permission from Aalto University, 2013 [29].

3.3. Alternative Scenarios for the Tonle Sap in 2040

The two research components described above provided new information about energy-, climate-, water- and food-related trends and pressures in the Tonle Sap area. To help understand future development from a nexus approach, these were integrated using a scenario formulation process that made use of the results from both research components. A scenario process was selected because it did not seem reasonable to create just one possible view about the ways the water, energy and food security are likely to develop in the future. Instead, the research findings were synthesised by creating four alternative scenarios for the Tonle Sap in 2040.

The starting point for the scenario formulation was energy-related, namely the decision whether or not to build more mainstream and tributary dams in the Mekong River Basin. This created two alternative "water and energy paths" for the scenario process: One with many Mekong dams (blue path) and one with only currently existing or under construction dams (green path). Next, the possible differences in the livelihood development and the implementation of government's policies [58] led to two different kinds of "societal development paths". Together, these two paths produced four alternative scenarios for the Tonle Sap by 2040: Major changes, Growing disparity, Green growth, and Stagnation (Figure 5).

All four alternative scenarios present different but nevertheless plausible futures for the Tonle Sap area. Importantly, they also have differing implications across the water-energy-food nexus at different scales (Figure 1). While the paths related to hydropower construction will impact energy security at regional scale, they also have direct implications for local water and food security. At the same time, the scenarios indicate that the future of the Tonle Sap depends very much on how socio-economic situation and livelihoods develop as a response to these external changes.

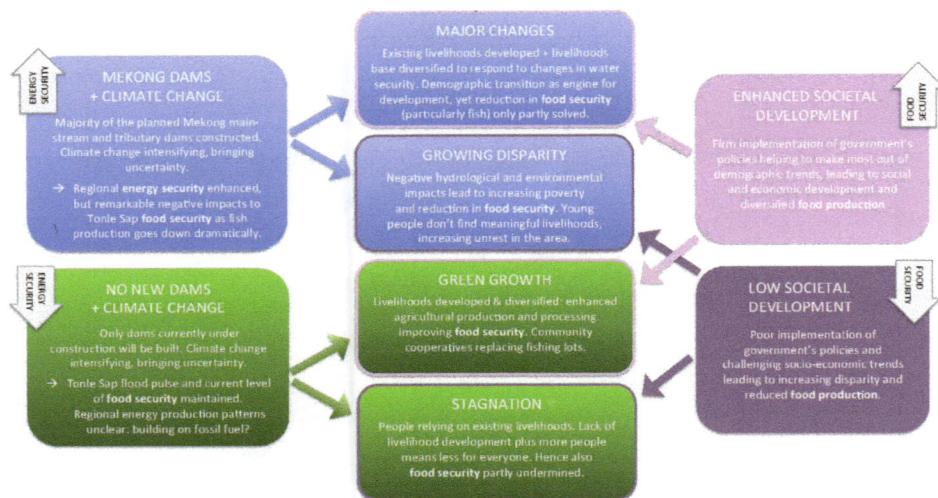

Figure 5. Four alternative scenarios created for the Tonle Sap in 2040 (centre), with two alternative water and energy paths on the left and two alternative societal development paths on the right. Grey arrows indicate how energy security and food security are likely to develop as part of different paths.

4. Discussion

In this article, our focus is two-fold. In practical terms, we assess water, energy, food and climate-related drivers in the Tonle Sap area. In methodological terms, we discuss the implications that using the water-energy-food nexus approach had in our study. The article provides a novel analysis of how the combination of external drivers (regional hydropower development and climate change) and internal drivers (demographic and socio-economic trends) is likely to affect the area's long-term development (four alternative futures). At a more theoretical level, the study provides an opportunity to discuss the pros and cons of a nexus approach in terms of both research and science-policy-stakeholder interaction, also making use of the authors' previous research experience in the area. Next, we will discuss both these aspects in more detail, together with the consideration of key limitations of our study.

4.1. Implications for the Tonle Sap Area: National Food Security in Danger

The Tonle Sap Lake area forms a critically important economic, social and environmental resource for the whole of Cambodia, yet is under threat from the possible changes that regional hydropower development will cause for its flood pulse. The possible changes in the flood pulse are expected to radically reduce the ecosystem productivity of the lake-floodplain system, leading also to remarkable social and economic impacts. Based on our analysis, we foresee that the most remarkable impacts are related to three, interlinked changes.

Firstly, the expected shifts in floodplain habitats [59] due to a changed flood regime can potentially be very destructive for existing floodplain ecosystems, also negatively impacting aquatic production. On the other hand, they can also lead to modest increases in agricultural areas in the upper floodplain, thus enabling potential increases in agricultural production. Secondly, increased dry season water levels are expected to lead to the destruction of the majority of the flooded forests surrounding the lake, having a notable impact on sedimentation processes, ecosystems and aquatic productivity. It would also impact environmental conservation and, hence, ecotourism, as large parts of the current conservation core areas would be permanently submerged.

Thirdly and most importantly, the hydrological and environmental changes put major pressure on the Tonle Sap's immense fish production. While the connection between Tonle Sap's primary production and fish production is still not very well understood [60,61], it is expected that the changes described in our results would lead to remarkable reductions in fish production. Such a radical reduction would naturally have negative implications for livelihoods in the Tonle Sap area, with people in the lower floodplain most affected. It would also have broader implications due to the critical importance of fish for food security: up to 80% of all animal protein consumption in Cambodia comes from fish and other aquatic animals [62], and Cambodian fisheries contribute considerably to regional food security thanks to fish migration and fish export [63,64]. Mitigating the impacts of reduced fish production on food security would require significant improvements in current food production, but this is challenging due to the limited amount of water and land available [65]. Consequently, the reduced fish production will lead to decreased self-sufficiency, increasing dependency on food imports and, ultimately, reduced food security.

In addition to these environmental pressures, the Tonle Sap area is seeing major demographic and socio-economic changes. Our analysis shows that almost 2 million people already live in the lake and its floodplains, with exceptionally large numbers of young people entering the work force, changing the livelihood dynamics. Given the dominance of agriculture, and the combination of management challenges and already heavy pressure on the area's natural resources, the Tonle Sap's future depends very much on what kinds of livelihood sources these young people are able to move to.

As illustrated by the four alternative futures (Figure 5), the future of the Tonle Sap area thus depends on both external drivers and internal changes in the demographic and socio-economic setting of the area. This puts pressure on the key actors, including the Cambodian government. In sum, we see that the Cambodian government and its line agencies should take simultaneous action at three levels. Regionally, the Cambodian government needs to engage into active dialogue with its upstream neighbours about the most sustainable ways to develop the Mekong River Basin to minimise the impacts on the Tonle Sap system. While the Mekong River Commission (MRC) provides the most obvious arena for such a dialogue, its inherent political challenges [66,67] mean that Cambodia needs to support the regional processes with proactive bilateral negotiations.

At a national level, firm implementation of the Cambodian government's development plans is the key for the Tonle Sap. On the positive side, Cambodia's Rectangular Strategy [58] already incorporates practically all relevant policies that are needed for the positive livelihood transformations to occur in the Tonle Sap within next decades. Yet implementing these ambitious plans requires close collaboration between the different ministries, active stakeholder participation, and, generally, good governance; all of these have traditionally been in rather short supply in Cambodia [68]. Finally, at a local level, the dominance of traditional livelihood sources and relatively slow livelihood transformations suggest that any livelihood development in the Tonle Sap area should use the existing livelihood structure as a foundation, and entirely new livelihood sources should be seen as complementary to this.

4.2. Methodological Implications: the Role of the Water-Energy-Food Nexus Approach

In terms of research, the nexus approach used in our study encouraged simultaneous use of several research methods, for the analysis of the interlinkages between the nexus themes cannot be done with one method alone. The nexus approach was also useful in highlighting the differing scales of the nexus themes; while the analysis focused on the local scale (the Tonle Sap), food security was examined at national scale, and energy security and climate change at regional scale as well.

However, our previous research activities in the Tonle Sap have already utilised similar kinds of approaches as well [31,34], their use being encouraged by Integrated Water Resources Management (IWRM), which is widely used in the Mekong [7,69] and shares many similarities with the nexus [70,71]. The nexus and IWRM can therefore be seen to encourage comparable analytical approaches, which are explicitly integrated, cross-sectoral and cross-scale. Compared to IWRM, we consider that the most important contribution of the nexus approach for our research was that it encouraged us to focus our attention on the linkages between water, energy, food and climate, which we achieved through the use of scenarios. The scenarios combined water, energy and food-related drivers, providing a synthesising framework for the entire study. The scenario formulation process brought two other notable benefits for our study. Firstly, it enabled the inclusion of the stakeholders' views from the very beginning, providing guidance and early feedback to our research. Secondly, it allowed the creation of alternative scenarios that facilitated discussion about the uncertainties related to our research results and also made the critical role of policy choices visible.

In terms of integrated planning and cross-sectoral collaboration, interesting findings emerge. When compared to our earlier research projects (that had a more explicit focus on water resources management), the simple fact that the nexus approach explicitly spelt out the key sectors and themes ensured more diverse engagement of stakeholders into the project workshops. The nexus thus enhanced cross-sectoral stakeholder dialogue by expanding the discussion from mere water resource management to the broader aspects of water, energy and food security; this link to cross-sectoral collaboration has been noted also by other authors [10,72,73]. In addition, the explicit focus on these themes created a common ground for stakeholders with different backgrounds. In this way, the nexus appeared as a kind of boundary object: the nexus as a concept is robust enough to establish a commonly agreed platform for discussion, yet flexible enough to allow different interpretations by various actors based on their differing views and interests [71,74,75]. This also opens up possibilities for new kinds of coalitions in transboundary contexts, alleviating dominant upstream-downstream tensions. For example, cross-country coalitions between energy ministries or fisheries authorities may help to counter riparian countries' often rather monolithic positions.

4.3. Limitations and Way Forward

Our study used a nexus approach to look at water, energy and food security in the Tonle Sap Lake. As the study was initiated already in July 2010, our approach was somewhat different when compared to other nexus approaches introduced later [1,6,10]. This meant that the majority of the actual research was done through two separate research components that focused only on one or two nexus themes, with the linkages between all of them being then considered only during the scenario formulation process. Also, while the nexus approach used did encourage the use of scenarios, scenarios are obviously not specific only to the nexus but could also be used with other approaches. In addition, the alternative scenarios created were methodologically quite close to conventional data-based forecasts, although building on the broader ideas of scenario formulation [53].

It must also be noted that in order to maintain the focus, we looked only at selected water-, energy and food-related themes and drivers. Such focus left out other important (and nexus-relevant) themes, including for example the possible flow impacts caused by regional irrigation development and land use changes. This highlights an important aspect of the water-energy-food nexus; given its broadness, it is practically unavoidable to leave out some less important but still relevant aspects from the nexus analysis. Such a finding relates to the conclusion by Guillaume *et al.* [14], who note that subsystems included and excluded from the nexus are case-specific and should be consciously scrutinised.

The research component focusing on livelihoods and food security relied heavily on the quantitative data available from Population Census 1998 and 2008. While ensuring very good coverage (the data includes every household and village in the Tonle Sap area), the datasets also have their limitations. Perhaps most importantly, the two Censuses recorded only main economic activity, meaning that the seasonal variation and general diversity of livelihood sources in the Tonle Sap were not properly captured [46]. This is likely to be particularly critical for fishing and related activities, which also form important secondary and tertiary livelihoods in the area [32]. On the other hand the Census results were generally in line with a household survey that considered also secondary livelihood sources [47]. The use of livelihood-related indicators (agriculture and fishing) as proxies for food security is not without problems either. Involvement in food-producing sectors does not directly correlate with food security, as there are also several other factors—including food imports and trade as well as other food sources—contributing to local food security. In addition, a large part of the Cambodian fish catch comes from privately owned fishing lots [76]; their contribution to food security remain invisible when looking only at livelihood-related indicators.

While our project aimed to enhance cross-sectoral collaboration, its primary focus was on research and analysis. This meant that our project did not include difficult policy decisions, nor the related politics of resource use. Such a non-contested setting is also likely to explain to a large extent why cross-sectoral discussions during the workshops occurred relatively easily. Yet, if the nexus is to actually influence policies, nexus processes will, and must, emerge also as contested policy arenas [77]. In such arenas, different actors are likely to use the vagueness and immaturity of the nexus concept to provide framings to advance their own agendas and safeguard their own interests. Indeed, if the nexus is to facilitate integrated planning and policy-making, the political nature of the nexus must also be understood and made visible, instead of it being hidden behind the "technical veil" [78].

5. Conclusions

This article presented the findings from a research project that looked at the water-energy-food nexus in the Tonle Sap Lake of the transboundary Mekong River between 2010 and 2013. Our study shows that the nexus is a very relevant approach in the Mekong River Basin, as the nexus components form the basis for social and economic development in the riparian countries and are closely interlinked. Our study concludes that the Tonle Sap Lake is a nexus hotspot within the Mekong, where local, national and regional drivers of water, energy and food security meet and merge in multiple ways.

Our results confirm that while the Tonle Sap flood pulse forms the basis for local and national food security, it is at the same time highly vulnerable to the flow changes that Mekong hydropower development is likely to cause in the near future. The results also show that with the time scale used in the model (up to the year 2042), the impacts from hydropower development are likely to be much more severe than those from changing climate. Such impacts are likely to lead to radical reductions in ecosystem productivity and, consequently, in the Tonle Sap fisheries. Such changes would create major socio-economic pressure on the Tonle Sap area, particularly when coupled with on-going demographic stress.

The enhanced regional and national energy security from hydropower developments (particularly in Laos) is thus likely to reduce food security at both local (Tonle Sap area) and national (Cambodia) levels. Our findings therefore illustrate that the nexus has a strongly spatial dimension, which in the context of the Tonle Sap means a transboundary dimension as well. For example, the decisions related to energy production are done at very different scales and locations compared to the scales and locations where the food security related impacts of such decisions are occurring. Such a multi-scale context emphasises also the geopolitical aspects related to the nexus themes: riparian countries have differing interests towards nexus themes. It also puts pressure on the Cambodian government to simultaneously act at regional, national and local scales to ensure a sustainable future for the Tonle Sap.

At a more theoretical level, our study provided interesting findings on the applicability of water-energy-food nexus approach. We identified the two main dimensions of the nexus, with one promoting research and analysis on the nexus linkages, and the other facilitating integrated planning and cross-sectoral collaboration. In terms of research, the nexus approach helped to frame our research, encouraged the use of different research methods and facilitated a multi-scale analysis of water-, energy- and food-related drivers. However, we noted that these aspects cannot really be considered nexus-specific, as they had already been present in our earlier research that built on IWRM. As a result, the nexus' main contribution to our research was that its focus on the linkages between water, energy and food encouraged us to find new ways to link these themes together, leading ultimately to a set of alternative scenarios for the Tonle Sap.

More broadly, the most important added value of the nexus approach was its contribution to cross-sectoral collaboration and science-policy-stakeholder interaction. We argue that this is because the nexus view is simultaneously more focused and broader than for example that of IWRM. It is more focused because the nexus approach explicitly spells out the themes and sectors that are to be included in the process and in this way helps to engage the relevant actors. Yet the nexus view is also broader, as it expands the discussion from mere water resources management—where energy and food security are largely considered external drivers—into a more general debate about water, energy and food security.

In this way, the nexus also emerges as a useful boundary concept that is both flexible and robust enough to engage the relevant sectors into a shared planning and policy platform. This makes the nexus a fascinating approach also in transboundary contexts, as it has a potential to alleviate dominant upstream-downstream tensions through new kinds of sectoral coalitions between the ministries in different riparian countries. At the same time it is clear that the lack of coordination between the nexus-related sectors and actors is not merely due to lack of appropriate approaches; the nexus is, after all, not the first approach calling for policy coherence. Instead, the ultimate challenge with integrated planning and policy-making lies in the politics related to natural resource use. If the nexus is to facilitate cross-sectoral collaboration, it should therefore address the inherently political nature of that collaboration and notice that the nexus itself is strongly political as well. Addressing the political aspects related to the nexus is particularly important in transboundary river basins, where the riparian countries often have very differing views on water, energy and food security.

Acknowledgments: The research presented in this article was done in the Exploring Tonle Sap Futures project under the regional Exploring Mekong Region Futures program funded by AusAID. We thank all project partners and funders, in particular the Tonle Sap Authority (TSA), Supreme National Economic Council (SNEC) and the Commonwealth Scientific and Industrial Research Organisation (CSIRO). Special thanks to Sophort So and Tony Hell for their collaboration and for Pech Sokhem and Hannu Lauri for their contribution in the implementation of the project. The comments provided by our colleagues at Aalto University's Water & Development Research Group, in particular by Joseph Guillaume, improved this article remarkably. Thank you for Hannu Lauri and Timo Räsänen for the help with the figures in the supplementary material. Thank you also for the constructive comments provided by our four reviewers. For writing the article, Marko Keskinen and Aura Salmivaara received funding from the Academy of Finland funded project NexusAsia (269901), and Matti Kummu from the Academy of Finland funded project SCART (267463).

Author Contributions: Marko Keskinen led the research project that this study is based on, and coordinated the writing process of the article. He was also mainly responsible for writing the Introductory, Discussion and Conclusions sections as well as parts related to alternative scenarios. Matti Kummu carried out the hydrological analysis and wrote parts related to it, while Paradis Someth carried out the analysis of local development in the Tonle Sap Lake and wrote parts related to it. Aura Salmivaara carried out the livelihoods analysis and, together with Marko Keskinen, the parts related to it. All authors contributed to the final writing up of the article.

Conflicts of Interest: The authors declare no conflict of interest.

References

1. Hoff, H. *Understanding the Nexus*; Stockholm Environment Institute: Stockholm, Sweden, 2011.
2. Waughray, D. *Water Security: The Water-Food-Energy-Climate Nexus*; Island Press: Washington, DC, USA, 2011.
3. Water-Food-Energy-Ecosystem Nexus. Available online: http://www.unece.org/env/water/nexus.html (accessed on 20 February 2015).
4. Bonn2011 Conference: The Water, Energy and Food Security Nexus—Solutions for a Green Economy. Available online: http://www.water-energy-food.org/documents/bonn2011_policyrecommendations.pdf (accessed on 3 September 2015).
5. World Economic Forum. *Global Risks 2011*; World Economic Forum: Cologny, Switzerland, 2011.
6. Bazilian, M.; Rogner, H.; Howells, M.; Hermann, S.; Arent, D.; Gielen, D.; Steduto, P.; Mueller, A.; Komor, P.; Tol, R.S.J.; *et al.* Considering the energy, water and food nexus: Towards an integrated modelling approach. *Energy Policy* **2011**, *39*, 7896–7906. [CrossRef]

7. Bach, H.; Bird, J.; Jonch, C.T.; Morck, J.K.; Baadsgarde, L.R.; Taylor, R.; Viriyasakultorn, V.; Wolf, A. *Transboundary River Basin Management: Addressing Water, Energy and Food Security*; Mekong River Commission: Vientiane, Laos, 2012.

8. Asian Development Bank. *Thinking about Water Differently: Managing the Water-Food-Energy Nexus*; Asian Development Bank: Manila, Philippines, 2013.

9. European Union. *Confronting Scarcity:Managing Water, Energy and Land for Inclusive and Sustainable Growth*; Overseas Development Institute (ODI), European Centre for Development Policy Management (ECDPM) and German Development Institute/Deutsches Institut für Entwicklungspolitik (GDI/DIE): London, UK; Brussels, Belgium; Bonn, Germany, 2012.

10. Flammini, A.; Puri, M.; Pluschke, L.; Dubois, O. *Walking the Nexus Talk: Assessing the Water-Energy-Food Nexus in the Context of the Sustainable Energy for All Initiative*; Food and Agriculture Organisation of the United Nations: Rome, Italy, 2014.

11. Beck, M.; Villarroel, W.R. On water security, sustainability, and the water-food-energy-climate nexus. *Front. Environ. Sci. Eng.* **2013**, *7*, 626–639. [CrossRef]

12. Middleton, C.; Allouche, J.; Gyawali, D.; Allen, S. The rise and implications of the water-energy-food nexus in Southeast Asia through and Environmental Justice Lens. *Water Altern.* **2015**, *8*, 627–654.

13. Allan, T.; Keulertz, M.; Woertz, E. The water-food-energy nexus: An introduction to nexus concepts and some conceptual and operational problems. *Int. J. Water Resour. Dev.* **2015**, *31*, 301–311. [CrossRef]

14. Guillaume, J.; Kummu, M.; Eisner, S.; Varis, O. Transferable principles for managing the nexus: Lessons from historical global water modelling of central Asia. *Water* **2015**, *7*, 4200–4231. [CrossRef]

15. Soliev, I.; Wegerich, K.; Kazbekov, J. The costs of benefit sharing: Historical and institutional analysis of shared water development in the Ferghana valley, the Syr Darya basin. *Water* **2015**, *7*, 2728–2752. [CrossRef]

16. Kibaroglu, A.; Gürsoy, S.I. Water-energy-food nexus in a transboundary context: The Euphrates-Tigris river basin as a case study. *Water Int.* **2015**, *40*, 1–15. [CrossRef]

17. Belinskij, A. Water-energy-food nexus within the framework of international water law. *Water* **2015**, *7*, 5396–5415.

18. Collins Dictionary: Nexus. Available online: http://www.collinsdictionary.com/dictionary/english/nexus (accessed on 25 August 2015).

19. Global Water Partnership (GWP). *Integrated Water Resources Management*; Global Water Partnership: Stockholm, Sweden, 2000.

20. Keskinen, M. *Bringing Back the Common Sense? Integrated Approaches in Water Management: Lessons Learnt from the Mekong*; Aalto University: Helsinki, Finland, 2010.

21. Food and Agriculture Organisation of the United Nations (FAO). *An Innovative Accounting Framework for the Food-Energy-Water Nexus—Application of the MuSIASEM Approach to Three Case Studies*; Food and Agriculture Organisation of the United Nations: Rome, Italy, 2013.

22. Räsänen, T.; Joffre, O.; Someth, P.; Thanh, C.; Keskinen, M.; Kummu, M. Model-based assessment of water, food, and energy trade-offs in a cascade of multipurpose reservoirs: Case study of the Sesan Tributary of the Mekong River. *J. Water Resour. Plan. Manag.* **2014**, *141*. [CrossRef]

23. Jalilov, S.M.; Olli, V.; Marko, K. Sharing benefits in transboundary rivers: An experimental case study of Central Asian water-energy-agriculture nexus. *Water* **2015**, *7*, 4778–4805. [CrossRef]

24. International Water Association (IWA); International Union for Conservation of Nature (IUCN). *Nexus Dialogue on Water Infrastructure Solutions: Building Partnerships For Innovation in Water, Energy and Food Security*; International Union for Conservation of Nature (IUCN) and International Water Association (IWA): Gland, Switzerland; London, UK, 2012.

25. Arias, M.E.; Piman, T.; Lauri, H.; Cochrane, T.A.; Kummu, M. Dams on Mekong tributaries as significant contributors of hydrological alterations to the Tonle Sap floodplain in Cambodia. *Hydrol. Earth Syst. Sci.* **2014**, *18*, 5303–5315. [CrossRef]

26. Kummu, M.; Tes, S.; Yin, S.; Adamson, P.; Józsa, J.; Koponen, J.; Richey, J.; Sarkkula, J. Water balance analysis for the Tonle Sap Lake-floodplain system. *Hydrol. Process.* **2014**, *28*, 1722–1733. [CrossRef]

27. Keskinen, M.; Kummu, M.; Käkönen, M.; Varis, O. Mekong at the crossroads: Next steps for impact assessment of large dams. *AMBIO* **2012**, *41*, 319–324. [CrossRef]

28. Grill, G.; Lehner, B.; Lumsdon, A.E.; MacDonald, G.K.; Zarfl, C.; Reidy Liermann, C. An index-based framework for assessing patterns and trends in river fragmentation and flow regulation by global dams at multiple scales. *Environ. Res. Lett.* **2015**, *10*. [CrossRef]

29. Keskinen, M.; Kummu, M.; Salmivaara, A.; Someth, P.; Lauri, H.; Moel, H.d.; Ward, P.; Pech, S. *Tonle Sap now and in the Future? Final Report of the Exploring Tonle Sap Futures Study*; Aalto University: Espoo, Finland, 2013.

30. Smajgl, A.; Ward, J. *The Water-Food-Energy Nexus in the Mekong Region: Assessing Development Strategies Considering Cross-Sectoral and Transboundary Impacts*; Springer: Berilin, Germany, 2013.

31. Mekong River Commission and Finnish Environment Institute (MRCS/WUP-FIN). *Research Findings and Recommendations: Final Report—Part 2*; Mekong River Commission and Finnish Environment Institute: Vientiane, Laos, 2007.

32. Keskinen, M. The lake with floating villages: Socio-economic analysis of the Tonle Sap lake. *Int. J. Water Resour. Dev.* **2006**, *22*, 463–480. [CrossRef]

33. Kummu, M.; Sarkkula, J. Impact of the Mekong River flow alteration on the Tonle Sap flood pulse. *J. Hum. Environ.* **2008**, *37*, 185–192. [CrossRef]

34. Keskinen, M.; Chinvanno, S.; Kummu, M.; Nuorteva, P.; Snidvongs, A.; Varis, O.; Västilä, K. Climate change and water resources in the lower Mekong River basin: Putting adaptation into the context. *J. Water Clim. Chang.* **2010**, *1*, 103–117. [CrossRef]

35. Varis, O.; Kummu, M.; Keskinen, M.; Sarkkula, J.; Koponen, J.; Heinonen, U.; Makkonen, K. Integrated water resources management on the Tonle Sap Lake, Cambodia. *Water Sci. Technol.: Water Supply* **2006**, *6*, 51–58. [CrossRef]

36. Keskinen, M. Institutional cooperation at a basin level: For what, by whom? Lessons learned from Cambodia's Tonle Sap Lake. *Nat. Resour. Forum* **2012**, *36*, 50–60. [CrossRef]

37. Grumbine, R.E.; Xu, J. Mekong hydropower development. *Science* **2011**, *332*, 178–179. [CrossRef]

38. Lauri, H.; de Moel, H.; Ward, P.J.; Räsänen, T.A.; Keskinen, M.; Kummu, M. Future changes in Mekong River hydrology: Impact of climate change and reservoir operation on discharge. *Hydrol. Earth Syst. Sci.* **2012**, *16*, 4603–4619. [CrossRef]

39. Koponen, J.; Lauri, H.; Veijalainen, N.; Sarkkula, J. *HBV and IWRM Watershed Modelling User Guide*; Mekong River Commission (MRC): Vientiane, Laos, 2010.

40. Jarvis, A.; Reuter, H.; Nelson, A.; Guevara, E. SRTM 90m Digital Elevation Database v4.1. Available online: http://www.cgiar-csi.org/data/srtm-90m-digital-elevation-database-v4-1 (accessed on 1 February 2010).

41. Joint Research Centre. *Joint Research Centre 2000 Database*; European Commission: Brussels, Belgium, 2003.

42. Food and Agriculture Organization of United Nations (FAO). *World Reference Base (WRB) Map of Soil Resources. Land and Water Development Division*; FAO: Rome, Italy, 2003.

43. Mekong River Commission (MRC). *Hydrometeorological Database of the Mekong River Commission*; Mekong River Commission: Vientiane, Lao, 2011.

44. Koponen, J.; Kummu, M.; Sarkkula, J. Modelling Tonle Sap Lake environmental change. *Verh. Int. Ver. Limnol.* **2005**, *29*, 1083–1086.

45. Västilä, K.; Kummu, M.; Sangmanee, C.; Chinvanno, S. Modelling climate change impacts on the flood pulse in the lower Mekong floodplains. *J. Water Clim. Chang.* **2010**, *1*, 67–86. [CrossRef]

46. Salmivaara, A.; Kummu, M.; Varis, O.; Keskinen, M. Socio-economic changes in Cambodia's unique Tonle Sap Lake area: A spatial approach. *Appl. Spat. Anal. Policy* **2015**, *2015*. [CrossRef]

47. Ward, J.; Poutsma, H. *The Compilation and Summary Analysis of Tonle Sap Household Livelihoods*; Commonwealth Scientific and Industrial Research Organisation (CSIRO): Canberra, Australia, 2013.

48. Chandler, D.P. *A History of Cambodia*; Westview Press: Boulder, CO, USA, 1993.

49. National Institute of Statistics (NIS). *Labour and Social Trends in Cambodia 2010*; National Institute of Statistics (NIS): Phnom Penh, Cambodia, 2010.

50. Someth, P.; Chanthy, S.; Kummu, M.; Keskinen, M. *Irrigation and Hydropower Development in the Catchment and Floodplain of the Tonle Sap Lake*; Aalto University: Espoo, Finland, 2012.

51. Foran, T.; Ward, J.; Lu, X.; Leitch, A.; Smajgl, A. *Excerpts from the Compilation of Scenarios Developed during the Regional and Local Studies*; CSIRO Ecosystem Sciences: Canberra, Australia, 2011.

52. Foran, T.; Ward, J.; Kemp-Benedict, E.J.; Smajgl, A. Developing detailed foresight narratives: a participatory technique from the Mekong region. *Ecol. Soc.* **2013**, *18*. [CrossRef]

53. Schwartz, P. *The Art of the Long View: Planning for the Future in an Uncertain World*; Doubleday: New York, NY, USA, 1996.

54. Van Notten, P. Scenario Development: A Typology of Approaches. In *Think Scenarios, Rethink Education, Schooling for Tomorrow*; Organisation for Economic Co-operation and Development: Paris, France, 2006.

55. Heikinheimo, E. *Four Scenarios for Cambodia's Tonle Sap Lake in 2030—Testing the Use of Scenarios in Water Resources Management*; Aalto University School of Engineering: Espoo, Finland, 2011.

56. McCarthy, J.J.; Canziani, O.F.; Leary, N.A.; Dokken, D.J.; White, K.S. *IPCC Report on Climate Change 2001: Impacts, Adaptation, and Vulnerability*; Intergovernmental Panel on Climate Change: Geneva, Switzerland, 2001.

57. National Institute of Statistics (NIS). *General Population Census of Cambodia 2008: Final Census Results—Figures at a Glance*; National Institute of Statistics (NIS): Phnom Penh, Cambodia, 2008.

58. Royal Government of Cambodia (RGC). *Rectangular Strategy for Growth, Employment, Equity and Efficiency—Phase III*; Royal Government of Cambodia: Phnom Penh, Cambodia, 2013.

59. Arias, M.E.; Cochrane, T.A.; Kummu, M.; Lauri, H.; Holtgrieve, G.W.; Koponen, J.; Piman, T. Impacts of hydropower and climate change on drivers of ecological productivity of Southeast Asia's most important wetland. *Ecol. Model.* **2014**, *272*, 252–263. [CrossRef]

60. Lamberts, D.; Koponen, J. Flood pulse alterations and productivity of the Tonle Sap ecosystem: A model for impact assessment. *J. Hum. Environ.* **2008**, *37*, 185–192. [CrossRef]

61. Holtgrieve, G.W.; Arias, M.E.; Irvine, K.N.; Lamberts, D.; Ward, E.J.; Kummu, M.; Koponen, J.; Sarkkula, J.; Richey, J.E. Patterns of ecosystem metabolism in the Tonle Sap Lake, cambodia with links to capture fisheries. *PLoS ONE* **2013**, *8*. [CrossRef]

62. Hortle, K.G. *Consumption and the Yield of Fish and other Aquatic Animals from the Lower Mekong Basin*; Mekong River Commission: Vientiane, Laos, 2007.

63. Ziv, G.; Baran, E.; Nam, S.; Rodriguez-Iturbe, I.; Levin, S.A. Trading-off fish biodiversity, food security, and hydropower in the Mekong River Basin. *Proc. Natl. Acad. Sci.* **2012**, *109*, 5609–5614. [CrossRef]

64. Chea, Y.; McKenney, B. *Fish Exports from the Great Lake to Thailand: An Analysis of Trade Constraints, Governance, and the Climate for Growth*; Cambodia Development Resource Institute: Phnom Penh, Cambodia, 2003.

65. Orr, S.; Pittock, J.; Chapagain, A.; Dumaresq, D. Dams on the Mekong River: Lost fish protien and the implications for land and water resources. *Glob. Environ. Chang.* **2012**, *22*, 925–932. [CrossRef]

66. Dore, J. An agenda for deliberative water governance arenas in the Mekong. *Water Policy* **2014**, *16*, 194–214. [CrossRef]

67. Grumbine, R.E.; Dore, J.; Xu, J. Mekong hydropower: Drivers of change and governance challenges. *Front. Ecol. Environ.* **2012**, *10*, 91–98. [CrossRef]

68. Asia Development Bank. *Cambodia: Country Governance Risk Assessment and Risk Management Plan*; Asia Development Bank: Metro Manila, Philippines, 2012.

69. Mekong River Commission (MRC). *IWRM-based Basin Development Strategy for the Lower Mekong Basin*; Mekong River Commission: Vientiane, Laos, 2011.

70. Benson, D.; Gain, A.K.; Rouillard, J.J. Water Governance in a Comparative Perspective: From IWRM to a "Nexus" Approach? *Water Altern.* **2015**, *8*, 756–773.

71. Juvonen, H.-M. *Nexus for What? Challenges and Opportunities in Applying the Water-Energy-Food Nexus*; Aalto University: Espoo, Finland, 2015.

72. Bach, H.; Glennie, P.; Taylor, R.; Clausen, T.J.; Holzwarth, F.; Jensen, K.M.; Meija, A.; Schmeier, S. Cooperation for water, energy and food security in transboundary basins under changing climate. In Proceeding of Second Mekong River Commission Summit and International Conference, Ho Chi Minh City, Viet Nam, 2–5 April 2014.

73. Global Water Partnership. *Beijing Nexus Dialogue Symposium*; Global Water Partnership: Stockholm, Sweden, 2014.

74. Star, S.L.; Griesemer, J.R. Institutional ecology, "translations" and boundary objects: Amateurs and professionals in Berkeley's museum of vertebrate zoology. *Soc. Stud. Sci.* **1989**, *19*, 387–420. [CrossRef]

75. Mollinga, P. *For a Political Sociology of Water Resources Management*; Universität Bonn: Bonn, Germany, 2008.

76. Johnstone, G.; Puskur, R.; Declerck, F.; Mam, K.; Mak, S.; Pech, B.; Seak, S.; Chan, S.; Hak, S. *Tonle Sap Scoping Report-CGIAR Research Program on Aquatic Agricultural Systems*; Consultative Group for International Agricultural Research: Montpellier, France, 2013.

77. Foran, T. Node and regime: Interdisciplinary analysis of water-energy-food nexus in the Mekong region. *Water Altern.* **2015**, *8*, 655–674.
78. Allouche, J.; Middleton, C.; Gyawali, D. Technical veil, hidden politics: Interrogating the power linkages behind the nexus. *Water Altern.* **2015**, *8*, 610–626.

![water logo] *water*

MDPI

Article

Chinese State-Owned Enterprise Investment in Mekong Hydropower: Political and Economic Drivers and Their Implications across the Water, Energy, Food Nexus

Nathanial Matthews [1],* and Stew Motta [2]

[1] CGIAR Research Program on Water, Land and Ecosystems, Colombo 10120, Sri Lanka
[2] CGIAR Research Program on Water, Land and Ecosystems, Vientiane 01000, Lao PDR;
stewart.motta@gmail.com
* Correspondence: n.matthews@cgiar.org; Tel.: +94-11-288-0000

Academic Editors: Marko Keskinen, Shokhrukh Jalilov and Olli Varis
Received: 18 May 2015; Accepted: 26 October 2015; Published: 6 November 2015

Abstract: Over the last decade, Chinese State-Owned Enterprises have emerged as among the most active investors in Mekong Basin hydropower development. This paper uses a political economy analysis to examine the forces that drive Chinese State-Owned Enterprises to invest in hydropower in the Mekong Basin. We focus our analysis on the Lancang (Upper Mekong River) in China and in the Greater Mekong Subregion (GMS), with an emphasis on Cambodia. The analysis reveals how powerful political and economic forces from within China and the GMS influence the pace, location and scale of investments in hydropower. These forces include foreign exchange reserves, trade packages and foreign direct investment, and political alliances. Combining the political economy and nexus approaches, we conclude that although policies from China recognize interconnections across the nexus, political and economic forces craft narratives that downplay or disregard these nexus interconnections and trade-offs. This in turn, influences how trade-offs and interconnections in hydropower development are managed and recognized in both local and transboundary contexts, thereby, creating potentially significant negative impacts on livelihoods, food security and the environment.

Keywords: China; state-owned enterprise; hydropower; water-energy-food nexus; greater mekong subregion; political economy; transboundary

---

1. Introduction

This paper examines the political and economic drivers surrounding Chinese State-Owned Enterprises' (SOE) activities in Mekong Basin hydropower, and the implications of these forces across the water, energy and food nexus, hereafter the nexus. The interconnections between water, energy and food have more recently been grouped together as *"the nexus"* [1–3]. Policy makers, international organizations, NGOs, academics and farmers, however, have been grappling with how to manage the interdependencies and trade-offs in the development of natural resources and transboundary rivers for decades [4,5]. One value of the nexus perspective is that it focuses attention on the interdependencies, choke points and trade-offs, thereby helping to understand leverage points and possible solutions [1–3].

The nexus concept takes on different manifestations depending on the context, scale and geography where it is examined [1–3]. For the purpose of this paper, we analyze the nexus between rivers (water), hydropower development (energy), and fish (food). Hydropower dams in the Mekong Basin provide an ideal case study to unpack nexus connections, interdependencies and trade-offs. Although the direct benefits of dams in the Mekong are often disputed, they are regarded by many states as a reliable and cheap source of electricity, and an important revenue stream [6]. There have also been well documented socioeconomic and environmental impacts that revolve around fisheries' losses, which are a key source of food security and livelihoods for millions of people [7].

Smajgl and Ward [8] provide a detailed assessment of how development-directed investments in the wider Mekong Region impact across water, energy and food. Keskinen *et al.*, (this issue) also demonstrate nexus connections in the Mekong by showing how hydropower will impact the region's annual floods, thereby affecting the flood-pulse of Tonle Sap, which will likely cause degradation of fisheries. In this paper, we aim to provide a link to these studies and others by illuminating the political and economic drivers of hydropower in order to better understand why some dams are built despite their well-known impacts at both transboundary and local scales. We demonstrate that the pace of Chinese SOEs' involvement in hydropower construction and the location of investments is heavily influenced by powerful political and economic forces. These forces emerge from both within China and the Greater Mekong Subregion (GMS). We argue that although China purports to recognize nexus interdependencies and trade-offs in its policies, the political and economic forces that drive Chinese SOEs' to build hydropower dams in the GMS Basin offer little space for the consideration of their impacts across the nexus. In this way, the political and economic realities in which hydropower is developed, and in which SOEs operate in the Mekong, often recreate and entrench the original silo approaches to development that nexus policies and dialogues are aiming to address.

The remainder of the paper is divided into four sections. Section 2 provides a brief overview of the methods used in the study, and of the political economy approach. Section 3 highlights the growth of renewable energy in China. This section also examines the policies and actions that are being deployed to respond to the challenges created by rapid economic growth that is driving Chinese SOEs' involvement along the Lancang-Mekong and in the GMS. The Lancang-Mekong is chosen because it is a transboundary river undergoing rapid hydropower development by SOEs. This section additionally critically examines the *"win-win"* narratives that emerge from China and SOEs surrounding hydropower development on the Lancang-Mekong. These narratives frame or ignore risks and uncertainties across the nexus in ways that buttress and endorse central, top down decision-making and the agendas of powerful actors. Section 4 examines the political and economic drivers of Chinese SOEs' rapid development of hydropower projects throughout the GMS with a focus on Cambodia and the implications of this development across the nexus. Cambodia is chosen as a focus because it is one of China's key allies in the GMS and increasingly a destination for Chinese investment. Finally, Section 5 concludes by reviewing how political and economic factors have shaped the application of nexus policies in the GMS and the negative implications this has had and will potentially have on water, energy and food.

2. Method

This paper uses a broad global political economy approach to analyze the key political and economic forces that drive and shape the ways water, food and energy are managed by Chinese SOE in hydropower development in the Mekong Basin. This political economy approach helps to explore the constellation of interests, motivations and histories that shape the institutions, governance and contestations across the nexus.

Political economy has been usefully employed to analyze or compliment nexus issues and approaches at varying scales. Allouche *et al.* [2], for example, highlight the importance of political economy considerations when examining nexus issues by arguing that the nexus does not adequately engage with the global political economy of energy and food. They argue that policy documents tend to promote economic and technological solutions that ignore the inequalities across the political economy. The nexus' recognition of trade-offs may also be used by decision-makers to justify negative environmental impacts as an unavoidable cost to ensuring water, food and energy security. Foran [9] recognizes the need for *"more vigorous thinking around the political economy of energy, water and food"*. Using a political economy approach coupled with historical and institutional analysis helps to uncover how power and inequality play out in the nexus [9]. Dupar and Oates [10] also use global political economy considerations to argue that the nexus approach must take account of rights-based approaches and transparent negotiation to ensure that resources are not commodified in ways that justify ignoring the environmental costs of their consumption. The importance of rights-based approaches including transparency, participation and fair negotiation in the Mekong setting has been well documented [11–13].

Political economy has also been used at more regional and local scales to examine nexus trade-offs and drivers. Middleton and Allen [14] use a political economy lens to demonstrate that the politics inherent in the nexus require that nexus approaches must consider both bottom-up analysis as well as historical, political and economic frames in order to understand the drivers and distribution of trade-offs. Political economy analysis of nexus issues is also complimented by an environmental justice lens [13]. Environmental justice helps to draw attention to the winners and losers of trade-offs. As we examine below, hydropower narratives that are driven by political and economic forces can strengthen powerful actors while excluding other stakeholders such as local people and the environment.

Political economy analysis helps to illuminate the power and economics embedded within water management and hydropower development. Mitchell's [15] analysis of the political economy of Mekong Basin development focuses on the political aspects of decision making and how the benefits and impacts of decisions have uneven distributions. Importantly, Mitchell identified the emergence of conflict between basin-wide coordination, promoted by the MRC (Mekong River Commission) and its donors, and the individual national agendas. As will be explored in this paper, this conflict continues today and is shaped by Chinese SOE investment. Political economy can also help to highlight the different manifestations of the nexus. Verhoeven's [16] analysis of the nexus in the highly political transboundary Nile context reminds us that the nexus is a human construct and the interconnections within it are understood in a multitude of ways and scales by different social and ethnic groups across societies, and this can result in contestation between the nexus components.

In addition to the political economy approach and a broad literature review, this paper draws from 25 informal interviews carried out from 2012 to 2014. The authors used both elite and normal interviewing to triangulate data with literature reviews and observation. Interviewees were identified by their prominent positions within organizations relevant to hydropower development and nexus issues. Interviews included senior and junior policy makers, employees of Chinese SOEs, bureaucrats, consultants, NGO staff and academics. The interview type was semi-structured using small sets of open ended questions. The questions focused around case studies and key events associated with nexus issues specifically linked to Chinese SOEs and hydropower development. The majority of interviews were carried out face-to-face in the Mekong Region, with some conducted by telephone and Skype. Several interviewees were interviewed multiple times. English was used for the majority of interviews, but some interviews and clarifying questions were done in Mandarin.

The authors take note that there are many different configurations of SOEs from China involved in hydropower development in the Mekong Basin and each have their own characteristics. We do not wish to brand all these SOEs or the Chinese government as always disregarding or downplaying the impacts of their investments and activities across the nexus. Rather, we are aiming to provide some insights into the forces that shape Chinese SOEs' decision-making and subsequent involvement in hydropower across the Basin. Our hope is that these insights might provide lessons for understanding the implications of hydropower investment in other regions. Throughout the article, we use the word China to refer to the Chinese government in a broad sense and Chinese SOEs to refer to the larger SOE hydropower developers that are representative of the main developers in the Basin. We have intentionally avoided naming many of the SOEs that we studied in this research in order to avoid any attribution to the people we interviewed. Where information came primarily from literature reviews we mention specific SOEs by name, but this does not indicate that these are the only SOEs to which the findings apply. Finally, it is worth noting that this paper focuses mainly on qualitative analysis because hydropower is deeply entwined with politics, economics and security issues across the nexus.

3. The Nexus, Hydropower and Chinese SOEs' Activities on the Lancang-Mekong

The Chinese government is well aware of the need to balance the trade-offs in the nexus as clearly shown by the first paragraph from the main targets of the 12th Five Year Plan (FYP):

We will maintain farmland reserves at 1.818 billion mu (approximately 121,260,600 hectares). We will cut water consumption per unit of value-added industrial output by 30%, and increase the water efficiency coefficient in agricultural irrigation to 0.53. Non-fossil fuel resources will rise to 11.4% of primary energy consumption. Energy consumption per unit of Gross Domestic Product (GDP) will decrease 16% and CO₂ emissions per unit of GDP will decrease 17% [17].

While China has policies that attempt to address all three aspects of the nexus, a significant focus of its policies are on energy [17,18]. This is primarily due to China's prioritization of unimpeded economic growth, which is currently dependent on energy intensive industries. Stresses on water and food security due to air, water, and soil pollution are all viewed within the context of emissions from energy intensive industries, and constitute concerns shared by both domestic and international communities [17,18].

China's Ministry of Environmental Protection was not given ministerial level status until 2007, and remains comparatively weak. The environmental and climate policies that do exist can be seen more as by-products of China's energy security worries. Energy security worries are also linked to concerns over social unrest due to widespread and persistent environmental degradation including air quality issues in the capital, and the policy responses to address these concerns initiated under the Hu Jintao administration [19]. China's ability to meet energy demands domestically is further threatened by the agricultural and industrial sectors' massive water and energy demands [20]. The energy intensive economy has also propelled China to become the world's leading carbon emitter, which increasingly places pressure on China's administration to play a more leading role in climate change goals as an emergent superpower within the context of global geopolitics.

Environmental policies and pollutant regulation may be viewed as an obstacle and a nuisance to development in China. This is especially true at the local levels where officials are valued and promoted primarily based on GDP increases. Renewable energy, however, is a priority pillar industry for the government's national development plan [21]. Renewable energy and especially hydropower is seen by Chinese scholars and the government not as an impediment to development, but as a solution to the challenges it faces across the nexus and is the dominant goal of energy security [21]. In addition, renewable energy is perceived as an important sector for China's economic expansion, trade, and technological enhancement [22].

While there have been massive investments into all three main sources of renewable energy in China (hydropower, wind and solar), many of the investments are geared towards GDP increases at the local and corporate levels, and lack the planning and collaboration necessary to contribute effectively to the country's macro-economic targets. Despite China being the world's largest investor in wind and solar power, 23% of the turbines and 28% of the panels are not connected to the grid, and therefore do not actually contribute to national power production [23]. If around a quarter of the billions of dollars of renewable energy projects are not actually contributing energy to the grid, it may be argued that these industries are not being driven by environmental policy alone. Although policy might be aimed at energy security, the investment does not necessarily contribute to the expected energy goals, an argument that this article will return to with regards to the Mekong Basin.

China has been unable to shift its economy to less energy intensive production higher up the value chain, which will make it increasingly difficult for renewable energy targets to be met. The renewable energy sector only contributed to around 1% of China's total energy consumption (not including hydropower) in 2010. Renewable energy is projected to contribute 15% of the energy supply by 2020 (See Table 1). Although there has been large investment in the solar and wind sectors, China's inability to shift the economy to less energy intensive production means that wind and solar energy will contribute only minimally to the overall energy supply. The vast majority of renewable energy supply that is needed to meet China's energy targets will come from hydropower production. The majority of which exists in the west and southwest of China on transboundary rivers.

Table 1. Key energy targets enacted in China's 12th FYP (2011–2015).

Energy	2010 Target	2010 Realized	2015 Target	2020 Target
Hydropower	190	272.6	290	420
Non-fossil fuel share in energy supply	10%	9.4%	11.4%	15%
Non-fossil fuel share in energy supply (excluding hydropower)	1%	1.2%	Unknown	3%

Source: FYP 2012 [17].

China's domestic hydropower expansion strategy is partially a response designed to alleviate the current stressors across the nexus. To legitimize this strategy, the Chinese government and its SOEs have employed a number of domestic and regional narratives that couch the impacts of hydropower on the transboundary Lancang River as a *"win-win"* for both China and the GMS [24]. These narratives have been developed and employed over decades of domestic dam construction [24]. For example, the Chinese government has couched its rigid policy stance on hydropower in a rhetoric of regional *"peace"*, highlighting the *"win-win"* outcomes for downstream riparian states [24,25]. Although China still rarely discusses the negative impacts of dams and their consequences within the nexus, it has recently been visibly more communicative with the Lower Basin states such as Thailand, Laos, Cambodia and Vietnam through the MRC [26].

Liebman [27] argues that in communicating with the GMS, China has used narratives that frame the benefits of hydropower as being shared between governments to avoid appearing as a hegemon within the region. Examples of this position can be seen in Chinese political speeches. During a speech at the MRC in 2010, H.E. Song Tao, Vice Minister of Foreign Affairs of the People's Republic of China stated that *"hydropower development of the Lancang River can improve navigation conditions and help with flood prevention, drought relief and farmland irrigation of the lower reaches"* and that dam construction would include *"equal consultation, stronger cooperation, mutual benefit and common development"* [28]. The speech made no mention of trade-offs across the nexus or how impacts would be mitigated.

In contrast to these narratives, studies show that China's cascade of dams on the Lancang is having a significant transboundary impact downstream by impeding vital sediment flows and reducing water levels by as much as 30% during the rainy season [29]. By ignoring or downplaying the fact that all energy production has trade-offs and that in the Mekong these are potentially significant and transboundary in nature, these narratives reinforce and empower decision-making that ignores nexus connections. These narratives not only legitimize silo decision-making, they also delegitimize civil society and environmental groups who attempt to place a grounded environmental justice lens into the nexus to speak for and illuminate who wins and loses from trade-offs [13]. As we will examine next, these narratives and political and economic drivers of Chinese SOEs' hydropower development extend past China's border into the GMS. We argue that the Chinese state is using its political and economic power to reshape science to legitimize its policy agendas and international and domestic strategies. This strategy and the political and economic forces that drive it has important implications for how SOEs address the trade-offs of hydropower development in both local and transboundary contexts. SOEs are able to use the ambiguous and contested discourses surrounding the nexus to further their political and economic interests, while disadvantaged actors are continually marginalized [9].

4. Chinese SOE Hydropower Expansion in the Greater Mekong Subregion and its Implications across the Nexus

The GMS is currently experiencing a surge in hydropower development, with more than 50 on-going large-scale dam projects (over 50 MW) being built and managed by Chinese companies, and many more in the proposal phase [30]. The distribution of large-scale Chinese dams in the region is as follows: Myanmar 30, Lao PDR 13, Cambodia 7, Vietnam 3, and Thailand with some Chinese dam projects, but none over 50 MW [30]. These projects are financed, developed, constructed, and contracted out primarily to Chinese SOEs. Sinohydro plays at least one of these roles in 30% of the large dam projects in the LMB, and handles the entirety of financing, developing, and building of five of the 13 large Chinese dams in the Lao PDR [30]. Large SOEs like Sinohydro often have higher capacity, scalability, greater political backing, more experience gained from domestic construction projects, and can usually build dams at a lower price than their competitors [31]. These large-scale projects usually require approval from the highest levels of government in both Beijing and in the LMB nations, and thus connections with the state serve as an asset. In the LMB, almost every major dam project has SOE involvement from well-known large enterprises, but Myanmar shows some contrast with much smaller and less well-known provincial level SOEs [30].

Beijing originally pushed its SOEs to go abroad and invest through the "*Going Global Policy*", which was proposed at the 5th plenary session of the 15th central committee in 2000. This policy acts as a roadmap for Chinese foreign investment. The ambitions of this policy were not articulated until the following FYP in 2005, when the Ministry of Foreign Trade and Economic Cooperation (MOFCOM), National Development and Reform Commission (NDRC) and Export-Import Bank of China policies began streamlining the process for investment to reach overseas markets [32]. From 2001 to 2005, China's ODI was $22.3 billion USD in total. As part of this policy, ODI rapidly increased to $18.7 billion USD in 2007 alone, $87.8 billion in 2012, $90.17 billion in 2013 and reaching a staggering $102.89 billion in 2014 [32–35].

China's "*Going Global*" strategy does not simply pertain to SOEs. China's overseas investments, trade, and aid must be viewed as a package rather than separate initiatives. ODI provides a conduit for China to invest its vast capital supplies and at the same time develop strategic partnerships and secure access to resources. Urban *et al.* [30] describe how these elements are packaged together in hydropower deals in the GMS:

> The Chinese practice is hence often to bundle aid, trade and investment by providing, for example, both investments and concessional loans for dam building and linking this to the export of electricity coupled with the import of Chinese manufactured goods and trade deals for Chinese firms. [30]

One example of this ODI policy, related to hydropower, is the Kamchay Dam in Cambodia, which was the first Build Operate Transfer (BOT) hydropower project in the country. Sinohydro is the primary SOE involved in the construction and operation of the Kamchay Dam project. The $280 million that was required to construct the dam, however, was part of a $600 million aid package given to Cambodia by China in 2006 [36].

Chinese investment in hydroelectric development in the GMS has expanded rapidly in the past decade. McNally *et al.* [37] argue that this investment should also be understood as exporting Chinese policies, expertise and ideologies, as discussed above, into developing countries. China has gone from having very few investments in Cambodia, Laos and Myanmar to becoming all three countries' largest foreign direct investor [38]. From 2006 to 2011, China invested over $6.1 billion into the three countries in their hydropower sectors alone [39] (See Figure 1 below). This figure is likely a low estimate as some concession agreements were signed but not counted during that time. Furthermore, it does not include the currently on hold Myitsone project in Myanmar valued at over $3.6 billion USD.

When Chinese ideologies and policies that disregard or downplay connections and trade-offs across the nexus, as outlined in the narratives above, are exported to the LMB, there are potentially significant consequences for livelihoods, food security and the environment across the region. An example of how these policies, expertise and ideologies are exported can be seen in how Beijing's outward investment policies are mirrored in Lao PDR's hydropower sector. Chinese corporations had just two projects completed as of February 2015, for a combined 210 MW. However, there is 1010 MW under construction, and 3892 MW that have MOUs signed but have yet to begin construction, including three out of six of the dams planned for the Mekong mainstream. If all of these projects are realized, the scale of China's installed capacity within the Lao PDR will have increased nearly 25 times between 2015 and 2030 [7]. We argue that China's lack of recognition of trade-offs across the nexus has driven many of these hydropower projects to disregard or downplay the evidence of their impacts on fisheries, livelihoods and the environment [7,40].

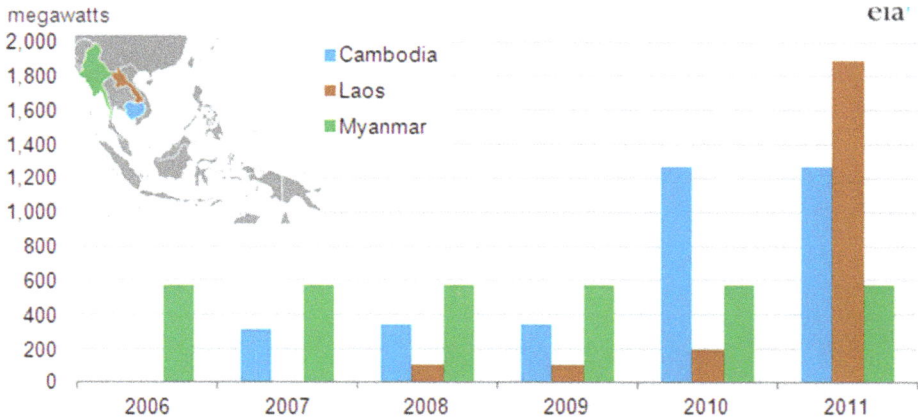

Figure 1. Cumulative hydroelectric capacity additions financed by China (2006–2011) (Source EIA, 2013 [39]).

The scale of the projects that the Chinese are pursuing is also increasing as shown by Myanmar's hydropower climate. Chinese companies signed the contract for the Yeywa dam in 2005, finishing the construction of the 790 MW project in 2011, making it the largest dam in the country. There are 46 dams planned from 2016 to 2030 with 12 of these over 1000 MW of installed capacity. Of these dozen 1000 MW plus projects, Chinese companies have 11, including the Mong Tong Dam, which is 7110 MW [41].

The potential and current negative implications of these investments across the nexus is significant [42–44]. Governments in Cambodia, Laos and Myanmar face major challenges in monitoring and regulating investment and its social and environmental impacts at this scale and speed [42–44]. The countries still have very young governments and judicial systems that are going through major institutional changes and improvements, but are currently not capable of holding large corporations accountable. In terms of transboundary impacts, different and often competing political and economic goals and historical and cultural issues compound opportunities for cooperation at the basin scale [45]. Transboundary water law also offers limited opportunities to challenge the transboundary impacts of upstream states on the water and food especially when legal frameworks across the basin states are based on different variables and rationales [46], and only Vietnam has signed up to the United Nations Watercourses Convention. An indication of the limitations and challenges of promoting good water governance and holding SOEs accountable can be seen in Transparency International's World Ranking indices as demonstrated in Table 2 [47–49].

Cambodia, Lao PDR, and Myanmar have portrayed large-scale hydropower development as essential for their own economic development and their citizens' livelihood improvement [50,51]. However, there have been very little independent data or studies that directly equate large-scale dam construction in these countries to the improvement of livelihoods. The more realistic driving force behind the domestic government support for these projects is the need for capital through foreign direct investment (FDI). This FDI comes with key revenue opportunities for national governments generated through energy export to neighboring countries, primarily Thailand, that have much higher energy demands [52–54].

Table 2. An overview of three international indices where China is heavily investing in hydropower in the GMS.

Country	Ease of Doing Business Rating (Out of 189)	Corruption Index (Out of 177)	Freedom of Press (Out of 179)
Cambodia	137th	160th	143rd
Lao PDR	159th	140th	168th
Myanmar	182nd	157th	151st

Chinese ODI in hydropower development in the GMS is often partly justified as supplying technology to developing countries or as an important strategy for China to increase its domestic energy supply and security. Despite this justification, China was a net exporter of energy to the region in 2010 [55]. In Lao PDR and Myanmar, the majority of the energy exported is destined for Thailand. There are no transmission lines from Cambodia to China, yet the Kingdom of Cambodia has recently begun an ambitious hydroelectric expansion with the help of Chinese SOEs.

Sinohydro completed construction of the country's first large-scale project, the Kamchay Dam, in 2011 [36]. Rather than having a vast array of small and medium projects (Lao PDR has more of this makeup), Cambodia is instead hosting a few large and controversial Chinese dam projects, including the Lower Sesan 2. These large-scale projects allow for greater injections of ODI into the Cambodian government and the scale and technical nature of the projects ensure that the Chinese companies involved will be central level SOEs.

Burgos and Sophal [56] argue that Hun Sen's relationship with China creates an opportunity for the country to develop its economy without conditional loans and grants from Western donors that may include calls for reform or restructuring of existing powerbases. Government-sanctioned projects give the Chinese SOEs political security and the ability to operate with greater autonomy in the country [55,56]. Examples of this can be seen in the Lower Sesan 2, where reservoir clearing began inside the concession area before dam designs were even approved [57]. The tree clearing at the Lower Sesan 2 was also illegal as it extended well beyond the borders defined in the concession agreement [58]. Many speculate that the central government and the SOEs are signing dam concession agreements in order to access the valuable timber reserves. The Ministry of Environment has the responsibility to hold SOEs responsible for producing and implementing Environmental Impact Assessments (EIAs) and Environmental Management Plan (EMPs). In the case of the Kamchay Dam, however, which was signed by Wen Jiabao and Hun Sen, construction went on for years without either an EIA or EMP [36].

Cambodia is China's most important political ally in the region and this relationship likely creates favourable conditions for Chinese SOEs operating in the country. In 2013, after ASEAN attempted to unite in its discussion with China over the South China Sea issues, Wang Yi, the Chinese foreign minister to Cambodia, refused to attend. Wang Yi explained in a statement to reporters that China is interested in protecting Cambodia from outside interests, hence his refusal to attend the ASEAN discussion [59]. As the main driver of the South China Sea tension is the potential undersea oil and gas reserves, it is possible to link China's interests in Cambodia back to China's domestic nexus. These examples demonstrate how hydropower investment can serve as a placeholder for larger political and economic issues that are not necessarily energy security or profit driven.

China is politically stronger in Cambodia than in the Lao PDR or Myanmar—the other two GMS nations where Chinese hydropower development is extensive. Chinese SOEs in Cambodia have a relatively large stake in the hydropower industry due to the scale of their projects and the lack of developer diversity in the sector. China has invested over $1.6 billion in just six projects, with four now operational, and two expected to complete construction in 2015 [58]. The Chinese Chamber of Commerce in Cambodia is dominated by hydroelectric SOEs and the industry will likely continue to expand. China's political and economic strategy within the region and ASEAN, and the Cambodian government's dependence on Chinese investment highlight some of the drivers behind hydropower decision-making. The analysis also shows that individual projects do not necessarily need to be profitable to be approved, but rather contribute to grander relationship goals that are key for both governments and their political and economic strategies.

Foreign Exchange Reserves as an Economic Driver of Investment

One of the issues around Chinese ODI meriting greater attention is the link between large-scale infrastructure projects and China's vast foreign exchange reserves (forex). China's forex has increased tenfold over the past decade, reaching $3.82 trillion at the end of 2013 [60]. This is by far the largest stockpile of forex in the world, three times that of the next largest held by Japan [61]. This large stock of forex threatens to cause the yuan to inflate, which in turn threatens China's export market and foreign investment. To prevent the yuan's inflation, China's central bank regulates fluctuations while purchasing large quantities of foreign reserves [62]. One of the main sources of foreign reserves are low-interest treasury bonds from the US and Japan, which can be as low as 2.8% over 10 years [61]. This huge amount of forex also makes China vulnerable to fluctuations in global forex markets, especially in instabilities of USD and JNY.

During the global economic recession both the U.S. and Japan responded to the crisis through monetary easing, or in other words, by aggressively running the printing presses, churning out cash, and devaluing their currencies. This concerns Beijing, which is sitting on huge sums of U.S. and Japanese forex while vast amounts of investment continue to flood into the country, threatening the yuan and therefore China's economic stability.

In order to alleviate these concerns, the Chinese banking system quickly pushes money out of the country to its SOEs in the form of ODI, and the easiest way to move large quantities of money outward is through large-scale infrastructure projects, such as hydropower dams. The Export-Import Bank of China (China Exim) is one of three policy banks that was established in 1994 along with the China Development Bank (CDB) and the Agricultural Development Bank of China. China Exim and CDB are the primary lenders for overseas investment in resource development. China Exim's mandate states that it is to *"assist Chinese companies with comparative advantages in their offshore project contracting and outbound investment, and promote international economic cooperation and trade"* [63]. In the last decade, China Exim's lending has increased dramatically and the bank is now the largest export credit agency in the world [64]. China Exim has been able to increase lending by 30%–40% year on year, fueling the ever expanding ODI under China's Going Global Strategy [64].

Long term payback systems are generally preferred by banks, which coincides well with large-scale hydropower projects where the Build Operate Transfer contract is usually a few decades in duration. Some of the large infrastructure projects carried out by SOEs might not be profitable, but this is inconsequential as the government has a mandate to spend money. Additionally, many of these concessional loans are attached to natural resource access [65]. Therefore, China's ODI into large-scale hydropower projects is not necessarily based on alleviating nexus pressures as its policies state, but can be seen as part of a strategy to ensure the country's greater economic security against global currency fluctuations that threaten the value of the RMB, and the depreciation of China's forex reserves.

5. Conclusions

This paper examined some of the key political and economic drivers behind Chinese SOEs' hydropower activities and the implications of these drivers on the recognition and management of nexus interconnections and trade-offs. Chinese SOEs are amongst the biggest players in global hydropower development and are heavily involved in dozens of projects across the Mekong Basin. The political economy approach used in this analysis reveals the forces that drive the pace, location and scale of this hydropower development both domestically in China and in the GMS. These forces include how hydropower projects are linked to bigger packages of trade and aid and FDI, which are important sources of revenue for developing countries such as Lao PDR, Myanmar and Cambodia. In addition, hydropower projects also provide a guise to access lucrative natural resources such as timber. Other key forces analyzed include the link between large-scale infrastructure projects and China's vast foreign exchange reserves, which facilitate SOEs to make hydropower investments abroad. The benefits of hydropower investments for both SOEs and domestic states are also linked to political alliances that have geopolitical dimensions.

When the political economy and nexus approaches are combined, the analysis reveals some of the key drivers that cause narratives surrounding hydropower to downplay or disregard trade-offs and interconnections across water, energy and food. These narratives, which position hydropower as a win-win for both upstream and downstream states, are in contrast to hundreds of scientific studies that demonstrate the substantial trade-offs of hydropower development. Hydropower dams in the Mekong Basin have been shown to have significant negative impacts on livelihoods, food security and the hydrology and ecosystems of the Mekong River and its tributaries [7,42–44].

Hydropower development by SOEs in the Mekong Basin also has potentially important transboundary impacts across water and food [42–44]. Domestic and transboundary water law currently has limited influence to address these impacts. Lower Basin states are challenged by limited capacity and often weak and underfunded institutions to negotiate or manage large-scale impacts across the nexus. Furthermore, diverse political and economic goals and important cultural and historical differences have impeded basin-wide cooperation surrounding mainstream dams despite the best efforts of the MRC [45].

Chinese SOEs' hydropower investments and activities are part of an increasing trend for both individual state and private sector led infrastructure financing and development that is expected to reach as much as four trillion dollars by 2017 [66–68]. The Asian Infrastructure and Investment Bank and the Silk Road Fund, for example, will soon bring an additional $140 billion dollars of investment for development, a large proportion of which will be aimed at hydropower. By using a nexus and political economy approach this study provides a useful link to studies that examine impacts of hydropower development on water and food through its analysis of the drivers and implications of silo decision-making. Further research is needed to consider the opportunities and challenges the enormous investment by SOEs and the private sector will have on the interconnected spheres of water, energy and food.

Acknowledgments: We thank the anonymous reviewers and academic editors for their constructive comments that significantly helped to improve the paper. We also thank Juliet Lu for comments on an early draft.

Author Contributions: Nathanial Matthews coordinated the writing process of the article. He was mainly responsible for writing the introduction, method and conclusion. Stew Motta led the analysis in section three and four. Both authors contributed to the final writing of the piece.

Conflicts of Interest: The authors declare no conflict of interest.

References

1. Hoff, H. *Understanding the Nexus*; Background Paper for the Bonn2011 Conference: The Water, Energy and Food Security Nexus; Stockholm Environment Institute: Stockholm, Sweden, 2011.
2. Allouche, J.; Middleton, C.; Gyawali, D. Technical veil, hidden politics: Interrogating the power linkages behind the nexus. *Water Altern.* **2015**, *8*, 610–626.
3. Allouche, J.; Middleton, C.; Gyawali, D. *Nexus Nirvana or Nexus Nullity? A dynamic Approach to Security and Sustainability in the Water-Energy-Food Nexus*; STEPS Working Paper 63; STEPS Centre: Brighton, UK, 2014.
4. White, G.F.; deVries, E.; Dunkerley, H.B.; Krutilla, J.V. *Economic and Social Aspects of Lower Mekong Development*; Mekong Committee: Bangkok, Thailand, 1962.
5. Scott, C.A.; Kurian, M.; Wescoat, J.L. The water-energy-food nexus: Enhancing adaptive capacity to complex global challenges. In *Governing the Nexus: Water, Soil and Waste Resources Considering Global Change*; Kurian, M., Ardakanian, R., Eds.; Springer: Dordrecht, The Netherlands, 2014.

6. Lee, S. Benefit Sharing in Environmental Governance: Beyond Hydropower in the Mekong River Basin. In *The International Handbook of Political Ecology*; Bryant, R., Ed.; Edward Elgar: London, UK, 2015; p. 189.

7. Ziv, G.; Baranb, E.; Namc, S.; Rodríguez-Iturbed, I.; Levina, S. Trading-off Fish Biodiversity, Food Security, and Hydropower in the Mekong River Basin. *Proc. Natl. Acad. Sci. USA* **2012**, *109*, 5609–5614. [CrossRef] [PubMed]

8. Smajgl, A.; Ward, J. *The Water-Food-Energy Nexus in the Mekong Region: Assessing Development Strategies Considering Cross-Sectoral and Transboundary Impacts*; Springer-Verlag: New York, NY, USA, 2013.

9. Foran, T. Node and regime: Interdisciplinary analysis of water-energy-food nexus in the Mekong region. *Water Altern.* **2015**, *8*, 655–674.

10. Dupar, M.; Oates, N. Getting to grips with the water-energy-food "nexus". Climate and Development Knowledge Network (London). Available online: http://cdkn.org/2012/04/getting-to-grips-with-the-water-energy-food-nexus (accessed on 10 July 2015).

11. Molle, F. *Irrigation and Water Policies in the Mekong Region: Current Discourses and Practices*; International Water Management Institute (IWMI) Research Report 095; IWMI: Colombo, Sri Lanka, 2005.

12. Hirsch, P. Water Governance Reform and Catchment Management in the Mekong Region. *J. Environ. Dev.* **2006**, *15*, 184–201. [CrossRef]

13. Middleton, C.; Allouche, J.; Gyawali, D.; Allen, S. The rise and implications of the water-energy-food nexus in Southeast Asia through an environmental justice lens. *Water Altern.* **2015**, *8*, 627–654.

14. Middleton, C.; Allen, S. Asia Pacific Sociological Association Conference Paper. Available online: http://carlmiddleton.net/wp-content/uploads/2014/12/MiddletonAllan_Nexus_FINAL.pdf (accessed on 11 July 2015).

15. Mitchell, M. The Political Economy of Mekong Basin Development. In *The Politics of Environment in Southeast Asia: Resources and Resistance*; Hirsch, P., Warren, C., Eds.; Routledge: London, UK, 1998; pp. 71–89.

16. Verhoeven, H. The Nexus as a Political Commodity: Agricultural Development, Water Policy and Elite Rivalry in Egypt. *Int. J. Water Resour. Dev.* **2015**, *1–15*, 360–374. [CrossRef]

17. China's 12th Five Year Plan (FYP). Available online: http://www.britishchamber.cn/sites/default/files/full-translation-5-yr-plan-2011-2015.doc (accessed on 20 November 2014).

18. Zhang, Q.; Xia, Q.; Clark, C.K.L.; Shu, G. Technologies for Efficient Use of Irrigation Water and Energy in China. *J. Integr. Agric.* **2013**, *12*, 1363–1370. [CrossRef]

19. Air pollution in China: A hazy future? Available online: http://sustainability-ranking.channelnewsasia.com/news-a-hazy-future.html (accessed on 5 November 2015).

20. Grumbine, E.; Xu, J. Recalibrating China's environmental policy: The next 10 years. *Biol. Conserv.* **2013**, *166*, 287–292. [CrossRef]

21. Schuman, S.; Lin, A. China's Renewable Energy Law and Its Impact on Renewable Power in China: Progress, Challenges and Recommendations for Improving Implementation. *Energy Policy* **2012**, *51*, 89–109. [CrossRef]

22. Wang, F.; Yin, H.; Li, S. China's Renewable Energy Policy: Commitments and Challenges. *Energy Policy* **2010**, *38*, 1872–1878. [CrossRef]

23. Ma, B.; Zhou, S. Smart-grid policies: An international review. *Wiley Interdiscip. Rev.: Energy Environ.* **2013**, *2*, 121–139.

24. Magee, D. New Energy Geographies: Powershed Politics and Hydropower Decision Making in Yunnan, China. Ph.D. Thesis, University of Washington, Seattle, WA, USA, 2004.

25. Xinhua. China to Build Huge Power Station on Lancang-Mekong River. Available online: http://english.peopledaily.com.cn/200201/20/eng20020120_89013.shtml (accessed on 13 May 2014).

26. Mekong River Commission (MRC). Mekong River Commission and China Boost Water Data Exchange. Available online: http://www.mrcmekong.org/news-and-events/news/mekong-river-commission-and-china-boost-water-data-exchange/ (accessed on 15 October 2014).

27. Liebman, A. Trickle-down Hegemony? China's "Peaceful rise" and Dam Building on the Mekong. *Contemp. Southeast Asia* **2005**, *27*, 281–304. [CrossRef]

28. MRC. Remarks by H.E. Song Tao, Vice Minister of Foreign Affairs of the People's Republic of China. Available online: http://www.mrcmekong.org/news-and-events/speeches/first-mrc-summit-5/ (accessed on 16 May 2014).

29. Kummu, M.; Varis, O. Sediment-related Impacts Due to Upstream Reservoir Trapping, the Lower Mekong River. *Geomorphology* **2007**, *85*, 275–293. [CrossRef]

30. Urban, F.; Nordensvärd, J.; Khatri, D.; Wang, Y. An analysis of China's investment in the hydropower sector in the Greater Mekong Sub-Region. *Environ. Dev. Sustain.* **2013**, *15*, 301–324.

31. McDonald, K.; Bosshard, P.; Brewer, N. Exporting Dams: China's Hydropower Industry Goes Global. *J. Environ. Manag.* **2009**, *90*, S294–S302. [CrossRef] [PubMed]

32. Lee, J. State Owned Enterprises in China: Reviewing the Evidence. In *OECD Working Group on Privatisation and Corporate Governance of State Owned Assets*; OECD: Paris, France, 2009.

33. China's Outward FDI Reaches New Highs on Strong Growth in 2012–13. Available online: http://asiatoday.com.au/content/china%E2%80%99s-outward-fdi-reaches-new-highs-strong-growth-2012-13 (accessed on 5 November 2015).

34. Ministry of Commerce People's Republic of China (MOFCOM). Regular Press Conference of Ministry of Commerce on January 21 2015. Available online: http://english.mofcom.gov.cn/article/newsrelease/press/201501/20150100878729.shtml (accessed on 18 June 2015).

35. MOFCOM (Ministry of Commerce of the People's Republic of China). Brief Statistics on China Direct Investment Overseas in 2013. Available online: http://english.mofcom.gov.cn/article/statistic/foreigntradecooperation/201401/20140100464237.shtml (accessed on 2 June 2014).

36. Grimsditch, M. *China's Investments in Hydropower in the Mekong Region: The Kamchay Hydropower Dam, Kampot, Cambodia*; World Resources Institute: Washington, DC, USA, 2012.

37. McNally, A.; Magee, D.; Wolf, A.T. Hydropower and Sustainability: Resilience and Vulnerability in China's Powersheds. *J. Environ. Manag.* **2009**, *90*, S286–S293. [CrossRef] [PubMed]

38. Matthews, N.; Motta, S. *China's Influence on Hydropower Development in the Lancang River and Lower Mekong River Basin*; CPWF Mekong: Vientiane, Lao PDR, 2013.

39. EIA (Energy Information Administration). U.S. Energy Information Administration. Available online: http://www.eia.gov/todayinenergy/ (accessed on 2 June 2014).

40. Magee, D. The dragon upstream: China's role in Lancang-Mekong development. In *Politics and Development in a Transboundary Watershed*; Springer: Dordrecht, The Netherlands, 2012.

41. Electricite Du Laos (EDL). *Power Development Plan of Lao PDR 2015–2025*; Power Plant Development Department, Electricite Du Laos: Vientiane, Lao PDR, 2015.

42. Lu, X.X.; Li, S.; Kummu, M.; Padawangi, R.; Wang, J.J. Observed changes in the water flow at Chiang Saen in the lower Mekong: Impacts of Chinese dams? *Q. Int.* **2014**, *336*, 145–157. [CrossRef]

43. Siciliano, G.; Urban, F.; Kim, S.; Lonn, P.D. Hydropower, social priorities and the rural–urban development divide: The case of large dams in Cambodia. *Energy Policy* **2015**, *86*, 273–285. [CrossRef]

44. Orr, S.; Pittock, J.; Chapagain, A.; Dumaresq, D. Dams on the Mekong River: Lost fish protein and the implications for land and water resources. *Glob. Environ. Chang.* **2012**, *22*, 925–932. [CrossRef]

45. Suhardiman, D.; Giordano, M.; Molle, F. Scalar disconnect: The logic of transboundary water governance in the Mekong. *Soc. Nat. Resour.* **2012**, *25*, 572–586. [CrossRef]

46. Rieu-Clarke, A. Transboundary hydropower projects seen through the lens of three international legal regimes—Foreign investment, environmental protection and human rights. *Int. J. Water Gov.* **2015**, *3*, 1–24.

47. Greater Mekong Dams Database. *CGIAR Program on Water, Land and Ecosystem's Greater Mekong Program*; WLE: Vientiane, Lao PDR, 2015.

48. International Finance Corporation (IFC). Doing Business: Economy Rankings. Available online: http://www.doingbusiness.org/rankings (accessed on 2 June 2014).

49. Corruption Perceptions Index 2013. Available online: http://cpi.transparency.org/cpi2013/results/ (accessed on 2 June 2014).

50. Reporters without Borders (RWB). Press Freedom Index 2013. http://en.rsf.org/press-freedom-index-2013, 1054.html (accessed on 2 June 2014).

51. Sokheng, V. Phnom Penh Post. Available online: http://www.phnompenhpost.com/national/pm-hun-sen-ushers-new-dam (accessed on 2 June 2014).

52. Geheb, K.; West, N.; Matthews, N. The Invisible Dam: Hydropower and its narration in the Lao People's Democratic Republic. In *Hydropower Development in the Mekong Region: Political, Socio-Economic and Environmental Perspectives*; Matthews, N., Geheb, K., Eds.; Routledge: London, UK, 2014.

53. Matthews, N. Water Grabbing in the Mekong basin—An Analysis of the Winners and Losers of Thailand's Hydropower Development in Laos. *Water Altern.* **2012**, *5*, 392–411.

54. Kattelus, M.; Rahaman, M.M.; Varis, O. Myanmar under Reform: Emerging Pressures on Water, Energy and Food Security. *Nat. Resour. Forum* **2014**, *38*, 85–98. [CrossRef]

55. Zhai, Y.; Jude, A.J. Energy Sector Integration for Low-carbon Development in the GMS: Towards a Model of South-South. In *Greater Mekong Subregion: From Geographical to Socio-Economic Integration*; Shrestha, O.L., Chongvilaivan, A., Eds.; Institute of Southeast Asian Studies: Singapore, Singapore, 2013; p. 216.

56. Burgos, S.; Sophal, E. China's Strategic Interest accessed in Cambodia: Influence and Resources. *Asian Surv.* **2010**, *50*, 615–639. [CrossRef]

57. Cambodia's Hydropower Development and China's Involvement. Available online: https://www. internationalrivers.org/files/attached-files/cambodia_hydropower_and_chinese_involvement_jan_2008.pdf (accessed on 5 November 2015).

58. Sovuthy, K. 2014. The Cambodia Daily. Available online: http://www.cambodiadaily.com/news/lawmaker-blasts-illegal-logging-in-s-treng-59494/ (accessed on 2 June 2014).

59. Sothanarith, K. Voice of America. Available online: http://www.voanews.com/content/cambodia-urges-closer-asean-ties-with-china/1734944.html (accessed on 2 June 2014).

60. Xinhua. Global Times. Available online: http://www.globaltimes.cn/content/841301.shtml#.UvxnMt9BH (accessed on 2 June 2014).

61. China's Forex Reserves Reach 3.82 Trillion USD. Available online: http://news.xinhuanet.com/english/china/2014-01/15/c_133046451.htm (accessed on 2 June 2014).

62. McKinnon, R.; Schnabl, G. China's exchange rate and financial repression: The conflicted emergence of the RMB as an international currency. *China World Econ.* **2014**, *22*, 1–35. [CrossRef]

63. Morrison, W.; Lebonte, M. China's Currency Policy: An Analysis of the Economic Issues. Available online: https://www.fas.org/sgp/crs/row/RS21625.pdf (accessed on 6 November 2015).

64. Islam, T.; Li, X. South-South Cooperation Policy and Practice by the Export Import Bank of China. *IOSR J. Econ. Financ.* **2014**, *3*, 9–19. [CrossRef]

65. Tan, X. World Resources Institute. Available online: http://www.wri.org/blog/2011/06/emerging-actors-development-finance-closer-look-chinese-and-brazilian-overseas (accessed on 2 June 2014).

66. Step Lightly: China's Ecological Impact on Southeast Asia. Wilson Center. Available online: http://www.wilsoncenter.org/publication/step-lightly-chinas-ecological-impact-southeast-asia (accessed on 2 June 2014).

67. Global Infrastructure Investment to Reach Four Trillion Dollars by 2017, Finds New Bain & Company Study. Available online: http://www.bain.com/about/press/press-releases/global-infrastructure-investment-to-reach-four-trillion-dollars-by-2017.aspx (accessed on 25 August 2015).

68. Merme, V.; Ahlers, R.; Gupta, J. Private Equity, Public Affair: Hydropower Financing in the Mekong Basin. *Glob. Environ. Chang.* **2014**, *24*, 20–29. [CrossRef]

Article

Water-Energy-Food Nexus within the Framework of International Water Law

Antti Belinskij

Law School, University of Eastern Finland, Joensuu Campus, P.O. Box 111, Joensuu FI-80101, Finland; antti.belinskij@uef.fi; Tel.: +358-46-920-9189

Academic Editor: Marko Keskinen
Received: 31 May 2015; Accepted: 11 September 2015; Published: 12 October 2015

Abstract: International water law, which regulates the uses of international watercourses that are situated partly in different States, is a highly topical sector of law. In 2014, two conventions covering the subject matter entered into force globally. At the same time, a water-food-energy nexus has become part and parcel of the development canon that emphasises the importance of the complex relationship between water, energy and food. In this article, it is discussed whether international water law supports the water-food-energy nexus approach, which aims to reconcile the different water uses in international basins. The analysis also covers the human rights to water and food from the nexus viewpoint. The legal regime of the Mekong River is used as an example of the possibilities and challenges of the nexus approach in international water law. It is concluded that despite its deficiencies international water law provides a very useful platform for the cooperation between States and different sectors that aim at guaranteeing water, food and energy security.

Keywords: international water law; water-energy-food nexus; transboundary cooperation; principle of equitable and reasonable utilisation; right to water; right to food; water uses; Mekong River

1. Introduction

International water law, which regulates the uses of international watercourses that are situated partly in different States, is a highly topical sector of law. In 2014, after a time when no general agreements were in existence in the field, two conventions covering the same subject matter entered into force globally: the 1997 UN Convention on the Law of the Non-navigational Uses of International Watercourses (UN Watercourses Convention) [1] and the 1992 United Nations Economic Commission for Europe (UNECE) Convention on the Protection and Use of Transboundary Watercourses and International Lakes (ECE Water Convention) [2]. The process of codification of international water law—dating back to the 1966 Helsinki Rules of the International Law Association (ILA Helsinki Rules) [3]—has thus culminated into binding international agreements, which are mutually compatible and complement each other [4–6]. The general principles of the two conventions, such as cooperation, equitable and reasonable utilisation and the no-harm rule, correspond to customary norms of international water law [7].

International water law has traditionally applied to inter-State relations concerning transboundary watercourses and not to the relationship between an individual and a State that defines the scope of international human rights law [8]. In recent years, however, human rights have also been increasingly discussed within the sphere of international water law. While the General Assembly of the United Nations recognised the right to safe and clean drinking water and sanitation as a human right in July 2010 [9], the 2004 ILA *Berlin Rules on Water Resources* (ILA Berlin Rules) [10] had already contained the right of access to water to meet every individual's vital human needs (Art. 17). In the legal literature, the discussion on the relationship between the right to water and international water law began in the 1990s [11].

At the same time as when the international water law conventions have entered into force, a water-food-energy nexus has become part and parcel of the development canon. The nexus approach emphasises the importance of recognising the complex relationship and interlinkages between the water, energy and food sectors and also aims to safeguard the human rights to water and food [12–14]. Hoff underlines the need for a nexus approach as follows:

Improved water, energy and food security can be achieved through a nexus approach—an approach that integrates management and governance across sectors and scales. A nexus approach can also support the transition to a Green Economy, which aims, among other things, at resource use efficiency and greater policy coherence. Given the increasing interconnectedness across sectors and in space and time, a reduction of negative economic, social and environmental externalities can increase overall resource use efficiency, provide additional benefits and secure the human rights to water and food. Conventional policy- and decision-making in "silos" therefore needs to give way to an approach that reduces trade-offs and builds synergies across sectors—a nexus approach. Business as usual is no longer an option [15] (p. 7).

In international water law, the nexus approach is currently being discussed within the regime of the ECE Water Convention. The Convention's program of work for 2013–2015 includes an assessment of the water-food-energy-ecosystems nexus that aims to improve the understanding of the interactions between water, food, energy and water-related ecosystems in international basins. Further, it intends to strengthen synergies and policy coherence between water, food and agriculture and land management sectors in the transboundary context [16]. In a transboundary setting, while water, energy and agricultural issues are often strongly interlinked in the international watercourses and aquifers [17], the possible frictions between the riparian countries and different interests make the nexus approach even more challenging than at the national level [14,18].

The legal regime of the Mekong River provides a concrete example of the challenges of a water-energy-food nexus approach in international river basins. As is well known, the Mekong River is one of the largest rivers in the world both according to its estimated length (4909 km) as well as its mean annual volume (475 km^3). The River supports a diverse and productive freshwater ecosystem and provides the basis of livelihoods for millions of people. While there are six riparian countries, Cambodia, China, Laos, Myanmar, Thailand and Vietnam, only four of them have signed the 1995 Agreement on the Cooperation for the Sustainable Development of the Mekong River Basin (1995 Mekong Agreement) [19]. This means that China and Myanmar are not among the members of the Mekong River Commission [20–22].

All the riparian countries of the Mekong River are going through periods of rapid changes and some are planning or have already realised large-scale water development projects. These include the construction of hydropower dams and irrigations projects, which on the one hand are important for the countries' economic development, but on the other hand may have remarkable negative impacts on ecosystems and thus on the livelihoods of people [20–22]. For instance, the hydropower dams in China have been suspected to cause a series of detrimental environmental impacts such as changes to the river's natural flood-drought cycle and to the transport of sediment on the Lower Basin of the Mekong. In addition, the hydropower dams in Vietnam in the Sesan River, which is a tributary of the Mekong, have had adverse effects on, e.g., fishing and river-side farming and thus, on the livelihoods of the communities in Cambodia [23,24].

In this article, it is discussed whether international water law supports the water-food-energy nexus approach, which aims to reconcile the different water uses in international basins. The principles of international water law (cooperation, equitable and reasonable utilisation, no-harm) will be studied from the nexus perspective following especially the methodology of the UNECE for the nexus assessment [13]. The analysis also covers the human rights to water and food from the nexus viewpoint. The legal regime of the Mekong River is used as an example of the possibilities and challenges of the nexus approach in international water law. Rather than providing complete answers on the relationship between international water law and the nexus approach, the article aims to add an important element to the nexus discussion, namely international water law.

The article consists of four sections. First of all, the relationship between the procedural features of the nexus approach and international water law is studied. Second, it is observed whether it is possible to reconcile different water uses in order for the nexus approach to be effective. Third, how the human rights to water and food—which the nexus approach aims to secure [15] (p. 7)—are regulated in international law is discussed. Fourth, the conclusions are presented at the end of the article.

In regard to the UN Watercourses Convention and the ECE Water Convention, the two important sources of the article, it must be noted that none of the Mekong countries is a party to the ECE Water Convention and only Vietnam is a party to the UN Watercourses Convention. Therefore, international customary law in addition to the 1995 Mekong Agreement largely determines their rights and obligations regarding international waters. The general principles of the UN and ECE conventions can be regarded as a source of international customary water law also in the Mekong River context.

In general, the role of international customary law is notably significant in international water law. Especially, the UN Watercourses Convention is reckoned to reflect the fundamental rules of customary law [17,24]. As McCaffrey points out:

For the most part, it (the UN Watercourses Convention) should be viewed not as an instrument that seeks to push the law beyond its present contours, but as one that reflects a general consensus as to the principles that are universally applicable in the field [7] (p. 261).

Also the non-formal ILA Helsinki Rules and ILA Berlin Rules are widely understood to reflect international customary water law, although some experts' approach towards those parts of the ILA Berlin Rules that seek to incorporate the requirements of international environmental and human rights law has been a bit cautious [25,26].

2. Procedural Features

2.1. Nexus Approach

The nexus approach is characterised by the UNECE by the following core features: participatory process, knowledge mobilisation, sound scientific analysis, capacity building, collective effort and benefits and opportunities [13]. From the viewpoint of international water law these are largely procedural features linked to the cooperation between riparian countries and different sectors. Cooperation provides an opportunity for early identification of possible disagreement between riparians and for means to prevent them escalating into conflicts [27].

According to the UNECE methodology, the participatory process of the nexus assessment in an international basin requires intersectoral cooperation between officials and experts from the countries sharing the basin. In addition, the consultation of various stakeholders such as local decision makers, planning authorities, practitioners, the representatives from different sectors and analysts is of utmost importance in order to ensure the responsiveness of the nexus approach to basin-specific needs and circumstances [13,18]. On the basis of the nexus assessment, the various interlinks between water, food and energy sectors should be illustrated thoroughly [28].

The methodology of the nexus assessment discussed within the regime of the ECE Water Convention includes the following steps: (1) identification of basin conditions and its socioeconomic context, (2) identification of key sectors and stakeholders to be included in the assessment, (3) analysis of the key sectors, (4) identification of intersectoral issues, (5) nexus dialogue and future developments, and (6) identification of opportunities for improvement. On a basin level, one can understand that the nexus cooperation requires a careful step-by-step approach that allows for enough time to build trust and deepen the intersectoral approach between States [13]. In this respect, international water law may offer a useful platform.

2.2. Principle of Cooperation

The principle of cooperation is one of the main features in international water law [29,30]. In the UN Watercourses Convention, it is stated that watercourse States shall cooperate on the basis of sovereign equality, territorial integrity, mutual benefit and good faith in order to attain optimal utilisation and adequate protection of an international watercourse (Art. 8.1). In a rather similar manner, the ECE Water Convention requires that the Riparian Parties cooperate on the basis of equality and reciprocity in order to develop harmonised policies, programs and strategies aimed at the prevention, control and reduction of transboundary impact and at the protection of the environment of transboundary waters or the environment influenced by such waters (Art. 2.6). While the ILA Helsinki Rules include numerous provisions implicitly requiring cooperation such as articles on equitable utilisation and on pollution prevention, the ILA Berlin Rules explicitly emphasise the duty to cooperate between basin States as the most basic principle of international water law (Art. 11). All in all, cooperation is seen as a logical extension of the principle of equitable and reasonable utilisation in international water law and is instrumental to full compliance with it [27,29].

The 1995 Mekong Agreement provides at the very start that the Parties agree to cooperate in all fields of sustainable development, utilisation, management and conservation of the water and related resources of the Mekong River Basin. These fields include, but are not limited to, irrigation, hydropower, navigation, flood control, fisheries, timber floating, recreation and tourism (Art. 1). The cooperation is characterised by the principles of sovereign equality and territorial integrity (Art. 4) in the same manner as in the UN Watercourses Convention. Thus, the perspective of the Mekong Agreement for the cooperation is rather wide and includes also water related resources such as energy and food. The development of the cooperation is at the heart of the Mekong Agreement since it is actually an agreement to agree on the further development of the regime [21].

In order to follow the participatory process in the nexus assessment of an international basin, both the UN Watercourses Convention and the ECE Water Convention include a number of subsequent provisions specifying the normative content of the cooperation between States. While the UN Watercourses Convention only suggests that watercourses States may consider the establishment of joint mechanisms for the facilitation of the cooperation (Art. 8.2), the ECE Water Convention requires that the Riparian Parties enter into bi- or multilateral agreements or other arrangements that provide for the establishment of joint bodies (Art. 9). According to the ILA Berlin Rules, basin States have the right to participate in the management of waters of an international drainage basin (Art. 10), and they should establish a basin wide or joint agency or commission with the authority to undertake the integrated management of international waters (Art. 64). Bi- and multilateral agreements and especially joint bodies offer a platform for the intersectoral cooperation between basin States [18,27]. Although it is unrealistic to expect that watercourse States could be compelled to establish strong and all-encompassing joint bodies from scratch, it is of the utmost importance that States enter into some kind of institutional arrangements to begin to build the mutual trust needed for effective cooperation in a transboundary context.

In the 1995 Mekong Agreement the Mekong River Commission provides an institutional framework for cooperation. It has three permanent bodies, namely the Council, Joint Committee and Secretariat as well as wide competence concerning the uses of waters in the basin (Arts. 11–12). The decisions of the Commission, however, have to be unanimous if not otherwise provided for in its rules of procedures (Arts. 20, 27, 34). While the structure of the Commission and the reliance of its decision-making on the governments have been reasonably criticised [21,24], the Commission nevertheless enables the Parties of the Mekong Agreement to develop their cooperation step-by-step to cover also intersectoral water-energy-food issues. The agreement does not specifically mention intersectoral cooperation but does not place any obstacles for that either.

Obviously, the efficiency and the geographical coverage of the Mekong Agreement is limited because Myanmar and China are not the Parties of the Agreement and thus not included in the actual cooperation in the Mekong River Commission. In addition, the Agreement does not include for regulating development on tributaries even though the tributaries provide almost a half of the flow to the Mekong River. China, especially, has a prominent role and massive plans for the development of the part of the Mekong River situated on its territory [20,21,31]. The construction projects of hydropower dams in China, for instance, have not included proper transboundary consultation and environmental impact assessment procedures [23].

However, the 1995 Mekong Agreement allows any riparian State to become a Party to the Agreement with the consent of the parties, and in 1996 China and Myanmar became so-called dialogue members of the Mekong River Commission. In 2002, China also signed an agreement with the countries of the Mekong River Commission on the provision of hydrological information on the Mekong River. Thus, although there are many challenges in the cooperation, the regime of the Mekong River is also an encouraging example of the development and management of an international watercourse despite several international conflicts in the area in recent decades [20,21,31].

The geographical scope of cooperation in international water law is a bit restricted from the nexus perspective. The UN Watercourses Convention refers to international watercourses, which are a system of surface waters and ground waters that constitute a unitary whole, parts of which are situated in different States (Arts. 1–2). In the ECE Water Convention, in contrast, transboundary waters are any surface or ground waters that mark, cross or are located on boundaries between two or more States (Art. 1) [27,29]. A more broadly expressed scope of international water law comprising all waters within a State in addition to international drainage basins can be found in the ILA Berlin Rules (Art. 1), but the extension of the Rules to cover national waters has also been subject to critique from the viewpoint of traditional international water law [25].

While it is safe to say that international water law regulates the use and protection of international basins, in regard to the security of water, energy and food there may be a requirement that cooperation should not be contained within river-basin boundaries [15] (p. 32). The nexus approach in international water law cooperation might necessitate that the water, energy and food security and the alternatives for the use of international water resources are taken into account on a wide geographical scale. International water law does not place any restrictions, however, on the widening of the geographical scope of transboundary cooperation, and the principle of equitable and reasonable utilisation actually covers, as will be analysed below, a wide range of water related and other factors that extend beyond international waters.

2.3. Knowledge Mobilisation

International water law offers some water-specific tools for the mobilisation of knowledge in international basins. According to the UN Watercourses Convention, watercourse States must exchange on a regular basis readily available data and information on the condition of the watercourse, and they need to employ their best efforts to comply with the request of non-readily available data by another watercourse State (Art. 9), whereas the ECE Water Convention states that the parties have to provide as early as possible for the widest exchange of information on issues covered by the Convention (Art. 6) [27,29]. The ILA Helsinki Rules recommend that basin States furnish relevant and reasonably available information to each other concerning international waters and their use (Art. XXIX), while the Berlin Rules call for the regular provision of all relevant and available information on the quantity, quality and state of international waters and the aquatic environment (Art. 56).

The ECE Water Convention, as well as the ILA Berlin Rules (Art. 29), requires that environmental impact assessment be applied within the authorisation regime of a planned measure (Art. 3 h), whereas the UN Watercourses Convention states that the notification of planned measures must include the results of any environmental impact assessment (Art. 12) [27,29]. The International Court of Justice has since stated in a case concerning pulp mills on the River Uruguay in 2010, that an environmental impact assessment can be considered a requirement under general international law where there is a risk that a proposed industrial activity may have a significant adverse impact in a transboundary context; in particular, on a shared resource [32]. Thus, international customary law indeed includes a requirement for an environmental impact assessment whenever there is a risk that a planned water use activity may have significant and adverse transboundary impacts.

The ECE Water Convention invites all the Parties concerned to perform scientific data gathering and analysis on the conditions of transboundary waters. First of all, the Parties must establish programs for monitoring the conditions of transboundary waters (Art. 6). Second, the tasks of joint bodies include, inter alia, data gathering and analysis to identify pollution sources, the elaboration of joint monitoring programs concerning water quality and quantity as well as the participation in the implementation of environmental impact assessments relating to transboundary waters (Art. 9). Third, the riparian parties have to undertake specific research and development activities in support of achieving and maintaining the water quality objectives and criteria that they have agreed to set and adopt (Art. 12) [27]. The ILA Berlin Rules, for their part, demand that a joint mechanism has authority over e.g., the coordination of their research programs and the establishment of networks for permanent observation and control (Art. 65).

In the 1995 Mekong Agreement, the regulations on the mobilisations of knowledge in the basin are rather general. In its institutional provisions it is stated that one of the functions of the Joint Committee is to regularly obtain, update and exchange information and data that is necessary to implement the Agreement (Art. 24 C). The Joint Committee also conducts appropriate studies and assessments for the protection of the environment and maintenance of the ecological balance of the Mekong River Basin (Art. 24 D), whereas the Secretariat maintains databases of information (Art. 30). While one of the most obvious deficiencies of the Agreement is that it completely lacks a provision for transboundary environmental impact assessment [21], the requirement for the assessment follows from international customary law, as is reflected by the International Court of Justice in the aforementioned Pulp Mills case.

When it comes to planned measures having possible transboundary impacts, the UN Watercourses Convention lays down rules on the process of information exchange and consultation. Accordingly, watercourse States must exchange information, consult and, if necessary, negotiate on the possible effects of planned measures on the condition of an international watercourse (Art. 11). The Convention includes specific rules on the consultations process concerning planned measures (Arts. 12–18) [29]. In addition, according to the ILA Helsinki Rules, a State should serve notice of any proposed construction or installation that would alter the regime of the basin in a way which might give rise to a dispute, while the Berlin Rules call for basin States to promptly notify other States or competent international organisations that may be affected significantly by a program, plan, project or activity and set out provisions on consultations (Arts. 57–58).

A good example of the participatory process and knowledge mobilisation under international water law is the Finnish–Russian transboundary cooperation that is based on the 1964 Agreement Concerning Frontier Watercourses [33]. The cooperation was originally motivated by the need to develop hydroelectric power sources and water protection. Other main topics addressed so far have concerned flood control and fisheries as well as log floating and transport [34] (p. 133). The cooperation is largely based on the collaboration between the two countries under the framework of the Joint Finnish–Russian Transboundary Water Commission. In practice, a wide range of expertise has been represented in the Commission, ranging from the management of natural resources including fisheries to hydropower production [35] (pp. 59–60). The 1991 Discharge Rule of Lake Saimaa and the River Vuoksi is the most significant individual outcome of the transboundary water cooperation between Finland and the Russian Federation.

3. Reconciliation of Different Water Uses

3.1. Nexus Approach

The nexus approach may help uncover the co-benefits as well as the external costs at the international level associated with the actions of different sectors. In the framework of international

Water **2015**, *7*, 5396–5415

water law, water cooperation can generate diverse and significant benefits such as economic growth, human wellbeing and environmental sustainability for cooperating States. Indeed, the nexus approach invites States to consider intersectoral implications beyond narrowly focusing on sharing water in transboundary water cooperation. However, there might be a need to reconcile water uses of different sectors after joint identification of opportunities for benefits and of solutions for capitalising on the synergies and addressing trade-offs in the interlinked pillars of the water-food-energy nexus [13,36]. Optimising and sharing the benefits of water uses between watercourse States is at the essence of international water law (Art. 5 of the UN Watercourses Convention).

The reconciliation of different uses of international water resources—including also groundwater [7]—is needed when there is a conflict of uses. This means that all reasonable and beneficial uses of all watercourse States cannot be fully realised. From the viewpoint of the water-food-energy nexus, for instance, there might not be enough international waters to fully meet all food, biofuel and hydropower production and domestic and environmental water needs. In that case, some adjustments or accommodations might be required in order to preserve sustainable utilisation and the equality of the right of each watercourse State. The equality of rights in international water law does not mean, however, that each State is entitled to an equal share of the uses or to identical quantitative portions of water [37] (p. 98). It rather means that the reconciliation of different water uses has to be based on the idea that each watercourse State is entitled to uses and benefits from a watercourse in an equitable manner [27] (p. 23).

The reconciliation of different water uses along the lines of the nexus approach requires thorough assessments based on detailed data and indicators. The nexus assessment of an international watercourse consists of different steps such as the identification of basin conditions, its socioeconomic context and economic sectors. Further on, different sectors as well as their intersectoral relationships and the synergies between them must be analysed. Water cooperation in the framework of international water law can generate significant benefits for States, and through the nexus approach these benefits can be considered in a broad intersectoral sense [13].

3.2. Principle of Equitable and Reasonable Utilisation

In international water law, the reconciliation of different water uses is based on the principle of equitable and reasonable utilisation and on the no-harm rule. The UN Watercourses Convention (Arts. 5–7) as well as the ECE Water Convention (Art. 2) obligate States to guarantee that the use of shared water resources is equitable and reasonable and does not cause significant harm to other States. These provisions are largely based on the ILA Helsinki Rules, according to which, each basin State is entitled, within its territory, to a reasonable and equitable share in the beneficial uses of the waters of an international drainage basin (Art. IV) [27,29]. The principle of equitable and reasonable utilisation is one of the cornerstones of the ILA Berlin Rules as well although the Berlin Rules have faced some criticism for emphasising the equitable and reasonable management of international waters and the no-harm rule (Art. 12) over the right to the beneficial uses of those waters [25,26].

The equitable and reasonable utilisation principle is the central piece of international water law when deciding on the utilisation of international watercourses between watercourse States and has been recognised as part of customary international law by the International Court of Justice in the judgment on the Gabčikovo-Nagymaros Project on the Danube River, which referred to an equitable and reasonable sharing of the resources of an international watercourse [8,21,27,38]. The ECE Water Convention provides that the Parties ensure that transboundary waters are used in a reasonable and equitable way by taking into account their transboundary character (Art. 2.2), whereas the UN Watercourses Convention, as well as the ILA Helsinki Rules (Art. IV), lays down more specific rules in this regard. According to the UN Watercourses Convention, watercourse States must in their respective territories utilise international waters in an equitable and reasonable manner with a view to attaining optimal and sustainable utilisation thereof and benefits therefrom by taking into account the interests of the watercourse States concerned (Art. 5) [5]. Thus, the general definition of the equitable utilisation makes it possible to take intersectoral nexus perspectives into consideration.

The no-harm rule is linked to the principle of equitable and reasonable utilisation in international water law and means an obligation for a State not to cause significant harm to other States when utilising an international watercourse in their territories. Basically, the no-harm rule lends itself to any activities causing or likely to cause transboundary impacts regardless of where an activity is located in the area of a State. The UN Watercourses Convention provides that States shall take all appropriate measures to prevent the causing of significant harm to other watercourse States and eliminate and mitigate such harm by having due regard for the principle of equitable and reasonable utilisation (Art. 7). The ECE Water Convention, for its part, requires States to take all appropriate measures to prevent, control and reduce any transboundary impact (Art. 2). The ILA Helsinki Rules clearly render the no-harm rule as subordinate to the principle of equitable utilisation by, e.g., stating that water pollution which would cause substantial injury to another State must be prevented in keeping with the principle of equitable utilisation (Art. X). The threshold of significant harm or transboundary impact as well as the sufficiency and appropriateness of the measures taken can be assessed in detail only on a case-by-case basis. A use of a watercourse that causes harm to other States does not necessarily mean that it is assessed as inequitable or unreasonable if all appropriate measures have been taken to minimise transboundary impacts [7,8,21,25,27].

The UN Watercourses Convention provides, largely following the ILA Helsinki Rules (Art. V), that the utilisation of an international watercourse in an equitable and reasonable manner requires taking into account all relevant factors and circumstances (Art. 6). According to the Convention, the factors include:

(a) geographic, hydrographic, hydrological, climatic, ecological and other factors of a natural character;
(b) the social and economic needs of the watercourse States concerned;
(c) the population dependent on the watercourse in each watercourse State;
(d) the effects of the use or uses of the watercourses in one watercourse State on other watercourse States;
(e) existing and potential uses of the watercourse;
(f) conservation, protection, development and economy of use of the water resources of the watercourse and the costs of measures taken to that effect; and
(g) the availability of alternatives, of comparable value, to a particular planned or existing use.

The ILA Berlin Rules add the sustainability of proposed or existing uses as well as the minimisation of environmental harm to these factors (Art. 13).

On the one hand, the factors relevant to equitable and reasonable utilisation of international watercourses are very much compatible with the water-energy-food nexus approach. For instance, the factors of the social and economic needs, potential uses of the watercourse, development and economy of use of the water resources direct towards the nexus approach as well as the requirement to take into account the availability of alternatives of comparable value. On the other hand, however, the needs of energy and food sectors are not particularly mentioned and the careful consideration of different factors requires thorough assessments. Thus, the development of the means for nexus assessments within the framework of international law is of the utmost importance from the nexus approach viewpoint.

The UN Watercourses Convention, as well as the ILA Helsinki Rules (Art. V) and the Berlin Rules (Art. 13), states that the weight to be given to each factor relevant to equitable and reasonable utilisation is to be determined by its importance in comparison with that of other relevant factors (Art. 6). It has to be taken into account, however, that circumstances pertaining to the factors relevant for equitable and reasonable utilisation are subject to change. The existing uses of a watercourse are only one of the factors that have to be taken into account when determining on the utilisation of an international watercourse. Therefore, while a use of transboundary waters may be regarded as equitable and reasonable at a given point of time, there might be a need to reverse such an assessment at a later stage [27]. Circumstances and thus the balance between different factors of equitable and reasonable utilisation may change or the nexus assessment, for instance, may provide new and more in-depth information on the factors relevant to equitable and reasonable utilisation such as the social and economic needs of the watercourse States and direct towards the reconciliation of different water uses. In the ILA Helsinki Rules, it is indeed stated that an existing reasonable use may continue in operation unless the factors justifying its continuance are outweighed by other factors leading to the conclusion that it be modified or terminated so as to accommodate a competing incompatible use (Art. VIII).

In accordance with the principle of equitable and reasonable utilisation and the no-harm rule, the goal of the 1995 Mekong Agreement is to optimise the multiple-use and mutual benefits of all riparians and to minimise the harmful effects that might result from natural occurrences and man-made activities. Further, the cooperation covers all fields of sustainable development, utilisation, management and conservation of the water and related resources of the Mekong River Basin (Art. 1). Parties have also agreed to make every effort to avoid, minimise and mitigate harmful effects from the development and use of the Mekong River Basin water resources or discharge of waters and return flows (Art. 7). While the equitable and reasonable utilisation is defined in the agreement basically only through procedural obligations to notify and consult the Joint Committee on water uses (Art. 5), the Mekong countries are subject to the principle of equitable and reasonable utilisation also under customary international law [24].

In the Mekong context, however, it has proved to be difficult to determine whether uses of the watercourse such as the construction of hydropower dams in China and Vietnam are indeed equitable and reasonable and whether all the efforts expected of a diligent State to avoid harmful effects to the interests of other countries have been taken. There have been some obvious shortages in data sharing and cooperation that have made the analysis of development difficult, if not impossible. In addition, the interpretations of the ambiguous substantive rules and principles of international water law depend on the local context and the States concerned [24]. When it comes to hydropower dams in the Mekong River Basin, their harmful effects on other basin States and the environment are easily overshadowed by potential benefits in energy production in the States where the dams are constructed [23].

3.3. Vital Human Needs

According to the UN Watercourses Convention, in the absence of an agreement or custom to the contrary no use of an international watercourse enjoys inherent priority over other uses (Art. 10.1). In addition, the ILA Helsinki Rules state that there are no inherently preferred uses over other uses (Art. VI). These provisions mean that the order of priority between different uses must be decided on a case-by-case basis by taking all the relevant factors into account [27,29]. However, in the event of a conflict between uses special regard is given to the requirements of vital human needs (Art. 10.2). Vital human needs must be taken into account as part of the social and economic needs of the watercourse States when deciding on the equitable and reasonable utilisation of international waters [5,37].

The special regard given to the requirements of vital human needs may lead to the actual priority of vital human needs over other uses in the event of a conflict between uses of international waters [8]. If there are no comparable alternatives such as using national water resources over the use of international waters to meet vital human needs, these needs must enjoy the highest priority in the light of the UN Watercourses Convention (Art. 10.2). The ILA Berlin Rules that comprise international as well as national water resources mention clearly that in determining an equitable and reasonable use States have to first allocate waters to satisfy vital human needs (Art. 14).

In the commentary of the International Law Commission, it is stated that in the UN Watercourses Convention vital human needs refer to water to sustain human life, including not only drinking water but also water required for the production of food in order to prevent starvation [37]. In the context of international law, drinking water usually means water that is used in addition to drinking for cooking, food preparation, personal hygiene and similar purposes to meet basic human needs, and there are different estimations on the amount of water needed for these purposes; ranging from 20 to 100 L per day [39]. The ILA Berlin Rules outline that vital human needs means waters used for immediate human survival that includes cooking and sanitary needs as well as water needed for the immediate sustenance of a household (Art. 2.20).

3.4. Community of Interests

The principle of cooperation and, in general, the procedural requirements of international water law cannot be emphasised too much when reconciling the uses of international water resources on the basis of equitable and reasonable utilisation and the nexus approach. Indeed, cooperation is seen as a logical extension of the principle of equitable and reasonable utilisation in international water law [7,29], and the UN Watercourses Convention requires that watercourse States participate in the use, development and protection of an international watercourse in an equitable and reasonable manner (Art. 5). In general, the determination of equitable and reasonable utilisation and of significant harm to other States requires close collaboration between States or third party intervention [24].

In the regime of the 1995 Mekong Agreement, a basin development plan could be used as a tool to reconcile different water uses. According to the Agreement, the aim of the basin development plan is to identify, categorise and prioritise the development projects and programs at the basin level. In this way the development of the full potential of sustainable benefits for all riparian States could be promoted, supported and coordinated and the wasteful use of Mekong River Basin waters prevented (Art. 2). The effective reconciliation of different water uses does not seem realistic in the Mekong context in the near future when taking into account the aforementioned contradictions in the uses of waters, the lack of cooperation between States in the Mekong River Basin as well as the relative weakness of the Mekong River Commission.

In general, international water law as well as the regime of the Mekong River provides a framework where different water uses can be reconciled in the light of the nexus approach. However, one must take into account that as opposed to absolute duties international water law creates due diligence obligations to comply with its provisions in the best and most rational way. The due diligence nature is reflected, for instance, in the duty to take all appropriate measures to prevent, control and reduce any transboundary impact (Art. 2 of the ECE Water Convention) [7,27]. Thus, international water law allows the nexus approach and can be seen to direct cooperation to that direction. Perhaps in the future the nexus assessment can be seen as part of the best international practice and due diligence obligations of international water law. Nevertheless, it is very difficult, if not impossible, to enforce reluctant States to undergo thorough nexus assessments and to accept the reconciliation of different water uses by means of international law.

In order for the nexus approach to be effective in the framework of international water law States should be guided by the concept of community of interests in their cooperation concerning international waters. While the claims of States over the uses of transboundary waters might have been based upon absolute doctrines of territorial sovereignty (upstream States) and territorial integrity (downstream States) in the past, international water law has been characterised by the doctrine of limited territorial sovereignty, which has evolved upon the principles of equitable and reasonable utilisation and of cooperation as well as the no-harm rule. The concept of community of interests would represent another step forwards and is based on the perfect equality of the riparian States in the uses of international waters [27,29]. From the nexus perspective the community of interests would require that riparian States identify long-term common interests and common means to guarantee water, food and energy security through cooperation and through the reconciliation of different water uses.

Riparian States should negotiate mutual gains agreements for international watercourses as put forward by Grzybowski *et al.* [17] to put the nexus approach into practice. The idea for the mutual gains agreements is that the development of joint opportunities outweighs the benefits of acting independently. The negotiations concerning, for example, significant hydropower developments in an international watercourse in the territory of one riparian State should include alternative scenarios that attempt to maximise the benefits of the project for other riparian States; for example, energy supply and flow regulation. These kind of negotiations, based on the identification and development of opportunities with reciprocal sharing of benefits, expand the cooperation between riparian States from only reflecting sovereignty and the legal principles of international law to planning and devising opportunities for mutual gains associated with cooperation from the viewpoint of the community of interests [17]. The nexus approach requires that instead of only concentrating on water demand and uses the mutual gains agreements should deal with water, energy and food strategies that aim to optimise benefits from all viewpoints [14].

4. Human Rights to Water and Food

The nexus approach aims to secure water, energy and food security in-line with human rights and basic human needs [15]. While water security means sustainable access to adequate quantities of water of an acceptable quality for basic human needs, also availability and access to water for other critical human and ecosystem uses can be included in it from the nexus perspective [15,40]. Energy security is defined by the International Energy Agency as the uninterrupted availability of energy sources at an affordable price [41], whereas for an individual it refers to access to clean, reliable and affordable energy services for cooking, heating, lightning, communication and productive uses [42]. Food security for its part is defined to exist when all people, at all times, have physical, social and economic access to sufficient, safe and nutritious food, which meets their dietary needs and food preferences for an active and healthy life [43]. The rights to water and food are two of the most essential human rights from the nexus perspective.

In July 2010, the General Assembly of the United Nations recognised the right to safe and clean drinking water and sanitation as a human right [9]. The recognition is a significant milestone in the debate that focuses on the rights related to access to water during the last decades [44]. In the resolution, the United Nations recognises the right to safe and clean drinking water and sanitation as a human right that is essential for the full enjoyment of life and all human rights; and calls upon States and international organisations to provide financial resources, capacity-building and technology transfer, through international assistance and cooperation in particular, to developing countries in order to scale up efforts to provide safe, clean, accessible and affordable drinking water and sanitation for all [9].

The General Assembly recalls in its resolution different human rights treaties and instruments that lend support to the right to water. These include, among others, the International Covenant on Economic, Social and Cultural Rights (ICESCR) and the International Covenant on Civil and Political Rights (ICCPR). Although water is not explicitly mentioned in these covenants, the right to water is usually derived from articles 11 and 12 of the ICESCR. According to articles 11 and 12 of the ICESCR, everyone has the right to an adequate standard of living including adequate food, clothing and housing and to the enjoyment of the highest attainable standard of physical and mental health. The right to water is seen to implicitly fall under the elements of the adequate standard of living [45].

Within the framework of international water law, the right of access to water is recognised in the ILA Berlin Rules in substantially the same way as in the General Assembly resolution. Accordingly, every individual has a right of access to sufficient, safe, acceptable, physically accessible and affordable water to meet that individual's vital human needs. States should ensure the implementation of this right on a non-discriminatory basis and progressively realise the right by various means (Art. 17). In regard to the right of access to water, the Berlin Rules are based on the national constitutions as well as human rights and water law instruments.

International water law also includes a concrete tool for the implementation of the human right to water in the form of the 1999 UNECE Protocol on Water and Health to the ECE Water

Convention [8]. According to the Protocol, the Parties must pursue the aims of access to drinking water and the provision of sanitation for everyone (Art. 6) and take all appropriate measures for the purpose of ensuring adequate supplies of wholesome drinking water as well as adequate sanitation (Art. 4). National and local targets for the standards and levels of performance are revised periodically (Arts. 6–7). In general, the UNECE Protocol on Water and Health represents a step that international water law is taking in the direction of the protection of the right to water also at the domestic level [8].

Unlike water, food is explicitly mentioned in Article 11 of the ICESCR as an element of the right to an adequate standard of living. The right to food or the right to adequate food is defined as a right to all nutritional elements that a person needs to live a healthy and active life and to the means to access them [46]. The right to food is realised when everyone has physical and economic access at all times to adequate food or means for its procurement [47].

In order to have access to food, a person must be able to either buy or produce their own food. The rights to food and water are interlinked in a way that the right to food necessitates the access to safe drinking water for personal and domestic uses such as food preparation and food production. In addition, the obligation to fulfill the right to food means that States have to strengthen people's access to and use of resources such as water for food production [46].

The special regard and preference given to the vital human needs in the UN Watercourses Convention (Art. 10.2) and ILA Berlin Rules (Art. 14) respectively combined with the resolution on the human right to water, the right of access to water in the Berlin Rules (Art. 17) as well as the right to food provide strong grounds for the priority of the indispensable water use to meet basic human needs over other uses. However, also in this regard the possible alternatives to the water uses (Art. 6 of the UN Watercourses Convention) must be taken into account. The priority of indispensable water use needed to meet basic human needs is very much in line with the nexus approach, which aims, as stated earlier, to secure the human right to water and food.

The relevant question is, however, whether international water law provides wider support than the traditional inter-State dimensions in regard to the regulation of international waters for the human rights to water and food. From the traditional point of view, the priority of indispensable water use in international water law proves only partial and indirect help for the protection of individual needs through inter-State claims that cannot deal with situations pertaining to exclusively domestic waters. Nevertheless, first of all the UNECE Protocol on Water and Health is redirecting this tradition through a strong emphasis on the access to water and sanitation and with its mainly domestic scope of application [8]. Secondly, the ILA Berlin Rules reflect the requirements of international environmental and human rights law in addition to traditional international water law by including waters within a State in their scope (Art. 1) and by the specific provision on the right of access to water (Art. 17). Thus, international water law is gradually taking steps in the direction of securing the basic human needs related to water uses in a wide geographical scope.

The 1995 Mekong River Agreement contains no specific provisions on how to guarantee the basic human needs for water and food in the River Basin. However, the Agreement regulates the maintenance of the flows on the mainstream from diversions, storage releases or other actions of a permanent nature (Art. 6). Accordingly, for instance, during the dry season the minimum monthly natural flow is to be maintained at the acceptable level through cooperation between Parties. The maintenance of flows, for its part, makes it possible that there is enough water for individuals to meet basic needs for water and food.

According to Article 2 of the ICESCR, States have to take steps by themselves but also through international cooperation, to the maximum of their available resources, with a view to achieving progressively the full realisation of the rights recognised in the Covenant. The means for this purpose include particularly the adoption of legislative measures. Thereby, the ICESCR also obligates that the rights to water and food are taken into account in the cooperation between watercourse States.

5. Conclusions

It is obvious that there are many challenges for the water-food-energy nexus cooperation in the regime of international water law. One may start by mentioning, for example, the general nature of the procedural and substantive provisions of international water law and its weakness in situations where riparian countries are not willing to cooperate. However, at the same time, international water law provides a very useful platform for the cooperation between States and different sectors aiming at guaranteeing water, food and energy security. It not only allows but also supports the nexus approach in the cooperation between States.

International water law and especially its two conventions (UN Watercourses Convention and ECE Water Convention) provide footsteps that riparian States can follow in their cooperation and when concluding bi- and multilateral water agreements. While the 1995 Mekong Agreement offers an encouraging example of the possibilities of international water law in enhancing cooperation, it also includes serious deficiencies from the nexus and holistic water management perspectives: not all of the basin countries are members of the Mekong River Commission, there are no specific provisions for development on tributaries and a transboundary environmental impact assessment is not required in the Agreement although it nowadays is a part of international customary law. All in all, the Agreement emphasises the sovereignty of States more than the UN and ECE conventions and international customary water law [21]. However, the Agreement also provides the tools and possibilities for the Basin countries to develop their cooperation in the direction of the nexus approach.

International water law offers an institutional framework for the nexus approach but does not offer concrete specifications for its procedural elements. It is up to States sharing the same watercourse to implement the main features of the nexus assessment (participatory process, knowledge mobilisation, sound scientific analysis, capacity building and collective effort [13]) into their cooperation within joint bodies or other arrangements. International water law provides a clear basis for the participatory process between States as well as knowledge mobilisation and analysis. In addition, capacity building and collective effort can be included in the cooperation.

The most difficult part of the nexus cooperation in the framework of international water law is the reconciliation of different water uses in a situation where there is not enough water to meet all the competing needs. International water law contains the general provisions on equitable and reasonable utilisation and on the minimisation of harmful transboundary effects. From the nexus perspective the most relevant factors of equitable and reasonable utilisation are the social and economic needs of the watercourse States concerned, the population dependent on the watercourse and the availability of alternatives for a particular planned or existing use. While existing uses of the watercourse is also one of the factors to be taken into account when deciding on equitable and reasonable utilisation, the nexus approach requires that States understand their common interests in enhancing water, food and energy security in a transboundary context and are willing to negotiate on the changes to existing use patterns.

The understanding of the community of interests of watercourse States can lead to fruitful negotiations and mutual gains agreements for international watercourses. If, however, watercourse States are not willing to cooperate, international water law does not contain any strong enforcement provisions to back up the compliance of its general norms. After all, very few international water disputes, for instance, have been dealt with in the International Court of Justice [17]. As for the Mekong context, the enforcement tools of the Mekong River Commission consist mainly of diplomacy, negotiation and persuasion [24]. Thus, in order to put the nexus cooperation in action in the framework of international water law, watercourse States have to understand its benefits to them. In this regard, the negotiations based on the nexus approach must include the different goals, actors and development alternatives that are attempting to maximise the long-term benefits of intersectoral cooperation between riparian States along the lines of the UNECE methodology for the nexus assessment of an international basin.

International human rights law as well as international water law supports the idea of the nexus approach to secure human rights to water and food. They lend clear support to the prioritisation of access to water to meet basic human needs over other water uses at both the international and national level. Therefore, when planning the uses of waters of international basins such as the Mekong, the rights to water and food must be guaranteed before allocating international waters for other purposes. At the same time, however, it must be taken into account that there might be alternative water resources such as national waters to be allocated for meeting these basic human needs. All in all, the traditionally strict divisions between international and national waters as well as inter-State and internal legal relations are gradually reshaping in the field of international water law by taking into account e.g., the emphasis on the access to water in the UNECE Protocol on Water and Health and the ILA Berlin Rules.

The general nature of the procedural and substantive requirements of international water law is a problem and an advantage at the same time from the nexus perspective. On the one hand, the rules on cooperation, on data gathering and analysis and on equitable and reasonable utilisation are ambiguous and concentrate on the uses and protection of international waters with no clear references to the energy and food sectors. On the other hand, international water law does not place any restrictions whatsoever on the consideration of energy and food sectors when the joint bodies of riparian countries cooperate and when different needs for water uses are reconciled. However, the thorough consideration of energy and food issues requires also viewpoints that go beyond the scope of traditional international water law.

The nexus approach to transboundary cooperation requires a long-term capacity and trust building between riparian States to create new opportunities through cooperation. It cannot be emphasised too much that it is of the utmost importance to create a reliable institutional structure as a basis for that cooperation. Cooperation is a step-by-step process that starts, for example, with the regular meetings of a joint committee and develops towards the in-depth nexus approach with intersectoral working and expert groups and public hearings. The essential building blocks of the cooperation are reciprocity and good faith between States as well as mutual benefits [27,30]. The nexus approach to the transboundary watercourses cooperation requires that riparian States perceive that the cooperation is in their common interests in the long term.

Of the two international water law conventions there are no institutional provisions in the UN Watercourses Convention, while the ECE Water Convention regulates on the Meeting of the Parties and the Secretariat with a view to be able to review the implementation of the Convention (Arts. 17–19). Thus, the ECE Water Convention is a living and dynamic convention [27] that offers a regime where the nexus approach can be developed further in the general framework of international water law.

The water sector provides an entry point for a nexus analysis and international water law is thus one of the starting points for the discussion on the nexus approach and law [18]. However, that discussion requires specific attention also from the viewpoints of energy and food law and national laws. It is a massive task to try to combine all the elements of the nexus approach and law and would require long-term cooperation between States and different organisations such as the UN, Food and Agriculture Organization of the United Nations (FAO) and UNECE. In addition to the coordinated and intersectoral discussions between the representatives of States, international non-governmental organisations such as the International Law Association could play an important role in enhancing the nexus approach in the context of international law.

Acknowledgments: The author have participated in the research project "Legal framework to promote water security" (WATSEC), financed by the Academy of Finland (268151). The author would like to thank the anonymous reviewers for their very valuable suggestions.

Conflicts of Interest: The author declares no conflict of interests.

References and Notes

1. *Convention on the Law of the Non-Navigational Uses of International Watercourses* (adopted 21 May 1997, entered into force 17 August 2014) UN Doc A/51/869.

2. *Convention on the Protection and Use of Transboundary Watercourses and International Lakes* (adopted 17 March 1992, entered into force 6 October 1996) 1936 UNTS 269.

3. International Law Association. *The Helsinki Rules on the Uses of the Waters of International Rivers, Adopted by the International Law Association*; International Law Association: London, UK, 1967; pp. 7–55.

4. Eckstein, G. (Ed.) Specially invited opinions and research report of the international water law project: Global perspectives on the entry into force of the UN Watercourses Convention 2014: Part one. *Water Policy* **2014**, *16*, 1198–1217.

5. Tanzi, A. *The Economic Commission for Europe Water Convention and the United Nations Watercourses Convention: An Analysis of their Harmonized Contribution to International Water Law*; United Nations Publication: Geneva, Switzerland, 2015.

6. McCaffrey, S.C. International Water Cooperation in the 21st century: Recent developments in the law of international watercourses. *Rev. Eur. Comp. Int. Environ. Law* **2014**, *23*, 4–14. [CrossRef]

7. McCaffrey, S.C. The contribution of the UN convention on the law of the non-navigational uses of international watercourses. *Int. J. Glob. Environ. Issues* **2001**, *1*, 250–263. [CrossRef]

8. Tanzi, A. Reducing the gap between international water law and human rights law: The UNECE protocol on water and health. *Int. Community Law Rev.* **2010**, *12*, 267–285.

9. United Nations. *Resolution Adopted by the General Assembly on 28 July 2010: The Human Right to Water and Sanitation*; United Nations: New York, NY, USA, 2010.

10. International Law Association: Berlin Conference (2004), Water Resources Law. Available online: http://internationalwaterlaw.org/documents/intldocs/ILA_Berlin_Rules-2004.pdf (accessed on 17 July 2015).

11. McCaffrey, S.C. A human right to water: Domestic and international implications. *Georget. Environ. Law Rev.* **1992**, *5*, 1–24.

12. Bizikova, L.; Roy, D.; Swanson, D.; Venema, H.D.; McCandless, M. *The Water-Energy-Food Security Nexus: Towards a Practical Planning and Decision-Support Framework for Landscape Investment and Risk Management*; International Institute for Sustainable Development: Winnipeg, MB, Canada, 2013.

13. United Nations Economic Commission for Europe (UNECE). *Methodology for Assessing the Water-Food-Energy-Ecosystems Nexus in Transboundary Basins*, ECE/MP.WAT/WG.1/2015/8. 15 June 2015.

14. Bach, H.; Bird, J.; Clausen, T.J.; Jensen, K.M.; Lange, R.B.; Taylor, R.; Viriyasakultorn, V.; Wolf, A. *Transboundary River Basin Management: Addressing Water, Energy and Food Security*; Mekong River Commission: Vientiane, Laos, 2012.

15. Hoff, H. *Understanding the Nexus*; Background Paper for the Bonn2011 Conference: The Water, Energy and Food Security Nexus; Stockholm Environment Institute: Stockholm, Sweden, 2011.

16. United Nations Economic Commission for Europe. *Report of the Meeting of the Parties on its Sixth Session, Programme of Work for 2013–2015*; Economic Commission for Europe: Geneva, Switzerland, 2015.

17. Grzybowski, A.; McCaffrey, S.C.; Paisley, R.K. Beyond international water law: Successfully negotiating mutual gains agreements for international watercourses. *Pac. McGeorge Glob. Bus. Dev. Law J.* **2010**, *22*, 139–154.

18. Lipponen, A.; Howells, M. Promoting Policy Responses on the Water and Energy Nexus Across Borders. In *Water & Energy*; Water Monographies II-2014; Oficina de Naciones Unidas de apoyo al Decenio «El agua, fuente de vida» 2005-2015/Programa de ONU-Agua para la Promoción y la Comunicación en el marco del Decenio, WCCE-World Council of Civil Engineers; Fundación Aquae: Madrid, Spain, 2014; pp. 44–55.

19. *Agreement on the Cooperation for the Sustainable Development of the Mekong River Basin* (entered into force 5 April 1995 by signature) 2069 UNTS 3.

20. Keskinen, M.; Mehtonen, K.; Varis, O. Transboundary cooperation *vs.* internal ambitions: The role of China and Cambodia in the Mekong region. In *International Water Security: Domestic Threats and Opportunities*; Pachova, N.I., Nakayama, M., Jansky, L., Eds.; United Nations University Press: Tokyo, Japan, 2008; pp. 79–109.

21. Bearden, B.L. Legal regime of the Mekong River: A look back and some proposals for the way ahead. *Water Policy* **2010**, *12*, 798–821. [CrossRef]

22. MacQuarrie, P.R.; Viriyasakultorn, V.; Wolf, A.T. Promoting cooperation in the Mekong Region through water conflict management, regional collaboration, and capacity building. *GMSARN Int. J.* **2008**, *2*, 175–184.

23. Pottenger, S.G. Biodiversity conservation v. *hydropower dams: Can saving the fish save the Mekong River basin?* *Pac. McGeorge Glob. Bus. Dev. Law J.* **2009**, *22*, 111–133.

24. Rieu-Clarke, A.; Gooch, G. Governing the tributaries of the Mekong—The contribution of international law and institutions to enhancing equitable cooperation over the Sesan. *Pac. McGeorge Glob. Bus. Dev. Law J.* **2010**, *22*, 193–224.

25. Salman, S.M.A. The Helsinki rules, the UN watercourses convention and the Berlin rules: Perspectives on international water law. *Water Resour. Dev.* **2007**, *23*, 625–640. [CrossRef]

26. Bogdanovic, S.; Bourne, C.; Burchi, S.; Wouters, P. ILA Berlin Conference 2004—Water Resources Committee Report, Dissenting Opinion. Available online: http://www.internationalwaterlaw.org/documents/intldocs/ila_berlin_rules_dissent.html (accessed on 17 July 2015).

27. United Nations Economic Commission for Europe. *Guide to Implementing the Water Convention*; United Nations Economic Commission for Europe: Geneva, Switzerland, 2013.

28. Rahman, M. Legal Knowledge Framework for Identifying Water, Energy, Food and Climate Nexus. Available online: http://ceur-ws.org/Vol-1105/paper3.pdf (accessed on 27 May 2015).

29. Rieu-Clarke, A.; Moynihan, R.; Magsig, B.-O. *UN Watercourses Convention User's Guide*; IHP-HELP Centre for Water Law, Policy and Science: Scotland, UK, 2012.

30. Leb, C. One step at a time: International law and the duty to cooperate in the management of shared water resources. *Water Int.* **2015**, *40*, 21–32. [CrossRef]

31. Eckstein, G. (Ed.) Specially invited opinions and research report of the International Water Law Project: Global perspectives on the entry into force of the UN Watercourses Convention 2014: Part two. *Water Policy* **2015**, *17*, 162–186.

32. International Court of Justice. *Pulp Mills on the River Uruguay (Argentina v. Uruguay), Judgment*; International Court of Justice: The Hague, Netherlands, 2010.

33. *Agreement between the Republic of Finland and the Union of Soviet Socialist Republics Concerning Frontier Watercourses* (signed 24 April 1964, entered into force 6 May 1965) 537 UNTS 231.

34. Kotkasaari, T. Transboundary cooperation between Finland and its neighbouring countries. In *Management of Transboundary Rivers and Lakes*; Varis, O., Tortajada, C., Biswas, A.K., Eds.; Springer: Berlin, Germany, 2008.

35. Kaatra, K. Outcomes of Vuoksi River cooperation and tasks between Finland and Russia since the 1960s. In *Creating a Peace and Ecology Lake Park in the Upriver of Bukhan River and the Cases of International River Cooperation*; Korea DMZ Council: Seoul, Korea, 2012.

36. United Nations World Water Assessment Programme. *The United Nations World Water Development Report 2015: Water for a Sustainable World*; United Nations Educational, Scientific and Cultural Organization: Paris, France, 2015.

37. International Law Commission. Draft articles on the law of the non-navigational uses of international watercourses and commentaries thereto and resolution on transboundary confined groundwater. In *Yearbook of the International Law Commission*; International Law Commission: Geneva, Switzerland, 1994.

38. International Court of Justice. *GabCikovo-Nagymaros Project (Hungary/Slovakia), Judgment*; International Court of Justice: The Hague, The Netherlands, 1998.

39. Special Rapporteur on the Human Right to Safe Drinking Water and Sanitation. Available online: http://www.ohchr.org/Documents/Issues/Water/FAQWater_en.pdf (accessed on 5 September 2014).

40. UN-Water. *Water Security the Global Water Agenda—A UN-Water Analytical Brief*; United Nations University: Tokyo, Japan, 2013.

41. International Energy Agency. Energy Security. Available online: http://www.iea.org/topics/energysecurity/ (accessed on 20 April 2015).

42. Energy for a Sustainable Future. Available online: http://www.un.org/chinese/millenniumgoals/pdf/AGECCsummaryreport%5B1%5D.pdf (accessed on 20 April 2015).

43. Committee on World Food Security. Available online: http://www.fao.org/cfs/cfs-home/en/ (accessed on 20 April 2015).

44. McIntyre, O. The human right to water as a creature of global administrative law. *Water Int.* **2012**, *37*, 654–669. [CrossRef]

45. Committee on Economic, Social and Cultural Rights. General Comment No. 15: The Right to Water (Arts. 11 and 12 of the Covenant). Available online: http://tbinternet.ohchr.org/_layouts/treatybodyexternal/Download.aspx?symbolno=E%2fC.12%2f2002%2f11&Lang=en (accessed on 7 April 2015).

46. Office of the United Nations High Commissioner for Human Rights. *The Right to Adequate Food*; Office of the United Nations High Commissioner for Human Rights: Geneva, Switzerland, 2010.
47. Committee on Economic, Social and Cultural Rights. General Comment No. 12: The Right to Adequate Food (Art. 11 of the Covenant). Available online: http://tbinternet.ohchr.org/_layouts/treatybodyexternal/Download.aspx?symbolno=E%2fC.12%2f1999%2f5&Lang=en (accessed on 7 April 2015).

water

MDPI

Article

Methods of the Water-Energy-Food Nexus

Aiko Endo [1],*, Kimberly Burnett [2], Pedcris M. Orencio [3], Terukazu Kumazawa [4], Christopher A. Wada [2], Akira Ishii [5], Izumi Tsurita [6] and Makoto Taniguchi [1]

[1] Research Department, Research Institute for Humanity and Nature, 457-4 Kamigamo-motoyama, Kita-ku, Kyoto 603-8047, Japan; makoto@chikyu.ac.jp

[2] University of Hawaii Economic Research Organization, University of Hawaii at Manoa, 2424 Maile Way Saunders Hall 540, Honolulu, HI 96822, USA; kburnett@hawaii.edu (K.B.); cawada@hawaii.edu (C.A.W.)

[3] Urban Disaster Risk Reduction Department, Catholic Relief Services Philippines (Manila Office), CBCP Building 470 Gen Luna Street, Intramuros, 1002 Manila, Philippines; pedcris.orencio@crs.org

[4] Center for Research Promotion, Research Institute for Humanity and Nature, 457-4 Kamigamo-motoyama, Kita-ku, Kyoto 603-8047, Japan; kumazawa@chikyu.ac.jp

[5] Yachiyo Engineering Co., Ltd., 2-18-12 Nishiochiai, Shinjuku-ku, Tokyo 161-8575, Japan; akri-ishii@yachiyo-eng.co.jp

[6] Department of Cultural Anthropology, Graduate School of Arts and Sciences, The University of Tokyo, 3-8-1 Komaba, Meguro-ku, Tokyo 153-8902, Japan; izumitsurita@gmail.com

* Correspondence: a.endo@chikyu.ac.jp; Tel.: +81-75-707-2477; Fax: +81-75-707-2509

Academic Editors: Marko Keskinen, Shokhrukh Jalilov and Olli Varis

Received: 29 May 2015; Accepted: 16 October 2015; Published: 23 October 2015

Abstract: This paper focuses on a collection of methods that can be used to analyze the water-energy-food (WEF) nexus. We classify these methods as qualitative or quantitative for interdisciplinary and transdisciplinary research approaches. The methods for interdisciplinary research approaches can be used to unify a collection of related variables, visualize the research problem, evaluate the issue, and simulate the system of interest. Qualitative methods are generally used to describe the nexus in the region of interest, and include primary research methods such as Questionnaire Surveys, as well as secondary research methods such as Ontology Engineering and Integrated Maps. Quantitative methods for examining the nexus include Physical Models, Benefit-Cost Analysis (BCA), Integrated Indices, and Optimization Management Models. The authors discuss each of these methods in the following sections, along with accompanying case studies from research sites in Japan and the Philippines. Although the case studies are specific to two regions, these methods could be applicable to other areas, with appropriate calibration.

Keywords: water-energy-food nexus (WEF); integrated tools; integrated indices; benefit-cost analysis (BCA); optimization management models; integrated maps; ontology engineering; physical models; interdisciplinary; transdisciplinary

1. Introduction

1.1. Background

The concept of the water-energy-food (WEF) nexus emerged in the international community in response to climate change and social changes including population growth, globalization, economic growth, urbanization [1], growing inequalities, and social discontent. These issues are putting more pressure on water, energy, and food resources, presenting communities with an increasing number of trade-offs and potential conflicts among these resources that have complex interactions. It is estimated that the global population will grow to 8 billion by 2025, 10 billion by 2050, and 11 billion by 2100 [2]. In terms of globalization, the traded percentage of food produced has grown globally from 10% in

1970 to 15% in 2000 [3]. The World Population Prospects estimates that 54% of the global population lives in urban areas, and that this proportion is likely to rise to 66% by 2050 [4]. Currently, some 1.1 billion people in the developing world do not have access to a minimal amount of clean water [5], and 1.2 billion still live in extreme poverty [6].

At the same time, the global water cycle is changing in response to warming caused by climate change, though the effects are expected to vary across areas and seasons, with some exceptions [7]. The demands for water, energy, and food are estimated to increase by 40%, 50% and 35%, respectively, by 2030 [8]. The interlinkages between these areas further complicate the matter of addressing their growing demands. Addressing the WEF nexus in a sustainable manner has therefore become one of the most critical global environmental challenges of our time.

1.2. The Study Context

This article presents the key methodological results from the Water-Energy-Food Nexus (WEFN) project by the Research Institute for Humanity and Nature (RIHN) [9]. The objective of the project is to maximize human-environmental security (*i.e.*, minimize risk) in the Asia-Pacific region by choosing policies and management structures that optimize WEF links, including water-energy (water for energy and energy for water) and water-food (water for food) connections. We base our approach on the view that it is important for transformative, sustainable solutions to maximize human-environmental security and decrease vulnerability by optimizing the linkages within the WEF clusters. We will take a regional perspective to tackle these global environmental problems around the Pacific Ocean, where the Asian monsoon dominates hydro-meteorological conditions. The populations that live under these natural circumstances face an elevated risk of negative impacts due to natural disasters, while also benefiting from abundant ecological goods and services. Thus, there are trade-offs and conflicts within WEF resources, as well as among the region's various resource users.

Figure 1 shows the dynamics of the WEF nexus under RIHN's WEFN project, which focuses on groundwater, spring water, and surface water for the water cluster; geothermal, micro-hydropower and shale gas for the energy cluster; and fishery, aquaculture, and agricultural production for the food cluster. A primary challenge of this undertaking is to examine the interlinkages between groundwater and fishery production, in terms of the hypothesis that the flow of nutrients from the land to the ocean affects the coastal ecosystem. This suggests that water use for producing and consuming food and energy on land might affect fishery production in coastal zones. To examine this theory, the authors present the four case study areas of RIHN's WEFN project: Obama City, Otsuchi Town, Beppu Bay, and Laguna de Bay (Table 1).

Table 1. Case study areas of RIHN WEFN project.

	Otsuchi			Obama			Beppu			Laguna de Bay		
	W	E	F	W	E	F	W	E	F	W	E	F
for W	/	—	—	/	P	—	/	—	—	/	—	—
for E	H	/	—	Gr	/	—	H/G/Gr	/	—	H	/	—
for F	F	—	/	F	P	/	F	—	/	F/A	—	/

Notes: H: micro-hydropower; F: fishery production; P: pumping; G: geothermal energy; Gr: ground *heat exchanger system*; A: agriculture production.

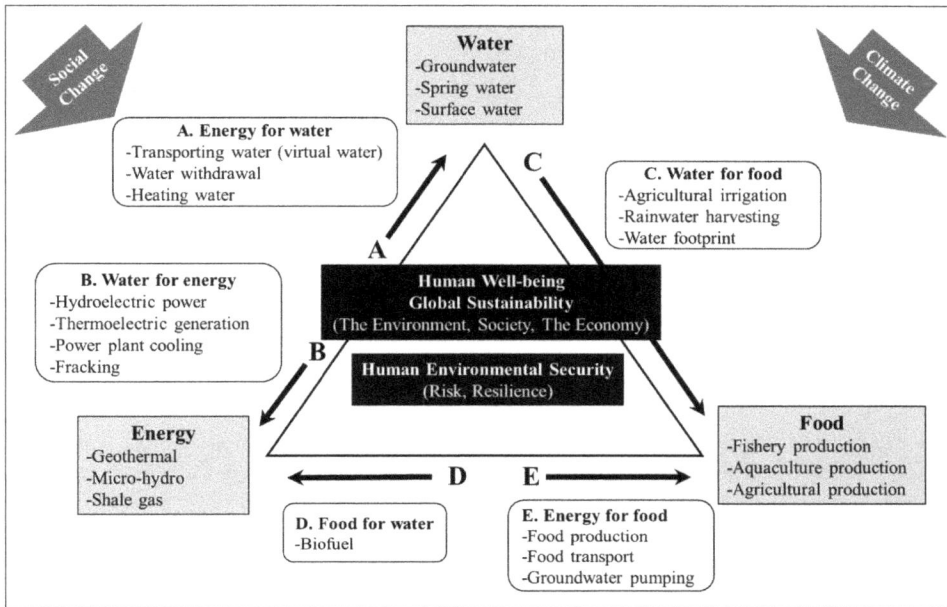

Figure 1. Dynamics of the WEF nexus under Research Institute for Humanity and Nature Water-Energy-Food Nexus (RIHN WEFN) project. Source: Authors modified from [10].

1.2.1. Obama City

Obama City is a coastal city of about 33,000 in central Wakasa District, at the mouth of the Kita River on the Sea of Japan in southwest Fukui Prefecture, Japan. The coastline around Obama City and Wakasa Bay is steep, irregular, and very scenic. The mixture of cold and warm currents in Wakasa Bay provides fertile fishing grounds. While fishing used to be the main industry, now the city is mostly supported by tourism. Groundwater has always been an important resource for the Obama area for domestic, municipal, industrial, and limited agricultural uses; in addition, groundwater has a cultural and historical significance. For example, it is used to melt snow in winter since it maintains a constant temperature throughout the year. It also provides a steady supply of nutrients to the nearshore region via submarine groundwater discharge (SGD) into the sea. Several commercially important fish species appear to thrive in the low-salinity, high-nutrient environment that SGD provides such as sea bass, oysters, and crabs.

1.2.2. Otsuchi Town

Otsuchi Town is located in Iwate Prefecture in northeast Japan, and was one of the most adversely affected towns following the March 11, 2011 Tohoku earthquake and tsunami that devastated the northeast part of the island of Honshu. The catastrophe obliterated the town harbor and all low-lying areas, inundating about half the city. The two important river basins include the Otsuchi and Kotsuchi Rivers. With an economy based on commercial fishing and to a lesser extent small-scale agriculture, Otsuchi lost nearly all its fishing boat, and the tsunami completely destroyed the town's sea farm industry. Through death and displacement, the local population fell from about 16,000 in 2011 to an estimated 12,000 in 2014. The flowing spring water wells played important roles before and after the tsunami as resources for drinking and other human security purposes. However, most of these springs have been covered in order to raise the ground level, thus cutting off local people's access. In addition, the national and local governments are planning to build 14-meter dikes along the coast to protect the town from future tsunamis.

1.2.3. Beppu Bay

In 2014, Beppu City's population was 120,000 and has been falling since its peak in 1981. The population of people 65 or older account for 35% of all inhabitants, and the number of children has decreased. Regarding topography, an alluvial fan is gently spreading from west to east. Hot springs originating at Mt. Tsurumi in western Beppu flow to the urban sectors. The hot spring water has a higher temperature in the mountainous areas than in the lowlands. Oita Prefecture ranked first in terms of the total amount of hot spring discharge points and spring water sources, and Beppu City has ranked first in the overall quantity of discharged hot spring water in the prefecture. The Hiya, Shin, Hirata, Haruki, Sakai, and Asami Rivers flow into Beppu Bay, and household and hot spring wastewater run into these rivers.

1.2.4. Laguna de Bay

Laguna de Bay is the largest freshwater lake in the Philippines and supports more than 35,000 fishermen. The lake consists of four bays in the west, center, south, and east. Calamba City and the urban municipality of Los Baños are the project's target sites, located in the south bay in Laguna Province in the southern part of Manila. The country's Department of Environment and Natural Resources classified the lake as class C freshwater, which means that it is generally designated for fish culture and propagation. In addition, the lake can be used for Class II recreation (no water-contact tourism activities) and Class I industrial water supply (for manufacturing purposes). Laguna de Bay has been extremely stressed by rapid urbanization and industrialization, which contributed to an increase in water demands for other uses such as agriculture, household consumption, tourism, and operating hydroelectric plants. Competition and unregulated use by multiple actors has been causing the quantity of water to decline, and its quality to degrade.

2. Methods for Analyzing the WEF Nexus

In this paper, we follow a framework for the interdisciplinary and transdisciplinary co-creation of new knowledge based on the concepts of co-design and co-production [11], the goal being to link the ideas and actions of numerous stakeholders from various sectors. We consider different temporal and spatial scales, including vertical and horizontal dimensions, to achieve sustainable development based on the international research platform Future Earth 2025 Vision, which prioritizes eight key focal challenges. The WEF nexus is one of them, and Future Earth's mission states *"Deliver water, energy, and food for all, and manage the synergies and trade-offs among them, by understanding how these interactions are shaped by environmental, economic, social and political changes."*

Following this concept, the project was designed consisting of five teams that carried out the following (see the project's structure in Figure 2):

(1) Biophysical measurements and analyses using space satellites, geothermic, and hydrogeological techniques (the water-energy nexus team);

(2) Biophysical measurements and analyses using geochemical, coastal oceanographic, geophysical, hydrologic, and ecological methods including isotopic tracers (the water-food nexus team);

(3) Social measurements and examinations of WEF relationships using stakeholder analyses, social network analyses, and community surveys, based on sociology, economics, anthropology, psychology, and behavioral science approaches (the stakeholder analysis team);

(4) Environmental governance, science in/for society, and co-design/co-production strategies emphasizing the integration of local-national scale stakeholders, and regional scale stakeholders (the science in/for society team); and

(5) The interdisciplinary team.

The interdisciplinary team conducted the research presented in this article with a mission to: (1) identify research problems; and (2) determine the methods and/or create new discipline-free methods [12] based on synthesizing and harmonizing team-based production, collected from individual scientists in different disciplines from each team in order to assess human environmental security. In addition, the team further developed these approaches to incorporate non-scientific/-disciplinary views on the analysis.

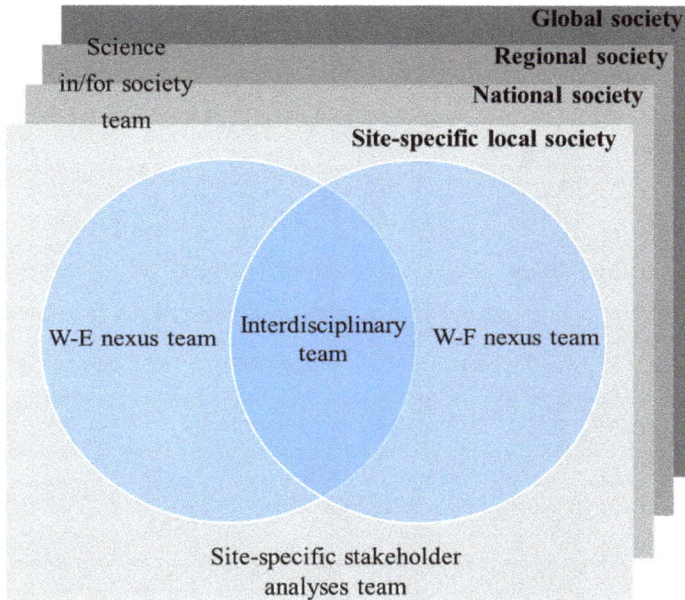

Figure 2. The structure of RIHN WEFN project.

We classified these methods as qualitative or quantitative (see the taxonomy in Table 2) for interdisciplinary and transdisciplinary research approaches, emphasizing the interaction between different scientific disciplines. The methods for interdisciplinary research can be used to unify a collection of related variables, visualize the research problem, evaluate the issue, and simulate the system of interest. Qualitative methods are generally used to describe the nexus in the region of interest, and include primary research methods such as Questionnaire Surveys, as well as secondary research methods such as Ontology Engineering and Integrated Maps. Quantitative methods for studying the nexus include Physical Models, Benefit-Cost Analysis (BCA), Integrated Indices, and Optimization Management Models.

The authors discuss each of these methods in the following sections, along with accompanying case studies from research sites in Japan and the Philippines. Although the case studies are specific to two regions, we see that these methods can apply to other areas with appropriate calibration. All methods discussed here are transdisciplinary in that they begin by engaging stakeholders in order to identify the appropriate research question. Then, they are used to design the scientific approach to collect appropriate data in order to parameterize and develop models. In turn, this allows researchers to answer the policy or management question of interest.

Table 2. Water-energy-food (WEF) methodology and taxonomy.

Type of Data		Functions	Interdisciplinary Research Approaches				Trans-Disciplinary Research Approaches
Primary	Secondary	Methods	Unification	Visualization	Evaluation	Simulation	
			Qualitative Methods				
√	√	Questionnaire Surveys	√	√	√	–	√
–	–	Ontology Engineering	√	√	√	√	√
√	√	Integrated Maps	√	√	√	√	√
			Quantitative Methods				
√	–	Physical Models	√	√	√	√	√
√	√	Benefit-Cost Analysis	√	√	√	–	√
√	√	Integrated Indices	√	√	√	√	√
√	√	Optimization Management Models	√	√	√	√	√

Source: Endo, A.; Orencio, P.M.; Kumazawa, T. and Burnett, K.

3. Qualitative Methods to Describe the Nexus

Although the fundamental relationships between water, energy and food have been used to operationalize concepts such as security, no single approach has been deemed suitable for every situation. A variety of factors including but not limited to scales, populations, institutions, and socio-economic conditions are central to deciding which approach should be used for an integrative and interdisciplinary analysis [13]. In developing countries, security can be defined simply as access to basic needs [14] or entitlements [15]. Following this concept, one way to evaluate security is by looking at the convergence of the core properties of water, energy and food systems; such attributes include access, availability, utility, and stability at the individual and household levels [16].

An analysis of how the core characteristics of water, energy, and food systems interact, whereby individual or household needs are compromised, usually requires specific measurements. However, most global frameworks developed to jointly analyze the three systems are not intended to be used at the local or regional levels because they do not incorporate the proper temporal and spatial scales [16]. Hence, studying the place-based interactions of each system at various spatial scales is valuable for decision-making. In the sub-sections below, we share our experiences using three qualitative methods to analyze water, energy and food systems: Questionnaire Surveys, Ontology Engineering, and Integrated Maps.

3.1. Questionnaire Surveys

We used Questionnaire Surveys in the case of Laguna de Bay (see above) to assess the WEF nexus. We based our methodology on the approach suggested by Strasser *et al.* [17], who stressed that basin-level information would be ideal for an accurate appraisal and can be gathered through a questionnaire that screens the nexus resources.

Questionnaire Surveys contributed to a nexus assessment that aimed to address the question of how the population's security is affected when various natural and social hazards disrupt the linkages among the three systems. Energy plays an important role in understanding human security within these connections. However, in this study, we paid particular attention to the relationship between water and food because the degradation of the water-food system has affected the socio-economic conditions of people in the study area in a multitude of ways; for example through pollution [18], competition for access to limited resources [19] and fish operations [20,21].

We designed the Questionnaire Survey based on a set of concepts that underlie water, energy, and food systems e.g., [22–24], namely availability, access, utilization, and management. The collected information will be used to develop the integrated index, which will support decision-making related to the inter-relationships of water and food. The Questionnaire Survey consists of four sections. The first section deals with demographic characteristics, while the second section looks at household access to and utilization of food and water resources. It also aims to determine each household's choice of food and water sources, and what types of activities are employed to manage major sources. The third section focuses on socio-economic activities, including an in-depth assessment of each household's main sources of income. The fourth section examines risk management, identifying available mitigation systems at the household level.

Prior to distributing the survey questionnaires, we gathered feedback on the survey's content and duration through field-testing. We then created a final web-based questionnaire using Google Forms to facilitate a paperless survey [25].

We carried out the field activities in March 2015 in targeted households of the two towns. In Calamba ($N = 258$), the households came from three agricultural barangays ($n = 85$), three barangays in sub-basins ($n = 87$) and three barangays in lakeshore areas ($n = 86$), respectively. In Los Baños ($N = 202$), the total number of samples came from households in two agricultural ($n = 73$) and two lakeshore ($n = 129$) barangays.

An assessment of household heads' sources of livelihood show that the income of the studied households in Calamba and Los Baños comes from various water-related activities. Farming and forestry contributes to 15% and fishery-related work around 20% of their livelihood sources. There is a strong dependence on water for food production, yet only around 15% recognize the government policies on water management systems currently in place. One can attribute this problem to a lack of management and access to information or financial systems.

We found the use of Questionnaire Surveys to screen and gather pertinent information on the inter-relationships of different nexus resources at the local level to be promising. The Questionnaire Survey was especially useful in incorporating the local people's general outlook on their level of economic, food, and livelihood security when various shifts occur in terms of the quality and quantity of the water-food nexus. Consequently, this provides the information necessary to make decisions and thus optimally manage local nexus resources. However, we have to acknowledge that the quality of the survey instrument always affects data resulting from this approach. Such quality includes the steps undertaken to develop the tool [26] and their limited spatial and temporal applications [27].

3.2. Ontology Engineering

Scholars have proposed methods and frameworks supporting interdisciplinary and transdisciplinary approaches in the field of team science. However, the methods and research designs for reasonably, effectively assessing the processes and outcomes of team science have not been sufficiently developed [28]. How do we facilitate collaboration using interdisciplinary and transdisciplinary approaches? Defila and Di Giulio [29] stated that the existence of many different frames, or definitions of the problem, suggests a need to develop shared goals and language, while Defila *et al.* [30] showed that those who achieved a synthesis also succeeded in identifying a common language and a collective theoretical basis.

Ontology Engineering is one of the base technologies in semantic web technology, where the Internet is used to create a knowledge base that computers can deal with directly by means of adding metadata, (*i.e.*, semantic information for computers, as annotations to information resources on the World Wide Web) [31,32]. An ontology consists of concepts and relationships that are used to describe the target world. It provides common terms, concepts, and semantics by which users can represent the contents with minimum ambiguity and interpersonal variation of expression. Construction of a well-designed ontology presents an explicit understanding of the system.

An ontology can deal with a model, an indicator system, or an analytical framework rather than a case itself. The main steps using an ontology are: (1) Sharing the definition of a term; (2) Sharing the relationship between items; (3) Sharing the relationship between models/indicator systems/analytical frameworks; and (4) Sharing the relationship between a defined term and a metadata item. The relationship between the WEF nexus model (Figure 1) and the ontology of sustainability science/social-ecological systems (SS–SES; please see below) corresponds to the relationship between the model and the ontology. Figure 3 illustrates these relationships.

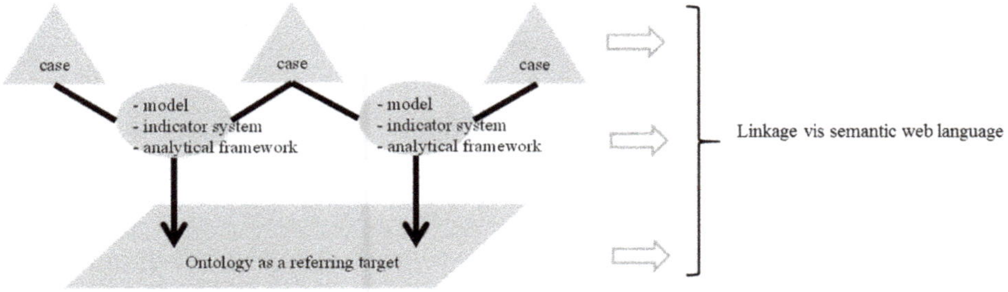

Figure 3. Mutual relationships structured by cases, models and ontology.

We used Ontology Engineering in Obama City and Beppu City (see above), building on a set of questions defined for those contexts. Figure 4 displays a conceptual map focusing on water, generated from the SS-SESs ontology [33,34]. This map shows that agriculture, fisheries and human life are the hubs that connect water, energy and food. By exploring the causal chains in the conceptual map, we can identify the question to be analyzed.

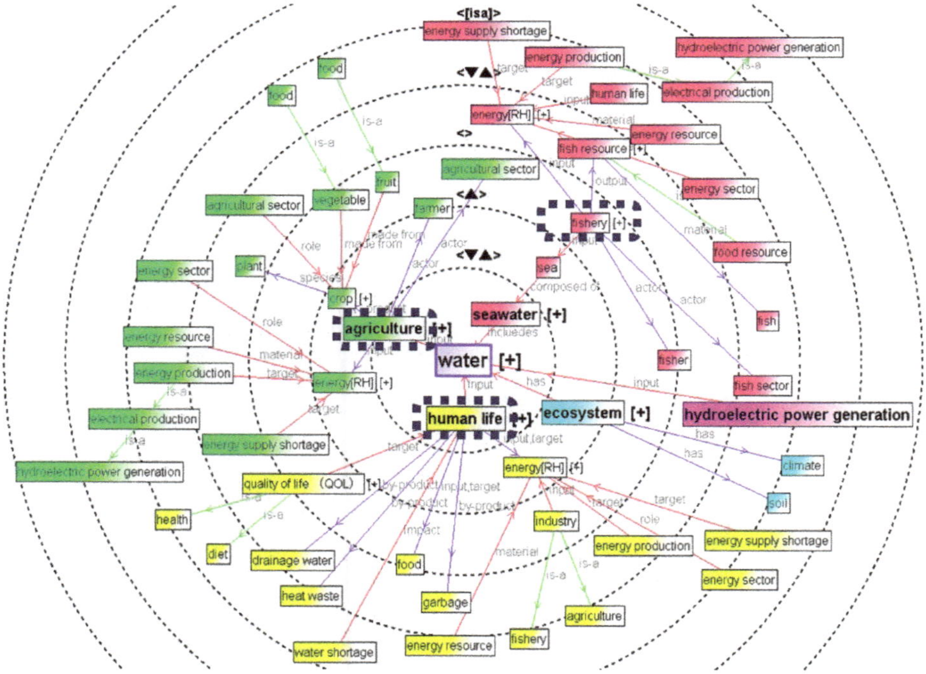

Figure 4. Conceptual map focusing on water.

Combining the exploratory questions and the ontology dealing with the nexus should facilitate our ability to establish the issue for further examination and share it among a variety of researchers or stakeholders.

3.3. Integrated Maps

An Integrated Map is an overlay of various single maps, and it can be used as a method to support the implementation of synthesized policies between the land and the sea. In contrast to sectoral management and monodisciplinary research approaches (which often focus on a single ecological system), an integrated map informs policies capable of restoring and maintaining the interdependence between the land and the sea.

Creation of an integrated map brings many benefits. Firstly, it can be used to incorporate individual research results into maps as integrated methods for interdisciplinary research approach to enhance mutual understanding between members. Secondly, it can be used to unify data, information and knowledge on maps to visualize and disseminate the current status of environment and utilization in river basins and coasts to stakeholders. Thirdly, an integrated map can facilitate the identification of key nexus issues, such as the impact that nutrient flows have for coastal ecosystem. Fourth, Integrated Maps can be used as a transdisciplinary method, engaging stakeholders and policy-makers to discuss through an integrated map how to implement integrated management of land and coastal areas. For an example of an Integrated Map created for Beppu Bay, see Annex.

It is possible to create a site-specific Integrated Map at the local level to visualize the current conditions of water, energy and food resources, as well as resource users. However, it would be challenging to create an Integrated Map at the national or global level. In addition, an Integrated Map shows a static condition, not future scenarios, which limits the map's utility to demonstrate inter-scale, inter-generational and inter-area circumstances.

4. Quantitative Methods to Examine the Nexus

Along with three qualitative methods, we took a combined approach using four quantitative methods: Physical Models; BCA; Integrated Indices; and Optimization Management Models. We used various kinds of quantitative methods, especially by the W-E and W-F nexus teams to analyze the interlinkages of the water-energy and/or water-food systems; this helped us understand the complexities of WEF nexus systems.

Furthermore, we presented the specific per-site results in several forms but normalized them to allow for direct comparison with other results at different project locations in the Asia-Pacific region. This makes it possible to decide on optimal policies regarding the sustainable management of water, energy, and food, not only for project members, but also for stakeholders.

4.1. Physical Models

A physical model simulates a biological or ecological system using mathematical formalizations of that system's physical properties. Such models are often used to predict the influence of a variety of factors on a complex system. In the context of water, we are often interested in how both exogenous (e.g., droughts, sea level rise, natural disasters) and endogenous (e.g., groundwater extraction, surface water diversion, water pollution) factors ultimately affect resource quality and availability over time.

To address questions about human-environmental security in the context of the WEF nexus, we developed a physical water model as part of RIHN's WEFN project based on a representation of water balance model called the similar hydrologic element response (SHER) model and a three dimensional variable density groundwater flow and transport model called the SEAWAT model. Using data collected daily, the SHER model can calculate water balance components such as groundwater recharge, river discharge, and surface runoff with precipitation and evapotranspiration. The SEAWAT model consists of the modular finite-difference flow model (MODFLOW) and the modular 3-D multi-species transport model (MT3DMS), which can simulate water and dissolved material transport by using the advection-diffusion equation [35].

Thus far, we have calibrated the physical model described above using data from Obama City. The primary issue currently facing Obama City is how to allocate groundwater among multiple uses, including for domestic purposes, melting snow (groundwater has a constant temperature throughout the year), as well as providing the necessary inputs to the nearshore fishery resources via submarine groundwater discharge. Results from this modeling exercise will allow us to consider how various future land use and climate change scenarios will affect water balance components such as recharge, runoff, and river discharge, in order to inform water management and land use decisions in Obama City. We intend to use these outcomes to help parameterize the economic models that we are developing to allocate groundwater in the most optimum way possible for these various uses over time.

Integrated physical models (such as those that measure water balance), and hydrological parameters (such as water exchange between rivers and groundwater, and groundwater discharge into the ocean) are useful methods for hydrologists. Material transport (including nutrients from the land to the ocean by rivers and groundwater) is important for fisheries. Hydrology, fisheries, and geochemical and biochemical information can be applied to this integrated physical model in an interdisciplinary way. To complete the transdisciplinary process, decision makers should be able to employ this model along with other stakeholders, such as scientists and business sectors, to decide on optimal policies for sustainable water and ecological management. This integrated physical model can deal with both the water-food and the water-energy nexuses. For the water-food nexus, we need additional data to link it with fisheries; however, we can obtain the basic components of water balance and geochemical parameters including chlorophyll-a with this model. For the water-energy nexus, the model can illustrate the connection between groundwater and subsurface temperature, which is important for heat pumping and geothermal energy.

Integrated physical models can simulate the balance between water, energy, and food production; therefore, simulations based on potential future scenarios can be useful for decision makers. However, the results of integrated model simulation without social and local knowledge may lead people to misconstrue the model's results if the numbers from simulations are unrealistic for political, economic and other reasons.

4.2. Benefit-Cost Analysis

BCA allows us to gauge an environmental project or investment by comparing an activity's economic benefits with its economic costs, usually over some fixed time horizon. BCA can be used to appraise a scheme's economic merit, or to compare the net benefits of competing projects. BCA examines potential changes to the ecosystem, with the objective of increasing social welfare.

One way to carry out BCA is to use benefit-cost models (BCMs), which assess the desirability of a proposed policy or project, either independently or ranked according to highest net benefit if selecting from a range of alternatives. BCMs can be used in the context of evaluating WEF nexus project to clearly consider the trade-offs in a particular region where one or more of the WEF elements will be utilized. The researcher begins the analysis by identifying all the potential benefits and costs of a particular action, regardless of whether he/she can quantify or monetize them. This step is usually conducted in partnership or consultation with stakeholders in the region of interest.

Once the time horizon for analysis is selected, we can calculate the net present value of quantifiable benefits and costs. Because nexus-related problems are typically dynamic, it is important to consider the timing and magnitude of benefits and costs. When monetized benefits largely exceed monetized costs, policy and management implications may be more straightforward than when monetized benefits fall short of monetized costs. In the latter case, the results can be interpreted to mean that if the non-monetized benefits are at least as large as the difference in shortfall between monetized costs and benefits, then the project or policy passes the benefit-cost test over the selected time horizon.

We developed a BCM to analyze the construction of a new dike between the Pacific Ocean and Otsuchi's coastline. In 2011, the Tohoku earthquake and tsunami completely destroyed a six-meter dike, and the Japanese government started to build a new 14-meter dike to replace the old one. While the project will be extremely costly in terms of construction expenses, the potential benefits associated with protecting lives and material property are also expected to be substantial. BCA is a method for evaluating the potential gap between these elements.

In terms of pros, the largest expected component will be avoided damages from another Tohoku-like event (future estimates for damage range from USD 50 million [36] to USD 210 million [37]). Other, much smaller potential benefits include ecotourism and tourism-related benefits that might be realized via the Itoyo Sanctuary Park, which is being planned to protect the nationally protected "itoyo" fish species.

On the cost side, the largest component will be the current and future construction expenditures involved in building the dike. In addition to construction costs, annual operation and maintenance expenses will be calculated and discounted for the appropriate time horizon. Similar costs will be collected for the Itoyo Sanctuary Park, though these are expected to be insignificant compared to the larger building costs of the dike. Another important component of the cost portion is ecological losses. Erecting the dike is expected to result in a 100% loss of the area's mudflat habitat, which would lead to economic losses in terms of the oyster, abalone, and seaweed fisheries. We calculated typical (or average) total annual profits from each of these fisheries and included them in the losses as a result of building the dike.

The WEF nexus is inherently about trade-offs. BCA enables researchers to provide decision-makers with information regarding the consequences of these trade-offs and to explicitly examine the net benefits of decisions in order to allocate scarce resources (such as water) toward food or energy. In addition, to improve understanding of the trade-offs, BCA makes the costs and benefits accrued over individual time periods transparent.

4.3. Integrated Index

A number of studies have shown that a mix of sociology, geography, and natural science is required to effectively analyze the relationship between people and their surrounding environment, e.g., [38–40]. Often, a key research objective aims to understand how people cope and develop, given prevailing social inequities and environmental stresses, which are typically area-specific [41].

Indicators are methods used to quantitatively describe and operationalize any system, no matter how inherently complex [42,43]. Indicators have served as an operational representation of various social and environmental characteristics, wherein measurable variables are used to create each indicator in order to quantify the system's overall characteristics in an analysis. To investigate and gauge the social system, MacGregor and Fenton [44] identified five indicator classes (informative, predictive, problem-oriented, program-evaluation, and target-delineation), which can be used to describe or model the different social sub-systems. Informative indicator types describe the social system, while predictive types model describe the social system's various components.

We employed indicators to assess the security of the human-environmental system in Laguna de Bay, using approach similar to the one presented in [43]. The objective was to identify indicators for measuring human-environmental security in Laguna, with a particular focus on the interconnection between food and water. This method is contextual and dynamic, and hence needs to be approached on a case-by-case basis. Common indicators for different environments are difficult to achieve because they should be strongly linked to the issue and objective for measurement and tailored specifically for the research area. Although not covered in this study, there is a strong interconnection between water and food with energy. The lake that supports the fisheries has been providing water that turns the turbines of the hydro-electric plants that supplement the region's energy supply. We believe this might entail a different approach for indicator development because energy operates on a different spatial scale.

Prior to selecting the indicator, we created a profile with help from local experts to establish area-specific information on physical and social characteristics. This profile included: the hazards that pose a risk to security of the social and natural environment; the susceptibility of the coupled system based on its intrinsic conditions (e.g., special needs populations, proximity to the hazard source); and the consequences based on the hazard effects.

An objective is formed and the approach to meet it is established within the target communities and their respective human and environmental values, the hazards that threaten their human values, and the means for mitigating their negative effects. We referred to the objective to develop the indicators.

The general dimensions for analyzing human-environmental security within the interconnections of water and food consisted of social, environmental, economic, governance, and risk components. For each dimension, specific indicators are identified based on prevailing indicators of the World Governance Index (WGI), the World Development Index (WDI), and the United Nations Development Programme's (UNDP) Human Development Index (HDI). Annex is an example of the indicators selected for the social dimensions to analyze the water-food nexus.

Based on our preliminary observations, an indicator-based assessment of the water-food nexus at the local level has allowed us to incorporate households' views and knowledge into our analysis of the effects of shifts from the water-food nexus to the human-environmental system. This is useful for minimizing bias assessment of social and environmental conditions at the study sites. However, we observed that this could also be challenging due to the unavailability of reliable and pertinent data. In this case, gathering responses at the household level and quantifying them based on a scale to gauge conditions has been substantial in the process of indicator-based assessment [26].

4.4. Optimization Management Models

While the BCM is an appropriate framework when the objective is to evaluate a project's independent or comparative desirability, a different approach is needed when the goal is to determine optimal allocation of a resource that has linkages to many other resources and may also cross physical, political, and administrative boundaries. In such situations, the Optimization Management Model provides one possible method to look at optimal resource allocation. Such a framework is invariant to the extent of transboundary interlinkages, as the objective is to maximize the net present value of total welfare. Once the optimal allocation is identified, however, the most effective way to incentivize behavior that approximates that social optimum can vary greatly depending on the particular situation.

We used the Optimization Management Model to study groundwater allocation problem in Obama City. Water has traditionally been thought of as a common property resource in the area, with publicly provided groundwater pumps and wells throughout the city. However, it is of interest how groundwater extraction influences the availability of SGD, as well as the fisheries. Such changes may lead to different management actions and policies regarding groundwater use in Obama City.

In order to address these questions, RIHN researchers developed an economic optimization framework that allows us to consider aquifer dynamics in response to a variety of different controls. The main model's control variables are pumping groundwater for domestic use and pumping groundwater for melting snow, since the groundwater remains at a constant temperature throughout the year. In response to these changes in pumping, the aquifer head level will be drawn down, thereby decreasing water pressure and subsequently changing the amount of SGD available in the nearshore. The optimization framework allows us to describe these linkages between groundwater pumping and the resulting dynamics of the aquifer, and optimize by choosing the benefit-maximizing levels of groundwater pumping for domestic and snow-melting uses during every period on the time horizon. For detailed explanation of the framework, see Annex.

While important trade-offs can be qualitatively identified using the theoretical optimization framework developed above, calculating the actual trajectories of optimal resource extraction and stock levels requires numerical methods. When the problem is relatively simple, applying a standard gradient method will often lead to the net present value (NPV) maximizing solution. However, as the complexity of the problem increases (for example, when the number of interlinked resources increases), a more complex, nonlinear programming algorithm may be required. Therefore, depending on the particular situation (e.g., the resource system's complexity, s strict or flexible project deadline, computational capability, data availability, the types of research/policy questions being addressed) then a BCM, optimization model, or a combination of the two may be most suitable.

An Optimization Management Model allows researchers to explicitly represent the interaction of natural resources, which is key to understanding trade-offs inherent in the WEF nexus. Decisions to draw down one resource often affect other resources, as well as the social welfare of the community of interest. For example, the decision to use groundwater for fisheries rather than agriculture depends on the production costs of both fish and agriculture, including energy. Economic optimization allows the researcher to determine how to allocate scarce resources over time, when doing so has consequences for the surrounding ecosystem and society.

5. Discussion

In this section, we discuss our experience on developing and using various integrated methods to address the WEF nexus. The main advantage of integrated methods is their ability to synthesize team-based production collected by individual scientists in different disciplines. Integrated methods are necessary to link the ideas and actions of numerous stakeholders from different sectors in light of distinct temporal and spatial scales. Both vertical and horizontal dimensions should be considered to reduce trade-offs and conflicts, and optimize the linkages within the WEF clusters.

In Table 3, we re-categorized and synthesized different methods to provide a critical reflection from an Ontology Engineering perspective. The methods we developed are shown in bold letters. We began by grouping methods according to how the target or objective of the research was expected to be achieved. A perspective-oriented approach uses information from a survey, analysis, or assessment, often in a forward-looking or predictive manner, in order to meet the research target. In other words, the method focuses on what could be understood. The state-oriented approach, on the other hand, aims to grasp the system in its current state; that is, it determines what should be understood about the target itself. We further sub-categorized these methods according to dimension/unit systems, and the integrated methods described in this paper each cover one or more of those dimensions (e.g., spatial, physical, monetary). In summary, we collected data from team-based production using each of the monodisciplinary approaches in order to create site-specific integrated methods.

Table 3. We re-categorized each method from the ontology engineering perspective.

How the Target World Exists	How to Recognize the Target	Dimension/Unit System	Individual Method	Integrated Method
Target system	√ Perspective-oriented √ What to be understood √ Format-oriented	Spatial	Map	Integrated Maps
		Physical	Physical Models	Integrated Physical Models
		Monetary	Cost Analysis	Benefit-Cost Analysis
			Benefit Analysis	Economic Optimization Models
		Non-unified unit	Indicator	Integrated indices
	√ State-oriented √ What to understand √ Content-oriented	Context-dependent	Specific	Comprehensive
			Questionnaire surveys	Interviewing

A remaining challenge is to develop integrated methods for linking the ideas and actions of various stakeholders from different sectors, while also considering distinct temporal and spatial scales, including vertical and horizontal dimensions (Figure 5). Ways of connecting local nexus issues within a community to broader national and global nexus issues (the vertical dimension) are often missing from site-specific case studies. At the same time, it is important to understand how an incident related to WEF resources and resource users in one case study area could affect other case study areas (the horizontal dimension). Finally, we should also consider how current events are likely to impact future WEF resources and resource users on a temporal scale.

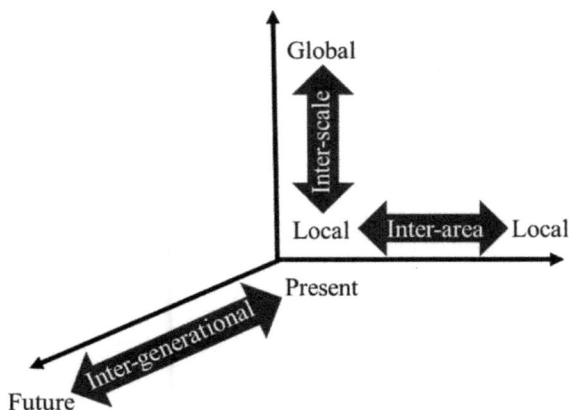

Figure 5. Relationships between targets on different spatial and temporal scales.

6. Conclusions

This article has provided a comparative analysis of various integrated methods that we used under the RIHN WEFN project in order to analyze the WEF nexus. We classified the integrated methods as qualitative and quantitative that contribute to both interdisciplinary and transdisciplinary research. Our approach employs the concepts of co-design and co-production to link the ideas and actions of numerous stakeholders from distinct sectors considering different temporal and spatial scales, including vertical and horizontal dimensions to achieve sustainable development.

Qualitative methods that we analyzed consisted of Questionnaire Surveys, Ontology Engineering and Integrated Map, while quantitative methods included Physical Models, BCA, Integrated Indices, and Optimization Management Models. We discussed our experiences using all of these methods based on case studies from research sites in Japan and the Philippines.

To take the approach of co-design and co-production through the project process, each method should be developed as a science-policy interface method, although each one has different uses at various stages. Ontology Engineering would be the most useful for designing the project during initiation stage to build a list of common concepts of term; the linkages of each term among stakeholders included researchers and practitioners. In addition, Ontology Engineering could be used at the policy planning stage to assess whether the policy/plan would cover all disciplines including natural sciences, social sciences and humanities, and sectors such as water, energy, and food (in order to address the key issues that are originally identified during the initiation stage). Questionnaire Surveys would be more useful for collecting information to analyze WEF interlinkages when few data exist; then, it would help to identify the key issues during the initiation stage. Integrated Maps can provide an opportunity to share knowledge showing actual conditions at a spatial scale among stakeholders during the policy planning stage. BCA and an Optimization Management Model would play important roles in clarifying trade-offs during the initiation stage, creating and providing policy options during the policy planning stage. Physical models could be quite essential to understand WEF nexus systems; if it were developed to clarify interlinkages between physical conditions of water, energy and food, as well as human activities by working with social scientists, then, it could be used to address the key issues more holistically during the policy planning stage. Using an Integrated Index can be a discipline-free method, which could incorporate and integrate each result with different disciplines, then evaluate trade-offs during the policy planning stage. At the same time, interdisciplinary team members themselves could be interpreters or coordinators for science-policy interface, using those approaches when they have a commitment to both science and society from the initiating stage.

From the perspective of spatial and temporal scales, although we covered spatial, physical and economic dimensions, our approach is somewhat limited in terms of vertical and horizontal elements, as well as on a temporal scale to address the WEF nexus. To address these challenges, it can be possible to use global data such as a global model to set our site-specific case studies within a global context on vertical spatial scale [45]. In addition, the creation of future scenarios further integrating each integrated method mentioned in this paper must be a challenge, however this will make it happen to analyze WEF nexus based on temporal scale [46]. While our case study areas focused on relatively small water bodies in Japan and the Philippines, we believe that all the methods can also apply in other contexts, including large transboundary rivers. The most important is first of all to understand the context and its specific characteristics and then recognize the key issues and related problems. Finally, it is key to select the most appropriate method(s) to analyze those issues. Overall, we conclude that developing integrated methods to link different scales and to achieve multi-dimensional targets is an important area for future research at both the case study level and in large river basin areas.

Acknowledgments: This research was financially supported by the R-08-Init Project, entitled *"Human-Environmental Security in the Asia-Pacific Ring of Fire: Water-Energy-Food Nexus"* at the Research Institute for Humanity and Nature (RIHN) in Kyoto, Japan. The authors are grateful to Mr. Shun Teramoto for his assistance. The authors acknowledge the academic suggestions for the paper provided by the reviewers and guest editors, especially Marko Keskinen.

Author Contributions: Aiko Endo and Izumi Tsurita reviewed WEF nexus practices and studies; Kimberly Burnett and Christopher Wada developed the BCA and Optimization Management Models; Pedcris M. Orencio conducted and analyzed the results of Questionnaire Surveys; Terukazu Kumazawa contributed to the section on Ontology Engineering as well as the discussion; Aiko Endo and Akira Ishii designed the Integrated Map; Makoto Taniguchi contributed to the section on Physical Models; Pedcris M. Orencio developed the Integrated Index with support from all authors. Although Aiko Endo and Kimberly Burnett conceived this paper, this research was conducted by an interdisciplinary team under the RIHN WEFN project performed this research. In order to assess human-environmental security, they aimed to develop integrated methods to synthesize and harmonize each project member's discipline and set of research skills.

Conflicts of Interest: The authors declare no conflict of interest.

Appendix

A. Integrated Map in Beppu Bay

The Basic Act on Water Cycle (enacted in 2014) and the Basic Act on Ocean Policy, (enacted in 2007) guide Japan's current water and coastal management. The former aims to re-establish the view that water is an integral part of water circulation and to promote integrated measures with respect to the water cycle. The latter broadly defines coasts as any areas where the land and sea interact. Water and coastal management thus requires an integrated approach.

It has gradually become clear that SGD plays an important role in the cycling of dissolved materials and is now viewed as one of the water cycle's invisible channels [47]. The map in Figure A1 overlays the actual conditions of use of the Hirata River, the Hiya River basin and Beppu Bay, with visually observed locations of spring water (in crossed pink lines).

Different bodies charged with overseeing specific targets oversee Japan's coastlines. The observed locations of spring water have been identified in both commercial port areas (in purple), which fall under the jurisdiction of the Coast Act (first enacted in 1956 and amended 1999), as well as common fisheries right areas (in light green), which fall under the Fishery Act. SGD has not been managed to date, because it occurs along the policy border between terrestrial and coastal areas. In order to clarify the dimensions where conflicts of interest emerge among stakeholders, effective administration will require interdisciplinary studies to reconsider the spheres and boundaries of water circulation handled by each ministry and agency, and the relationships between these actors.

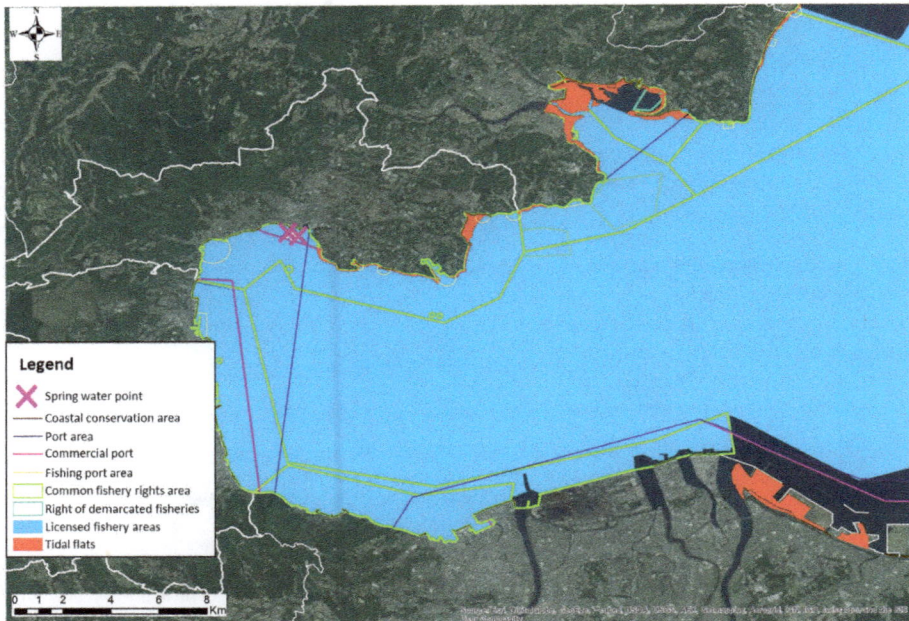

Figure A1. Beppu Bay—Multiple overlay map. Source: [48].

B. Social Indicators for Lagunade Bay

The Table A1 below shows selected indicators of the social component for the analysis of water and food nexus in Laguna de Bay. We identified the indicators based on the objective, within a specific boundary. Each boundary corresponds to a specific dimension and is referred to when distinguishing the indicators.

In general, social indicators are chosen based on their measurability when downscaled at the local level. The availability of real data for valuation at this level is one of the major concerns for establishing a quantitative measure. Nonetheless, household surveys that can be rapidly executed could be explored to develop the quantitative values. The use of the Likert scale to gauge household socio-economic data, with reference to a range of values, was very useful in the process of identifying, evaluating, and measuring indicators, e.g., [26]. This is an important component of metric development that is required to create a human-environmental security index, which, like the WGI, the WDI and the HDI, could portray the security of the coupled system in the research area.

Table A1. Selected indicators of the social component for the analysis of the water and food nexus in Laguna de Bay, Philippines.

Component	Indicators	Variables	Tentative Values
Social	Food sufficiency rates	% of protein needs sourced from fisheries	Total required dietary allowance (RDA) protein need per individual
		% of protein needs met	Total RDA of protein needed per individual
	Water sufficiency rates	% water demand supplied locally	80% of demand per end user is supplied

Table A1. *Cont.*

Component	Indicators	Variables	Tentative Values
	Health status	% of mortality rates of adults, women and children	Major cause of mortality
		access to hospital services	Standard hospital beds for size of population
		% occurrence of water-borne diseases	50% of children and female population are affected by water-borne diseases
	Change in the population	% population growth rate	Annual national growth rate
		% population density	Mean standard limit of population density
	Transportation	Passenger cars	Availability of passenger vehicles for public transport per standard population
	Communication	Mobile phone/TELCO subscriptions	80% of population has access to TELCO subscriptions

C. Description of the Optimization Management Model in Obama City

Figure A2 displays the general framework used to study Obama City's optimal groundwater management. The central resource of interest in the framework is the groundwater aquifer, illustrated by the blue box in the center of the flow chart. Aquifer volume $X_h(h)$ is a function of head level h, which will be changed via the controls (in red): groundwater quantity pumped for households q_H at a cost of $c_w(h)$ and the quantity of groundwater pumped to melt snow q_S, also at a cost of $c_w(h)$. Distribution costs for each use are given by c_{DH} and c_{DS}, respectively. The third control variable in the model is the quantity of fish caught q_F, harvested at a cost of $c_F(X_F)$. Aside from the aquifer state variable, this framework includes the fishery stock, given in orange by X_F, which is governed by the growth function $G(X_F, h)$. In this framework, the primary stock of groundwater directly affects the growth of the fish stock via SGD, which is a function of head level $SGD(h)$. Benefits for all uses of groundwater are shown in green: domestic benefits B_H, snow-melting benefits B_S, and fishery benefits B_F.

The framework allows researchers to understand the multiple trade-offs working against each other in the model. As more groundwater is used for one of the aboveground uses (domestic or melting snow), less is available to support fishery production in the nearshore region via submarine discharge. The optimal allocation of water will depend on the marginal benefits accrued under each of these uses, as well as the marginal costs for utilizing each one. These benefits and costs will change over time in response to their respective time paths, described in Equations (2) and (3). The objective is then to choose the optimal time paths of groundwater pumping for each use to maximize total NPV from all three uses of groundwater: domestic, melting snow, and fishery production. Equation (1) describes this goal mathematically.

$$\text{Max } (B - C) \text{ over time}$$
$$\text{subject to } dh/dt = \text{Recharge} - SGD(h) - (q_H + q_S) \tag{1}$$
$$dX_F/dt = G(X_F,h) - q_F$$

where

$$B = B_H(q_H) + B_S(q_S) + B_F(q_F) \tag{2}$$

$$C = [c_W(h) + c_{DH}]q_H + [c_W(h) + c_{DS}]q_S + [c_F(X_F) + c_{DF}]q_F \tag{3}$$

when the problem is transboundary in nature, there may be multiple benefit and cost functions for each end use, *i.e.*, the B and C-functions would be indexed not only by end-use but also by region or country. Although that complicates the mathematics, the underlying methodology remains the same. Once the optimal allocation is determined, payments for benefits or compensation for costs can be used to incentivize transboundary socially optimal outcomes. In this case, the commercial fishermen might give up a portion of their expected benefits to cover the cost to the service provider (the users of the aquifer) of ensuring the continued provision of the valuable SGD. This may require substituting groundwater use with costlier alternative freshwater sources (e.g., recycled wastewater).

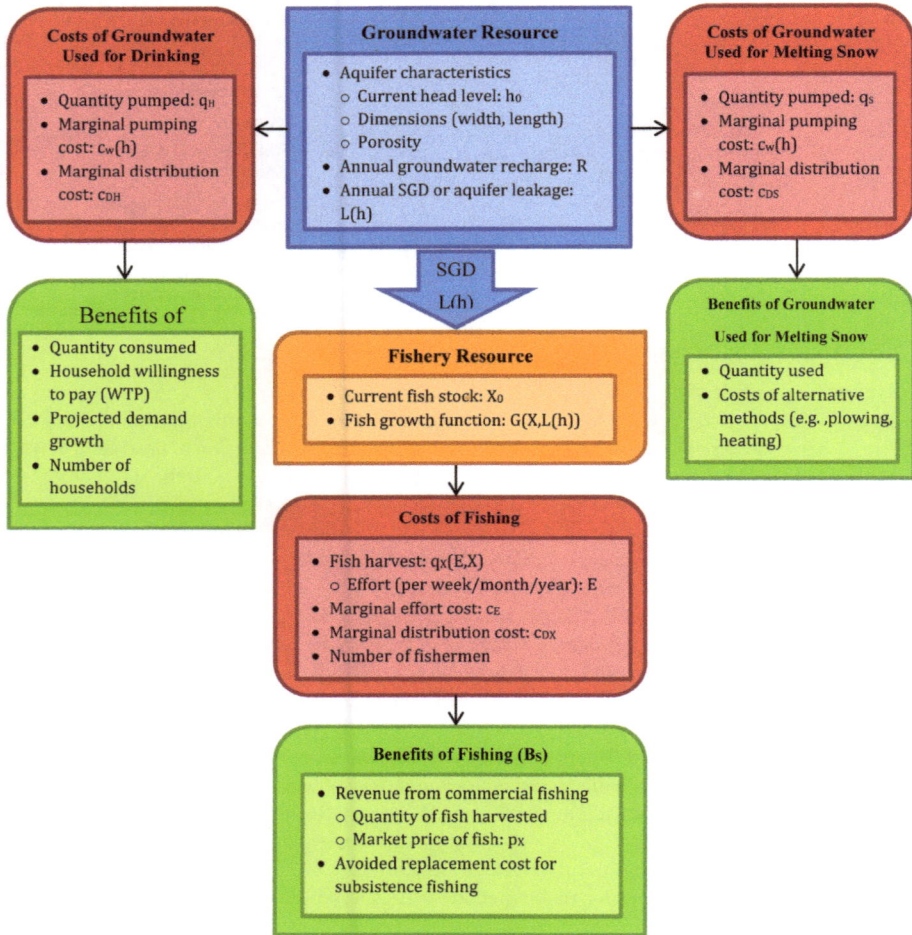

Figure A2. Groundwater optimization model for Obama City, Japan.

References

1. Hoff, J. Understanding the Nexus. In Proceedings of the Bonn 2011 Conference: The Water, Energy and Food Security Nexus, Bonn, Germany, 16–18 November 2011.
2. United Nations Department of Economic and Social Affairs, Population Division (UNDESA). *World Population Prospects: The 2012 Revision*; UNDESA: New York, NY, USA, 2013; p. 118.
3. Anderson, K. Globalization effects on world agricultural trade, 1960–2050. *Philos. Trans. R. Soc.* **2010**, *365*, 3007–3021. [CrossRef]
4. United Nations Department of Economic and Social Affairs, Population Division (UNDESA). *World Population Prospects: The 2014 Revision*; UNDESA: New York, NY, USA, 2014; p. 27.

5. United Nations Development Programme (UNDP). *Human Development Report 2013*; UNDP: New York, NY, USA, 2013; p. 203.

6. United Nations (UN). *The Millennium Development Goals Report 2013*; UN: New York, NY, USA, 2013; p. 60.

7. Intergovernmental Panel on Climate Change (IPCC). *Fifth Assessment Report*; IPCC: Geneva, Switzerland, 2014; p. 151.

8. United States National Intelligence Council (USNIC). *Global Trends 2030: Alternative Worlds*; US NIC: Washington, DC, USA, 2012; p. 137.

9. Taniguchi, M.; Allen, D.; Gurdak, J.J. Optimizing the water-energy-food nexus in the Asia-Pacific ring of fire. *EOS Trans. Am. Geophys. Union* **2013**, *94*. [CrossRef]

10. Scott, C.A.; Kurian, M.; Wescoat, J.L., Jr. The Water-Energy-Food Nexus: Enhancing Adaptive Capacity to Complex Global Challenges. In *Governing the Nexus: Water, Soil and Waste Resources Considering Global Change*; Kurian, M., Ardakanian, R., Eds.; Springer International Publishing AG: Gewerbestrasse, Switzerland, 2015; pp. 15–38.

11. Mauser, W.; Klipper, G.; Rice, M.; Schmalzbauer, B.; Hackmann, H.; Leemans, R.; Moore, H. Transdisciplinary global change research: The co-creation of knowledge for sustainability. *Curr. Opin. Environ. Sustain.* **2013**, *5*, 420–431. [CrossRef]

12. Keskinen, M. Bringing Back The Common Sense? Integrated Approaches in Water Management: Lessons Learnt from the Mekong. Ph.D. Thesis, Aalto University, Espoo, Finland, 3 September 2010.

13. Giampietro, M.; Aspinall, R.J.; Ramos-Martin, J.; Bukkens, S.G. *Resource Accounting for Sustainability: The Nexus between Energy, Food, Water and Land Use*; Routledge: London, UK, 2014; p. 270.

14. Hicks, N.; Streeten, P. Indicators of development: The search for a basic needs yardstick. *World Dev.* **1979**, *7*, 567–580. [CrossRef]

15. Sen, A.K. *Development as Freedom*; Oxford University Press: New York, NY, USA, 1999; p. 366.

16. Loring, P.; Gerlach, S.; Huntington, H. The new environmental security: Linking food, water and energy for integrative and diagnostic social-ecological research. *J. Agric. Food Syst. Community Dev.* **2013**, *3*, 55–61. [CrossRef]

17. Strasser, L.; Howells, M.; Alstad, T.; Rogner, H.; Welsch, M. Water-Food-Energy-Ecosystems Nexus: Reconciling Different Uses in Trans-boundary River Basins. In Proceedings of the an Informal Paper of the Second Meeting of the Task Force on the Water-Food-Energy-Ecosystem Nexus, Geneva, Switzerland, 8–9 September 2014.

18. Cariño, J.K. Integrated Water Resources Management: The Experience of the Laguna Lake Development Authority (LLDA), Philippines. In Proceedings of the First National Congress on Philippine Lakes, Southeast Asian Ministries of Education organization (SEAMEO)-Southeast Asian Resional Center for Graduate Study and Research in Agriculture (SEARCA), Los Baños, Laguna, Philippines, 25–28 November 2003.

19. Santos-Borja, L.C. Aquaculture Development and Management in Laguna de Bay. In Proceedings of the First Living Lakes African Regional Conference, Kisumu, Kenya, 27–30 October 2005.

20. The Current State of Aquaculture in Laguna de Bay. Available online: http://dirp4.pids.gov.ph/ris/dps/pidsdps0720.pdf (accessed on 26 May 2015).

21. Tan, R.L.; Alvaran, T.A.C.; Villamor, B.B.; Tan, I.M.A. Cost and return analysis of fishpen operation in Laguna de Bay and the economic implication of "zero fishpen policy". *J. Environ. Sci. Manag.* **2010**, *13*, 14–26.

22. Sullivan, C. Calculating a water poverty index. *World Dev.* **2002**, *30*, 1195–1210. [CrossRef]

23. Barrett, C.B. Measuring food insecurity. *Science* **2010**, *327*, 825–828. [CrossRef] [PubMed]

24. Designing Indicators of Long-Term Energy Supply Security. Available online: http://www.ecn.nl/docs/library/report/2004/c04007.pdf (accessed on 26 May 2015).

25. A Questionnaire Survey about Water and Food Sources at the Household Level. Available online: http://goo.gl/forms/WceqPbHspM (accessed on 3 March 2015).

26. Orencio, P.M.; Fujii, M. An Index to Determine Vulnerability of Communities in a Coastal Zone: A Case Study of Baler, Aurora, Philippines. *AMBIO J. Hum. Environ.* **2013**, *42*, 61–71. [CrossRef] [PubMed]

27. Orencio, P.M.; Endo, A.; Fujii, M.; Taniguchi, M. Using thresholds of severity to threats to and the resilience of human systems in measuring security. *Soc. Indic. J.* **2015**, in press.

28. Stokols, D.; Hall, K.L.; Moser, R.P.; Feng, A.; Misra, S.; Taylor, B.K. Cross-disciplinary team science initiatives: Research, training, and translation. In *The Oxford Handbook of Interdisciplinary*; Frodeman, R., Klein, J.T., Mitcham, C., Holbrook, J.B., Eds.; Oxford University Press: New York, NY, USA, 2010; pp. 471–493.

29. Defila, R.; di Giulio, A. Managing consensus in interdisciplinary teams. In *The Oxford Handbook of Interdisciplinary*; Frodeman, R., Klein, J.T., Mitcham, C., Holbrook, J.B., Eds.; Oxford University Press: New York, NY, USA, 2010; pp. 482–485.

30. Defila, R.; di Giulio, A.; Scheuermann, M. *Forschungsverbundmanagement. Handbuch für die Gestaltung Inter—Und Transdisziplärer Projekte*; Vdf Hochschulverlag an der ETH Zürich: Zurich, Germany, 2006; p. 118.

31. Mizoguchi, R. *Ontology Kougaku (Ontology Engineering)*; Ohmsha: Tokyo, Japan, 2005; p. 280. (In Japanese)

32. Mizoguchi, R. *Ontology Kougaku no Riron to Jissen (Theory and Practice of Ontology Engineering)*; Ohmsha: Tokyo, Japan, 2012; p. 248. (In Japanese)

33. Kumazawa, T.; Kozaki, K.; Matsui, T.; Saito, O.; Ohta, M.; Hara, K.; Uwasu, M.; Kimura, M.; Mizoguchi, R. Initial Design Process of the Sustainability Science Ontology for Knowledge-Sharing to Support Co-Deliberation. *Sustain. Sci.* **2014**, *9*, 173–192. [CrossRef]

34. Kumazawa, T.; Matsui, T. Description of social-ecological systems framework based on ontology engineering theory. In Proceedings of the 5th Workshop on the Ostrom Workshop (WOW5), Indiana, IN, USA, 18–21 June 2014.

35. United States Geological Survey (USGS). MODFLOW and Related Programs. Available online: http://water.usgs.gov/ogw/modflow/ (accessed on 3 March 2015).

36. United States Geological Survey (USGS). Pager: M 9.0, Near the East Coast of Honshu, Japan, 2011. Available online: http://earthquake.usgs.gov/earthquakes/pager/events/us/c0001xgp/#summary (accessed on 27 May 2015).

37. Topics Geo: Natural catastrophes 2011—Analyses, Assessments, Positions. 2012. Available online: http://www.munichre.com/site/corporate/get/documents_E-1152749425/mr/assetpool.shared/Documents/5_Touch/Natural%20Hazards/Publications/302--07225_en.pdf (accessed on 27 May 2015).

38. Cutter, S.L.; Mitchell, J.T.; Scott, M.S. Revealing the vulnerability of people and places: A case study of Georgetown County, South Carolina. *Ann. Assoc. Am. Geogr.* **2000**, *90*, 713–737. [CrossRef]

39. Boruff, B.J.; Emrich, C.; Cutter, S.L. Erosion hazard vulnerability of US coastal counties. *J. Coast. Res.* **2005**, *21*, 932–942. [CrossRef]

40. Adger, W.N. Vulnerability. *Glob. Environ. Chang.* **2006**, *16*, 268–281. [CrossRef]

41. Cutter, S.L.; Emrich, C.T. Moral hazard, social catastrophe: The changing face of vulnerability along the hurricane coasts. *Ann. Am. Acad. Political Soc. Sci.* **2006**, *604*, 102–112. [CrossRef]

42. Gallopin, G.C. Indicators and their use: Information for decision-making. In *Sustainability Indicators. Report of the Project on Indicators of Sustainable Development, Scope 58*; Moldan, B., Bilharz, S., Eds.; Wiley: Chichester, UK, 1997; pp. 13–27.

43. United Nations Economic Commission for Europe (UNECE). A proposed approach to assessing the water-food-energy-ecosystems nexus under the UNECE water convention. In Proceedings of the Discussion Paper of the First Meeting of the Task Force on the Water-Food-Energy-Ecosystems Nexus, Geneva, Switzerland, 8–9 April 2013.

44. MacGregor, C.; Fenton, M. Community values provide a mechanism for measuring sustainability in small rural communities in Northern Australia. In Proceedings of the Country Matters Conference Proceedings, Canberra, Australia, 20–21 May 1999; pp. 20–21.

45. Guillaume, J.H.A.; Kummu, M.; Eisner, S.; Varis, O. Transferable principles for managing the nexus: Lessons from historical global water modelling of central Asia. *Water* **2015**, *7*, 4200–4231. [CrossRef]

46. Keskinen, M.; Someth, P.; Salmivaara, A.; Kummu, M. Water-energy-food nexus in a transboundary river basin: The case of Tonle Sap Lake, Mekong River Basin. *Water* **2015**, *7*, 5416–5436. [CrossRef]

47. Taniguchi, M. Yusui (Spring water). In *Hito to Mizu II Hito to Seikatsu (Humans and Water II Humans and Lives)*; Akimichi, T., Komatsu, K., Nakamura, Y., Eds.; Bensei Publishing: Tokyo, Japan, 2009; pp. 79–103. (In Japanese)

48. Oita Prefectural Government. *Obtained Materials on the Areas for Licensed Fisheries in Beppu Bay 2013*; Oita Prefectural Government: Oita, Japan, 2013.

Article

A Methodology to Assess the Water Energy Food Ecosystems Nexus in Transboundary River Basins

Lucia de Strasser [1,*]**, Annukka Lipponen** [2,†]**, Mark Howells** [1]**, Stephen Stec** [3] **and Christian Bréthaut** [4]

[1] KTH Royal Institute of Technology, Stockholm SE-100 44, Sweden; mark.howells@desa.kth.se
[2] Environment Division, United Nations Economic Commission for Europe, Palais des Nations 8–14 avenue de la Paix, Geneva 10 CH-1211, Switzerland; annukka.lipponen@unece.org
[3] Central European University, Nador u. 9, Budapest 1051, Hungary; stecs@ceu.edu
[4] Institute for Environmental Sciences, University of Geneva, Boulevard Carl-Vogt 66, Geneva 1205, Switzerland; christian.brethaut@unige.ch
* Correspondence: lucia.destrasser@desa.kth.se
† The views expressed in this article are those of the authors and do not necessarily reflect the views of the United Nations Economic Commission for Europe or its Member States.

Academic Editors: Marko Keskinen, Shokhrukh Jalilov and Olli Varis
Received: 6 October 2015; Accepted: 11 January 2016; Published: 16 February 2016

Abstract: The "nexus" is a potentially very appropriate approach to enhance resource efficiency and good governance in transboundary basins. Until now, however, evidence has been confined to isolated case studies and the nexus approach remains largely undefined. The methodology presented in this paper, developed for preparing a series of nexus assessments of selected river basins under the Water Convention of the United Nations Economic Commission for Europe (UNECE), is a timely contribution to this ongoing debate. The nexus assessment of a transboundary basin has the objective of identifying trade-offs and impacts across sectors and countries and to propose possible policy measures and technical actions at national and transboundary levels to reduce intersectoral tensions. This is done jointly with policy makers and local experts. Compared to an Integrated Water Resource Management approach, the water energy food ecosystems nexus approach concurrently considers multiple sectors and their evolution. This offers the opportunity to better involve key economic sectors—energy and agriculture in particular—in the dialogue over transboundary water resource uses, protection and management.

Keywords: nexus; transboundary; methodology; participatory; water; energy; food; land; ecosystems

---

1. Introduction

The concept of regional, national and local integrated resource assessment and the links between resources and service supply chains have grown in understanding [1]. The term nexus has been used in a variety of contexts with the aim of advancing an understanding of how sectors are linked, and in turn to inform cross-sectoral governance coherence. On one hand, resources become scarcer as demand for them increases. Multiple uses of resources are increasingly at risk of becoming conflictual, undermining energy, water, food and environmental security [2–6]. On the other hand, the established "silos" approach to policy making (developing and implementing sectoral plans independently, without accounting for trade-offs and impacts across sectors) becomes more and more risky because spillover effects across sectoral policies become more expensive and unsustainable. In other words, the interlinkage (or "nexus") between sectors becomes stronger. This calls for coherent, responsible and consultative planning [7,8].

Despite its increasing popularity, there is no universal set of sectors to be analysed when the nexus is being studied. Depending on the context, the nexus framework has been used to include two or more sectors among energy, water, food, land, climate, environment, and ecosystems [9]. This lack of a clear definitions makes it difficult to establish what constitutes a good nexus analysis [10]. However, there is one characteristic that strongly defines it and makes it innovative, which is the shift from a sector- or resource-centric perspective to a multi-centric one [11]. At its simplest, the nexus is the complex of connections and interactions among water, food, energy, ecosystems and other related systems (or sectors) and the "nexus approach" to natural resource management takes into account these complex interactions, as such resembling the "multi-use sustained-yield" analytical framework applied to resources such as forests in the 1950s and 1960s.

In this paper, applying a "nexus approach" means taking into account the links and dynamics between resource systems to harmonize their outlook and management. The added value of a nexus approach compared to others (in particular, when it comes to water management, to the well-established Integrated Water Resource Management (IWRM), defined by the Global Water Partnership (GWP) as 'a process which promotes the coordinated development and management of water, land and related resources in order to maximize economic and social welfare in an equitable manner without compromising the sustainability of vital ecosystems' [12]) has been questioned [10,13]. However, traditional "integrated" approaches typically have limited analytical scope and often do not consider re-enforcing stresses or indirect links (such as, for example, climate change affecting water demand and in turn energy production) [14]. Therefore, a holistic approach that extends beyond IWRM can be recommended, for example, when assessing large infrastructure interventions where social, economic and environmental factors are of major importance [15].

A major criticism of the nexus approach is that the nexus concept itself is rooted in global considerations, such as increasing demands for water, energy and food, climate change and increased pressure on the environment (directly descending from the Gaia Hypothesis and "The Limits to Growth" [16] of the 1970s). Yet, little has been done to scale this understanding to a pragmatic local, national and regional planning approach [10].

The potential of using integrated assessments for the purpose of improving resource planning and management is becoming clearer, but more still needs to be done to make them respond to the actual needs of policy making [7]. Also, it has been observed that important aspects such as social factors can be overlooked in such assessments [17]. Indeed, an analyst aiming at understanding dynamics involving multiple sectors and resources should have the correct skill set to look beyond a purely physical systems analysis. At a governance level this can be essential: there may be overlaps in institutional mandates, lack of compatibility of geographical and political scales, differences in policy or enforcement culture, lack of consistency in regulations, or even power imbalances. Physical trade-offs, however, are a measurable result of sub-optimal management and provide evidence that intersectoral coordination is needed to improve policy coherence and resource efficiency [11].

From the perspective of the UNECE Convention on the Protection and Use of Transboundary Watercourses and International Lakes (Water Convention), which provides an overarching legal and institutional framework for transboundary water cooperation in the pan-European region, a multi-sector nexus approach might help improve transboundary cooperation on water resources. Not only does it explicitly involve key water users beyond drinking water supply, namely agriculture and energy production, as well as environment protection authorities, but by extending beyond the water management domain, it can encourage dialogue about broad benefits across sectors, potentially shifting focus from water allocation only. For instance, in the energy sector an understanding of water resource dynamics may be limited, hence energy planning could be improved with better consideration of water supply risks.

1.1. Intersectoral Policy Coherence in Transboundary Settings

Transboundary river or lake basins are extensive and affect many peoples' lives. In the world there are 263 such basins, covering almost half of the Earth's land surface, and some 40 per cent of the world's population lives within them [18]. Many transboundary issues that create friction between countries sharing the same water resources are intersectoral by nature. A typical example is given by the often conflicting water needs for hydropower production and irrigation purposes, both pulled by strong drivers such as employment, economic growth, energy and food security.

Recent work at country level, for example, in Mauritius [14] and California [19], indicates that intersectoral linkages (or nexi) are of material importance. Interestingly, sectoral policies can impose unintended consequences on other sectors even if there is no explicit competition for a single resource, because sectors are interlinked in a network [14]. Where feedbacks are better understood, sectoral policies can positively influence one another, reducing negative impacts or even generating co-benefits [20].

In transboundary settings, matters become even more complex [21]. The physical linkage of water makes riparian countries deeply interdependent, calling for policy coherence not only between sectors, but also across boundaries. For instance, if oil becomes too expensive in an upstream country, exporting/importing food may become unaffordable. This can cause an increase in subsistence agriculture and a less controlled water use, with potentially negative effects downstream. Such interdependency is accentuated when riparian countries are part of power pools where electricity is shared, or common markets for agricultural products.

1.2. The Nexus Work Under the UNECE Water Convention

Because of its potential in advancing transboundary cooperation, the "water-energy-food-ecosystems nexus" was selected as one of the thematic areas of work under the UNECE Water Convention for the 2013–2015 program of work [22]. A Transboundary River Basin Nexus Approach (TRBNA) methodology has been developed to support this work, which in practical terms involves carrying out a nexus assessment of selected basins. Results of this work have been published by the UNECE [23]. Up to now, the river basins assessed using this methodology are the following:

○ The Alazani/Ganykh shared by Azerbaijan and Georgia [24];
○ The Sava, shared by Bosnia and Herzegovina, Croatia, Montenegro, Serbia and Slovenia [25];
○ The Syr Darya, shared by Kazakhstan, Kyrgyzstan, Tajikistan and Uzbekistan [26];
○ The Isonzo/Soča, shared by Italy and Slovenia (not completed).

The nexus assessment of a basin aims at informing, supporting and promoting transboundary cooperation and assisting countries by:

○ Identifying interlinkages (trade-offs and impacts) across sectors and countries and incoherencies in governance;
○ Proposing actions to reduce negative impacts, minimise trade-offs and possibly take advantage of existing complementarities and win-win opportunities;
○ Providing evidence of benefits from improved cooperation at national and transboundary levels.

1.3. Scope of the Paper

The scope of this paper is to describe the TRBNA methodology and present findings from its application in three river basins: the Alazani/Ganykh, the Sava and the Syr Darya. The novelty of the presented methodology lies in the very fact that it aims at applying a pragmatic "nexus approach". It attempts to deal with the strong complexity linked with intersectoral analysis. Especially challenging is the case of transboundary settings, characterized by a diversity of stakeholders, the multiplicity of institutional settings and a variety of priority water needs.

By sharing key lessons learned from the UNECE nexus project, this paper aims at contributing to the current discussion on the nexus in general, and the practical value of applying a nexus approach in transboundary contexts in particular. After a step-by-step description of the methodology (Section 3) and the illustration of key lessons learned from the case studies (Section 4), the discussion part (Section 5) consists of a dissertation around the following questions:

○   How did using a nexus approach contribute to advancing the work of the UNECE?

○   What are the strengths and main limitations of this methodology to be taken into account for future applications?

2. Study Context

2.1. Cooperation and Benefits

Cooperation becomes crucial when resources become scarcer, costs are shifted across boundaries, or livelihoods and security of resources are threatened. It is therefore important to ensure that the management of shared water resources is both sustainable and coordinated across riparian countries, respecting international law. A case study on the Rhone demonstrated that increasing cooperation and involving a variety of actors in transboundary dialogue—principles of IWRM—do not guarantee, alone, a more coherent management of the river [27]. However, cooperation is also a necessary (although not sufficient) condition to maintain international relations, in turn necessary to establish agreements and share benefits "beyond the river" [28]. Good governance, including intersectoral coordination, participation of different stakeholders and the availability of transboundary legal frameworks, are important elements in putting the nexus approach into practice in transboundary basins [29].

Improved cooperation generates benefits that propagate across sectors at both national and transboundary level. Sharing water resources sometimes constrains the achievement of these objectives; their uncoordinated management can create tensions and undermine trust between countries, reducing opportunities for regional cooperation [28]. As a result, improved transboundary cooperation can greatly benefit riparian countries in many ways, also (but not only) in economic terms.

By broadening the perspective beyond water allocation, a nexus approach adds value to the IWRM approach in discussing the benefits of cooperation. It was observed that in transboundary contexts a dialogue focused on the value of water, the role of ecosystems and benefits of cooperation is less likely to get stuck on disputes over water resource allocation [30]. Moreover, increasing intersectoral coordination within and between riparian countries (including water management, energy and agriculture) opens opportunities for generating benefits, and also synergies, above the basin scale [15]. The nexus approach allows for a multi-sectoral dialogue that is in principle broader than the dialogue promoted with IWRM and that aims at discussing synergies out of the water management domain and beyond the basin scale.

2.2. Developing the TRBNA Methodology

The objectives of the Water Convention—namely promoting cooperation in the management of transboundary waters—influenced the development of the TRBNA methodology in many ways. First of all, despite the multi-centric nature of the nexus and the fact that it can be applied at different scales, water holds in this context an undeniable importance over the other resources, being the natural vector of transboundary impacts and the subject of transboundary cooperation. This results in the methodology having a certain emphasis on water as the entry point to the nexus (see Sections 3.1.5 and 5.2) and focusing on the basin scale like IWRM (even though specific components of the nexus are analysed at a different scales—notably the energy system, analysed at regional and national level). Nevertheless, since the nexus emerged as a concept rooted in the concepts of water, energy and food security, by nature it goes a step further to IWRM to improve multi-sectoral coordination and integration [31].

A high level of engagement with national administrations in the assessment resulted from the fact that countries are the constituency of the Water Convention (as Parties) and the UNECE (as Member States). The UNECE has engaged officially with the countries sharing the basins, while through the Water Convention's governing bodies the countries have contributed to shaping the assessment process. Stakeholder participation was therefore a pillar for the development of the methodology, being not only a useful tool to better analyse the nexus, but also a necessary step to officially validate results and ensure the policy relevance of the assessments.

In designing the participatory process of the TRBNA, high importance was given to the joint identifyication of the benefits of cooperation. This aims at ensuring that the assessment is relevant for the countries, in the sense that it takes into account national interests. Despite their differences, all countries aim at achieving or improving security in the supply of resources (social stability and equity) and economic stability or growth. Nevertheless, the pursuit of national interests should not prevent the use of shared watercourses from remaining equitable as well as reasonable, and a country's development should not cause significant harm to co-riparian countries, in line with the key obligations of the Water Convention, which are also the main principles of international water law [32].

By design, the nexus assessment goes only as far as to propose beneficial interventions and illustrate how they might improve inter-sectoral transboundary management Other activities can support it or complement it: risk assessments, cost and benefit analyses and integrated modeling efforts can build on its outcomes.

In its final form the methodology synthesizes elements from different approaches, notably the basin approach (inherent to the IWRM) [12]; the Climate, Land use, Energy and Water strategies (CLEWs) framework [33]; the nexus approach developed in the Food and Agriculture Organization of the United Nations [34] and a proposed approach to assess the governance aspects of the nexus [35]. The latter builds on the analytical framework proposed for the project "GOUVRHONE, Governance of the Rhône River from Lake Léman to Lyon", led at the University of Geneva [27], which has been inspired by the Institutional Resource Regime framework [36,37] developed for an analysis of regulatory frameworks through public policies, property rights and the interlinkages between these two legal corpus.

Attention was paid to the challenges of putting the nexus into practice at basin level and finding a balance between the technical and social approaches to the nexus. Consequently, key drivers to the development of the TRBNA were (1) the need to work on two parallel and coordinated lines of analysis: governance- and resource-based; and (2) the aim of pragmatically diagnose inconsistencies and suggest beneficial actions for consideration by the countries concerned, both in terms of administrative and infrastructural interventions.

3. Methodology. The Nexus Assessment of a Transboundary Basin

In order to systematically carry out nexus assessments, an ad-hoc terminology was developed to define nexus related concepts. This was particularly important to clarify differences with other approaches such as IWRM (to which most stakeholders involved in the participatory process are more familiar with) but also to be able to maintain a margin of comparability between the assessments to facilitate sharing of experiences.

As an example, "nexus issues" and "nexus solutions" were defined, respectively, as a problematic situation that affects more than one sector and an intervention that would benefit more than one sector (including interventions that reduce the pressure on ecosystems and the environment at large). Because of the transboundary focus of the assessments, both should have a transboundary dimension, involving or impacting more than one country. It should be noted that a "nexus solution" that affects two sectors may create a new "nexus issue" affecting a third sector. Nexus solutions can take various forms but in general terms they should contribute to improving overall resource use efficiency, improving resilience of socio-economic activities to external shocks (including climate change) and strengthening policy coherence (thereby minimizing negative externalities). A complete glossary can be found in the final publication of the project [23].

3.1. A Six Steps Process

The TRBNA methodology consists of six steps and this sequence is illustrated in Figure 1. In steps 1–3 the analysts prepare a desk study of the basin, which will be used as basis for steps 4–6, where stakeholders are actively involved and a more in-depth analysis of nexus interlinkages is made. The key instruments utilized at various steps of the process are described below.

The diversity of the basins to be assessed requires the methodology to be flexible enough to allow the analysts to consider a wide range of interlinkages and conditions, applying at the same time a simple and consistent framework. To allow for this flexibility, the assessment process has been designed to zoom-in from a broad socioeconomic diagnosis of the basin to its specific intersectoral issues and existing opportunities to mitigate them.

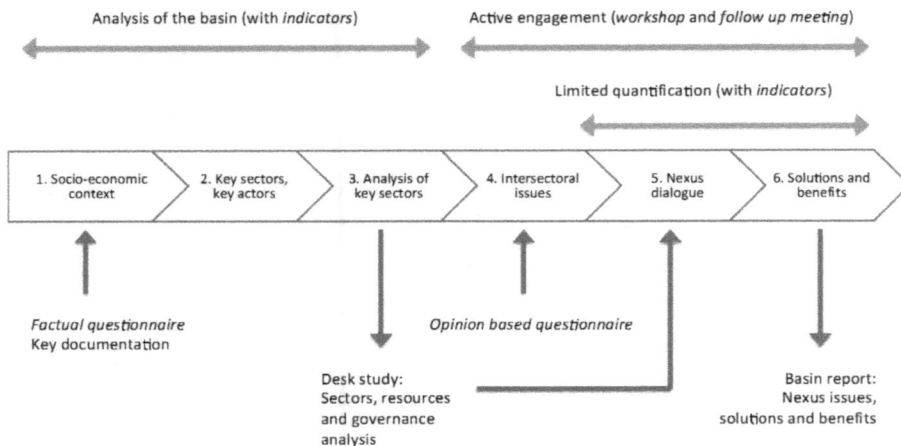

Figure 1. Schematic of the six steps with inputs and outputs.

○ Indicators—Steps 1,3,4,6 (see Appendix 1):

Three groups of indicators are used at different stages to substantiate the analysis of the basin:

(1) statistical and spatial screening indicators at country and basin levels;
(2) perspective indicators from the different sectors and countries (see "Opinion based questionnaire");
(3) basin-specific indicators of various kinds, to support the study of interlinkages.

○ Factual questionnaire—Step 1

Distributed to the participants to the workshop and local experts to collect basic information on the state and uses of resources as well as issues in the areas of water, energy, food/land and ecosystems.

○ Workshop—Steps 4,5,6 (see Appendix 2)

It includes several sessions where participants engage in the nexus assessment process directly, by discussing intersectoral and transboundary issues. On top of providing input to the assessment, this gives them ownership of the process and allows for direct confrontation of various sectors. Ad hoc material, reported in Appendix 3, was prepared to facilitate the discussion in working groups during the workshops. Sector-centric diagrams were used to facilitate discussion in sectoral groups (as part of Step 4); nexus diagrams were used to facilitate cross-sectoral dialogue (as part of Step 5) (see Figures A1 and A2).

○ Opinion based questionnaire—Step 4

Distributed, filled in and collected at the beginning of the participatory workshop to gather the opinions of stakeholders involved in the process and compare the different perspectives between sectors—water, energy, food/land and ecosystems—and countries on various issues. Issues that everyone agrees on and differences in perception are important nexus indicators.

○ Follow-up meeting—Step 6

Discussion with authorities on how the findings and solutions included in the assessment relate to policies or programs in the countries, and what could be done to address the identified intersectoral issues. It is a mean of verifying the real relevance of the assessment for policy development.

3.1.1. Step 1—Socio-Economic and Geographical Context

Step 1 aims at characterizing the basin conditions and its economic context and determining the level of dependency of riparian countries on the basin's resources.

(a) *State of energy, food, water and environmental security in the basin.* Emphasis is given to the needs of local populations living in the basin and in its riparian countries. The levels of poverty are established, important livelihoods and social issues understood. Access to and affordability of resources are the primary information to be collected, together with information on environmental issues in the basin. It should be noted that establishing a precise definition of "security" in each nexus areas (water, energy, food/land and environment/ecosystems) was not the focus of this project.

(b) *The relations that exist within the region, the basin and its riparian countries.* The basin is first of all understood in terms of its geographical and geopolitical aspects. The basin is most probably linked to national development plans in a way that affects the use of its resources. The basin may be valuable for specific economic activities taking place in its area, it may be an important transit route, its natural resources may be exploited for the benefit of external actors (e.g., transfer of water, mineral mining, *etc.*). This connection may be understood, at least in the cases of energy and agricultural production, in terms of dependency. For example, it may be interesting to determine how much each riparian country relies on energy produced using water from the basin.

(c) *Main strategic goals, development policies and challenges.* These exist at different scales: basin, country and region. A key strategy for promoting new technologies in irrigation, for example, will translate into changing water use, as will the goal of providing every household with safe drinking water. Strategic goals of riparian countries may affect the resources and population living in the basin also indirectly. For instance, a country can be investing less and less in agriculture and a basin with a high share of population employed in agriculture may experience migration or social change.

The necessary information may be derived from:

○ The factual questionnaire compiled by focal points from each riparian country
○ Key documentation on the basin and region such as socio-economic reports and environmental reviews. In this process, a basic set of reports was used, such as reports by various UN organizations (e.g., UNECE, FAO) and other international organizations (e.g., World Bank, Global Water Partnership) were typically included in this list, together with River Basin Management Plans)
○ Screening indicators at national and basin level. In this process, indicators from World Bank and FAO—Aquastat databases were widely used, however no fixed set of screening indicators was defined (see Appendix 1 and Section 5.2).

3.1.2. Step 2—Identification of Key Sectors and Key Actors

This step aims at identifying the key sectors to be included in the nexus assessment and the key actors to be involved in the assessment process (*i.e.*, workshop and further consultation).

(a) *Identifying the key sectors.* These are determined on the basis of the findings of Step 1 as the ones that play a major role in the basin's socio-economy and environmental protection. In general terms, sectors are resource users. They can be productive (e.g., industry) or just consumptive (e.g., households) (see Section 3.1.4. for a clarification on how the concepts of "sectors" and "resources" were interpreted). The water supply, energy production, agriculture and environmental protection sectors can be considered as a core set of key sectors, however some sub-sectors will be more relevant than others. For example, hydropower as a subsector of energy production, crop production (or even the production of a specific crop) as a subsector of agriculture and so on.

(b) *Identifying key organizations and other actors.* By taking active part in the workshops and in the consultation process, these stakeholders will not only share their knowledge but also offer an opportunity to include the findings from the assessment into actual plans and programs, playing a key role in the nexus assessment. Their identification should be informed or validated by the mapping of actors. It should be noted that in our terminology 'organizations' indicates formal actors such as River Basin Organizations and ministries; 'other actors' could include even individuals who have knowledge on and/or influence over the nexus and the study context.

3.1.3. Step 3—Analysis of Key Sectors

Understanding how the sectors use resources, their socio-economic value and what are the rules, plans and regulations associated with them is the objective of this step. The sectors considered are the ones defined in Step 2. The outcomes of Steps 1–3 constitute the core of the desk study, which informs the discussion on interlinkages and feeds into Steps 4–6.

(a) *Sectors and resource flows analysis*

Sectors need inputs in certain amounts and quality. For example, water is required in good quality for direct uses but also energy needs to be safe, available in sufficient quantity and clean. Similarly, different land types can be available for different uses. This drives various demands of resources that is satisfied by extracting and processing them. Mapping these resource flows is the start of an integrated system analysis, which could entail the use of sectoral models of the basin and countries. The level of detail that can be reached in this analysis depends on many factors, among which the most constraining are: data and time availability, number and complexity of sectors to include in the system, size of the basin. The minimum output would be a sketched schematic of resource needs by the different sectors, their outputs and impacts, at least semi-quantitatively or with indicative orders of magnitude.

(b) *Governance Analysis*

We define "governance" as "a system of responsibility and accountability involving formal and informal institutions that builds trust and capacity to cooperate in policymaking, decision-making and implementation of measures".

Conducting a governance analysis helps to gain a better understanding of the context in which the different sectors of activity operate. This multi-level context is composed by different elements. It includes formal rules that depend on public and private law; it entails varying consideration regarding the structure and mandates of public administration (such as varying degrees of centralized or self-organized configurations) and different combinations of actors and interlinkages that rely on formal and informal agreements. A governance analysis helps to generate a better understanding of the extent to which conditions are being met in order to achieve coherent (and sustainable) integration of different sectors (consumers) of resources and to identify its regulatory capacities at different levels.

In line with the objectives of the assessment, the governance analysis includes key sectors and it considers different scales: regional, national and local. Focuses of the governance analysis are the following aspects [23]:

1. Policy framework—strategies and other policy documents, instruments, *etc.*;
2. Legal and regulatory framework—rules and regulations;
3. Organizations and actors—mandates, responsibilities, administration.

While the basin is the appropriate level for consideration of traditional water resources management issues, other geographical scales are appropriate in relation to other sectors. For example, energy security is usually determined according to strict political boundaries. Moreover, cultures of decision-making and administration, and relationships among stakeholders may also be quite different from sector to sector, making comparisons difficult.

These (a) and (b) lines of work are complementary. On one hand, the resource analysis establishes availability and quality of the resources available, as well as the mechanisms (demands, supply, trade, *etc.*) that link them to their uses. On the other, the governance analysis understands how actors and rules determine the management of those resources.

Ad-hoc material to facilitate the discussion at the workshop is prepared as part of the desk study (see Appendix 3). The spatial dimension is important for assessing whether and where resource uses are less compatible. Hence, it may be useful to prepare, additionally to this material, basin maps of the basin displaying key aspects that will be discussed by the sectoral groups (see Step 4) such as protected areas, infrastructure, water bodies, *etc*. At this stage it is also possible to define a set of important drivers, such as policy directions, socio-economic trends and climatic trends. It will be useful to advance a preliminary list of these before the workshop, to inspire and facilitate dialogue.

3.1.4. Step 4—Intersectoral Issues

Step 4 takes place in the participatory workshop, which structure is synthesized in Appendix 2. This step is key in the participatory process because it defines how each sector will interface the others in the nexus dialogue. Here, intersectoral issues are explored from sectoral perspectives (before being jointly discussed and prioritized in Step 5). Participants are divided into thematic groups—water, energy, food/land and ecosystems—according to their expertise or area of interest. They are asked to discuss interlinkages from a sectoral perspective in a sort of brainstorming exercise, using sector-centered nexus diagrams and thematic maps of the basin (see Appendix 3). Information from the desk study is used to inform and facilitate the discussion (see Step 3). Key policies, sectoral plans and data sources are presented and validated by local actors, who also provide expert judgment for prioritization of issues.

An opinion based questionnaire is used to collect the different perceptions of sectors and countries. This contains statements that participants have to rank in terms of importance and personal perception. The questionnaire is anonymous but respondents have to specify their country and area of expertise (water, energy, food/land or ecosystems) so that comparisons between groups can be made. For example, one country could perceive energy security as a top concern while another not at all, or water resources could be described as "scarce" in one country but not in another. Answers can be grouped by country and by sector and compared to measure to what extent the groups agree or disagree.

In this work, the decision of maintaining an ambiguous definition of "sectors" was deliberate. The diagrams used to support intersectoral dialogue do not display key sectors, that we have seen can be various and depend on the basin. They display the four "components" of the nexus: "water", "energy", "food/land" and "ecosystems", which can be interpreted either as resources or sectors depending on the context of discussion. A limitation of using this diagram to illustrate intersectoral issues is that it emphasizes the physical aspect of resource flows across sectors over other important aspects (e.g., economic, legislative, *etc.*). Moreover, sectors consuming the different resources are not spelled out. At the same time, using a simple and intuitive diagram (that workshop participants are invited to modify if needed) allows for a less constrained dialogue. For the purpose of the assessment, it is important that all pressing intersectoral issues are reflected but there are no "wrong" arrows. For instance, industry could be part of "energy" (*i.e.*, as an extension of the energy industry) or "land" (*i.e.*, as a type of land use), depending on which option participants feel more comfortable with and by the type of industry discussed.

3.1.5. Step 5—Nexus Dialogue

Step 5 can be considered the core of the nexus assessment because it is the moment where intersectoral issues are discussed having all concerned sectors around the table. A shared understanding of the nexus is built on the basis of (1) an agreed picture of the basin conditions, national priorities (in terms of sectors or economic development) and environmental concerns and (2) sectoral perspectives on pressing intersectoral issues.

The interlinkages identified in Step 4 are jointly prioritized and combined into thematic "nexus storylines". Depending on the time available and the number of participants, this can be done in mixed groups or in an interactive plenary session. Typically, the storylines will evolve around the topics of water availability and water quality—in line with the fact that, in transboundary contexts, water is the natural entry point to the nexus dialogue.

Interlinkages (such as multiple uses of resources, negative impacts, trade-offs and dependencies between sectors) are discussed together with the existing obstacles to overcome them, to establish a shared understanding of intersectoral challenges—e.g., diverging objectives and priorities for development, gaps/overlaps of responsibilities and mandates, *etc.* Next, the relevant future tendencies (climate change, socio-economic trends) are identified jointly with participants and the effects that these will have on intersectoral issues are discussed.

3.1.6. Step 6—Solutions and Benefits

Following the discussion on intersectoral issues, possible solutions are discussed. A definition was made to limit candidate solutions, namely that they have to benefit at least two sectors and have a clear transboundary dimension. They can be of two kinds:

(a) Synergetic: when two or more sectors actually cooperate on actions and projects that create multiple benefits.

(b) Sectoral: when the action of one sector has side benefits on other sectors or at least minimizes the negative impact on other sectors.

Technical solutions as well as policy interventions are considered. To the first group belong infrastructural and operational interventions, technological innovation, *etc.* To the second, a broad range of potential interventions, from cooperation agreements to communications, implementation of economic instruments, change to existing policies or development of new ones, institutional arrangements, change in regulation, *etc.*

After the workshop, the analysts will set time to quantify, to the extent possible, the identified interlinkages and benefits associated with the discussed solutions. Quantification is made mainly to illustrate the importance of some of the identified priority issues and give an idea of what a possible fully integrated resource assessment of the basin could look at. The use of integrated modelling can help at this stage, for instance to estimate future water needs coming from the development of new hydropower, the impact of energy efficiency measures on water use and availability for ecosystems, the impact of food trade on domestic water consumption, and so on. However, the extent of the analysis basically depends on the availability of resources and the interest of involved policy makers.

The nexus assessments concludes with "potential beneficial actions", rather than "recommendations", due to the fact that no proper evaluation of the actions is made at this stage. Depending on the type of storylines and proposals emerging from the nexus assessment, follow-up analytical exercises could be set up to study the applicability of solutions, which can include risk assessments, cost and benefit analyses, integrated modelling of (climatic, socio-economic) scenarios, action planning, policies and plans for stakeholder engagement and other governance aspects.

A follow-up meeting with key stakeholders is needed to make sure that solutions are translated into feasible actions, ideally linked to actual policies or projects on the agenda of national governments or basin organizations. In the course of this meeting the results of the nexus assessment are presented and discussed. These results should point clearly at beneficial actions and benefits that have been identified.

4. Results

For the purpose of this paper some examples have been extracted from the basin assessments, both to illustrate the practical application of each step of the methodology and to provide a point of reference for the key lessons learned.

Examples from the application of Steps 1 and 2 are illustrated respectively in Appendix 4: "Highlights from the socio-economic analysis" and in Appendix 5: "Map of key organizations and actors". They contain comparable information from the three basin: Alazani/Ganykh, Sava and Syr Darya. Step 3 is illustrated with the "Analysis of the energy sector in the Sava basin" (Appendix 6). Examples relative to Step 4 and 5 can be found respectively in Appendix 7: "Intersectoral issues in the Syr Darya River Basin" and Appendix 8: "Nexus dialogue in the Alazani/Ganykh basin". Finally, Step 6 is illustrated in Appendix 9 with an example of "Solution and benefits from the Sava River Basin".

Lessons learned include a variety of considerations on the implementational aspects of the methodology. They are presented following the same step-by-step structure.

4.1. Analysis of the Basin—Steps 1 to 3

Step 1 helps to roughly establish key issues and concerns to be kept in mind during the whole assessment process (see Appendix 4). Resource security issues, strategies, policies and interests on the basin's resource base will either be the core of discussion in the participatory phase or hidden elements that will influence it. As an example, energy security issues and electricity capacity expansion are recurring elements, but they can play different roles in the nexus. In the Syr Darya energy insecurity is a clear driver to hydropower use and expansion (affecting directly seasonal water availability for irrigation); in the Sava, hydropower expansion is strategically relevant because it contributes to reach targets for reducing emissions and incrementing the share of renewables but creates environmental concerns in some areas. In the Alazani/Ganykh, despite strong national interest in expanding hydropower capacity, the particular geo-morphology of the river limits its development there. Energy insecurity in this case manifests itself indirectly, with the reliance of rural communities on fuel wood, contributing to deforestation and loss of forest related ecosystem services.

It is challenging to determine which sectors should be part of the nexus assessment at the early stage of Step 2. However, some initial boundaries need to be set up, not only to move on with the sectoral analysis but also to be able to communicate clearly to the stakeholders involved which sectors are considered as being "inside" the basin-specific, local nexus. These boundaries may be eventually revised and adjusted, after the workshop. In the UNECE nexus assessments, at least water, energy, ecosystems and agriculture were always considered as key sectors. Agriculture was often limited to food production (crops, livestock and fishing), while trade was taken into account only if important interlinkages would arise (Syr Darya). Other aspects of land use such as flood control (associated with the component "food/land" of the nexus) were also considered when relevant (Alazani/Ganykh, Sava).

It is also easily the case that too many actors are identified as relevant for the assessment. It is not always feasible to involve all of them in the process, and at the same time this would not necessarily guarantee a better outcome. In this case, priority should be given to ensure diversity and balance across sectors and countries. It is assumed that the set of stakeholders involved in the participatory process is sufficiently representative of all relevant sectors and interests. However, in practice, the choice of stakeholders can be influenced by many factors. In this project, the group of stakeholders involved in the process emphasized public administration and was complemented by a limited involvement of operators/state companies, civil society and academia (see Appendix 5). In terms of expertise, water was heavily represented. This should not be interpreted to infer inadequacy of the nexus approach to broaden stakeholder involvement, but rather as a feature of this application of the methodology in this particular framework. In fact, stakeholders should ideally be selected after having rigorously identified key actors (Step 2) but commonly the mapping of actors was completed after the workshop and time constraints in the process resulted in over-reliance on established networks. The application of the methodology revealed also that further refinements are necessary in order to be able to more fully assess the range of characteristics of stakeholder engagement, including the level of self-organization of stakeholder communities, their legitimacy and representativeness, the degree of cooperation, number and quality of opportunities for engagement, and access to review procedures, to name just a few.

The analysis of sectors and resource flows of Step 3 required to go beyond the mere account of resources availability, production and demands. For example, while the waters of the Syr Darya aliment extensive irrigation schemes in semi-arid areas, irrigation in the Sava is very limited, constituting less than the 1% of total water use in the basin. This does not mean, however, that the agricultural sector in the Sava is less relevant from a nexus perspective. Employing 5% to 10% of the population in the region (*i.e.*, all countries), the agro-industry is an important economic sector. Considering the low irrigation capacity and predictions showing a trend towards longer and hotter summers, the resilience of the agricultural sector to climate change and the future development of water demands for agriculture become important areas of investigation.

The experience from the governance analysis of the basins shows that comparisons across basins or countries need to be made with caution. Formal cooperation frameworks between the countries may vary, as well as their coverage in terms of sectors and issues. However, their presence does not necessarily translate into closer cooperation: despite the existence of various inter-state institutions in Central Asia, cooperation in natural resources management and trade remains difficult. Moreover, there are national but also subnational differences: for example, Slovenia does not have regional administration, while Bosnia-Herzegovina, mostly for historical reasons, has a special entity level with its own authorities that enjoys a high degree of autonomy.

It should be kept in mind that governance with respect to the nexus is not the same as governance in a sectoral context. In general, and sector-specific terms, governance has been studied extensively and standards for improving governance have advanced steadily in various fields, even if at different speeds. But governance requires a specific context, whether it be processes of policymaking and decision-making, or implementation and financing. Complex, multi-sectoral, multi-use frameworks are relatively undeveloped, amorphous and ad-hoc. Decision-making involving trade-offs between sectors are usually at a high political level. Consequently, the kinds of platforms, institutional arrangements, and development of practice over time that are characteristic of sectoral processes are largely absent in a nexus context. The most that can be done is to assess parallel governance contexts sector by sector (as can be seen in Appendix 6) and to propose governance principles in connection with nexus analysis as it evolves.

Lastly, an important aspect of a governance assessment is to understand whether there are important "undercurrents" with respect to specific sectors and uses. A well-functioning governance system will ensure transparency and help to resolve conflicts within a sector. If there is an imbalance in governance across sectors, the full range of interests and values will be represented to a different extent, and interactions through a nexus process might reveal weaknesses in the perceived consensus in a sector with relatively poor governance. Governance in the "weaker" sector may be improved thereby, but it should also be recognized that existing power structures may not welcome such changes.

4.2. Active Engagement—Steps 4 to 6

Approaching intersectoral issues starting from sectoral perspectives, as part of Step 4, ensures that all components of the nexus manage to contribute to the subsequent nexus dialogue (see Appendix 7). In particular, a challenge is to go beyond the "security nexus" of energy, food and water and put the fourth dimension of "ecosystems" on the same level. When discussing development objectives, it is sometimes difficult to keep environmental priorities in the discussion. This is because environmental needs are commonly seen as a constraint to the expansion of "productive" sectors. Discussing ecosystem services proved to be quite successful to shift this view: interlinkages with other sectors not only include resource needs and the negative impact of other sectors on the environment, but also the positive contribution of ecosystem services. So, for example, in the Alazani/Ganykh, the water retention and land stabilizing service provided by the forest is considered, in the final nexus dialogue, as a key interlinkage that bridges energy, land and water policies (see Appendix 8).

The nexus dialogue of Step 5 gave very different results from case to case. In the case of the Alazani/Ganykh intersectoral linkages combined clearly in a storyline with clear causality links (see Figure A4). Although non-comprehensive of all relevant intersectoral elements (for example, renewable energy installations in the agricultural sector or water transfer to Baku), this collectively developed picture includes important indirect linkages that provided new important insights. In the Sava and Syr Darya, the nexus dialogue did not produce equally articulated storylines, but it served the purpose of prioritizing the most pressing intersectoral issues.

The search for "nexus solutions" in Step 6 varied according to the fact that riparian countries in each basin are at different stages of transboundary cooperation with each other. In the Alazani/Ganykh, Georgia and Azerbaijan are in the process of negotiating a bilateral agreement on the Kura/Ara(k)s river—of which the Alazani/Ganykh is a tributary [38] and are both developing new legislation on water management. In the workshop, participants were highly engaged in the discussion on nexus solutions to put forward ideas for further work that would contribute to these efforts.

In the Sava basin, the countries were already engaged in transboundary cooperation on a variety of topics within the working program and mandate of the International Sava River Basin Commission (ISRBC). Further integration of water policy with other policies, as well as further dialogue with key sectoral stakeholders, have been set in the Strategy on Implementation of the Framework Agreement of the Sava River Basin as specific objectives in the field of river basin management [39].The nexus assessment was motivated by this wish to broaden the existing engagement of stakeholders, in particular by better involving the agricultural and energy sectors in the dialogue over water management at basin level.

In the Syr Darya, where improving cooperation requires first its restoration, it was necessary to identify solutions that could be taken at national level first. National level benefits were clearly spelled out together with transboundary ones and emphasis was put on national goals that could be pursued without compromising transboundary relations, or even helping its recovery [23].

Follow-up meetings, not foreseen in the initial stages of the methodology development, proved to be a highly valuable addition. The one for the Sava assessment was organized at basin level in May 2015. This meeting allowed discussing the findings across sectors at transboundary basin level. Moreover, modelling efforts initiated during the assessment (the results of which can be found in the final publication of the project [23]) were presented to the countries to discuss opportunities to use them in follow-up actions. In the case of Alazani/Ganykh and Syr Darya, results were presented for discussion to stakeholders in country level meetings in the context of European Union (EU) Water Initiative's National Policy Dialogues. Despite the use of some modeling tools, integrated modeling is not a necessary step of the TRBNA but it is valuable in these meetings to illustrate the potential of taking action in a coordinated manner (see, for example, the case of Sava in Appendix 9). In order to validate results, it was particularly important to prioritize the use of official data and validate all assumptions with local experts.

The nexus assessment of a basin will inevitably reflect the issues and opportunities identified during the course of the participatory process, potentially overlooking or paying only cursorily attention to other important aspects. The role of facilitators at the workshop can be key in ensuring that no major intersectoral issues are neglected; however, the choice of focus of the assessment will ultimately be made by the stakeholders involved, potentially reflecting a deliberate intention to "not discuss" a certain topic or lack of expertise in a certain area within the group.

Finally, it should be noted that sectoral and national interests remain strong even when intersectoral and transboundary understanding is improved. Application of the nexus approach highlights the differences in governance across sectors. The quality of governance can therefore be compared, particularly at the points where various decision-making processes interact. Taking into account such differences in a transparent manner with stakeholder involvement could lead to policy responses to achieve greater balancing of interests in complex decision-making. This may require additional steps in the process, but they should be carefully designed and tested.

5. Discussion

5.1. Value of a Nexus Approach for Advancing Transboundary Cooperation in the Framework of the Water Convention

The development of this methodology and its various applications have involved a number of stakeholders with different views and expectations on what a nexus assessment should focus on and include as well how it should support policy making and improve coherence between sectoral policies. However, this methodology has been developed for the Water Convention, and the nexus assessments were carried out to support transboundary, intersectoral dialogue and to inform policy-making. Considering the actual contribution to the UNECE's work is therefore appropriate, even though the final beneficiaries are the countries.

The nexus assessments build on previous work under the Water Convention, namely the Second Assessment of Transboundary River, Lakes and Groundwaters [40]. This was a comprehensive description of transboundary water bodies and major transboundary issues in the European and Asian parts of the UNECE region. The nexus work is a step forward to discussing transboundary issues and opportunities of cooperation. The effort was parallel to another work item under the Water Convention: "Quantifying the benefits of cooperation" [22,41]. These two mutually benefited each other: on one hand, many benefits of cooperation can be found by applying a nexus approach; on the other, recognizing benefits of cooperation is valuable to motivate engaging into a nexus dialogue.

The response of the countries involved in the assessments was positive. They reviewed favorably the nexus assessment work: the methodology as well as the general conclusions and recommendations were endorsed by the seventh session of the Meeting of the Parties to the Water Convention (Budapest, Hungary, 17–19 November 2015). The Parties also decided that the nexus assessment will be continued as a part of the work programme 2016–2018 under the Water Convention [42]. Although the development of an assessment framework was among the most relevant achievements of the project, the assessments provided for a joint identification and analysis of the main intersectoral issues with the sector authorities of all riparian countries, and for a structured dialogue on them, leading towards discussion about possible improvements to the current resource management and transboundary cooperation. This process could contribute to setting common objectives, reviewing and possibly expanding scope of cooperation or institutional mandates, provide evidence of need for further cooperation, *etc.*

Ultimately, the TRBNA was useful to advance cooperation because it offered the opportunity to identify issues and opportunities jointly with national key actors across sectors and international partners. Even though many solutions are not new, the multiple sectors' participation allows for potentially new input, and the nexus perspective allows for a broader scope of considering and analysing negative and positive impacts as well as measures that could be taken.

In the TRBNA, the implementational aspects of nexus solutions was quite limited and only introduced at the end of the process (Step 6) and suggested as follow-up work. If possible, the thinking and dialogue should be prolonged to explore who (which sector, organization, *etc.*) is in a position to address the potential solutions identified and what concrete actions could be undertaken by which actor. Actions could be incorporated into ongoing or planned initiatives to support policy processes in the countries. For instance, in some basins the riparian countries are part of the European Union (EU) Water Initiative's National Policy Dialogues or there are regional organizations such as basin organizations or other joint bodies, possibly with multiple-sector representation that could provide a framework for identification of beneficial future activities. However, further developing the methodology in this direction will be essential to improve the applicability of results.

5.2. Strengths and Limitations of the TRBNA Methodology

Findings from three nexus assessments highlight several strengths and limitations of the TRBNA methodology identified against the aim of advancing transboundary cooperation. These can help assessing its suitability to further applications.

Strengths:

○   Using a highly participatory approach. This allows to focus on actual issues and priorities as well as to validate results from the sectoral and intersectoral analyses.

○   Searching for opportunities of cooperation beyond the water domain. This helps discussing the direct and indirect beneficial effects brought by cooperating in other areas (e.g., trading or establishing common objectives at regional level) to potentially broaden consideration and involvement of different sectors and interests in the current water cooperation frameworks.

○   Being flexible and adaptable. Focusing more on the "local nexus" by, for instance, avoiding a prescriptive use of indicators, it allows for the consideration of diverse intersectoral issues and cooperative options.

○   Substituting direct transboundary confrontation with intersectoral (nexus) dialogue. In contexts where transboundary dialogue is politically sensitive, this may be very useful if not necessary.

○   Using a resource-flow approach and a governance analyses in parallel. These two reinforce each other, respectively by providing evidence of physical trade-offs arising from the multiple use of finite resources and by identifying incoherences in the definition of policies or gaps in the institutional and legal frameworks.

Limitations:

○   Using sometimes ambiguous definitions and inconsistent indicators. While useful to adapt to different understandings and circumstances, this can create confusion when it comes to comparing results across basins.

○   The assessment approach as defined does not address a number of aspects that can be important, for example, financial constraints related to the applicability of solutions, administrative cultures and power imbalances. While already the scoping nature of the assessments and the resources available limited what could be covered, also the institutional set-up and priorities affected what was considered appropriate and relevant to include. Being outlined at a fairly general level, some solutions may be perceived as incomplete or even unclear by sectoral experts involved in the process.

○   Over-emphasizing and over-representing water over the other sectors and resources. This causes some important interlinkages to be discussed less, namely food/land-energy, energy-ecosystems and food/land-ecosystems.

Some aspects of the methodology could be improved by amending the existing structure with feedbacks from future case studies or by adapting it to the needs of the basin to be assessed. For instance, with regard to the above limitations:

○   Establishing a glossary of terms and a minimum set of screening indicators on the basis of further consultation with experts in intersectoral issues.

○   Extending Step 6 to include a better description of what a future analysis of each solution would entail.

○   The imbalance of water could be improved by relying less on established networks and carefully selecting key actors (*i.e.*, applying more rigorously Step 2) or at least by better involving those actors in the review of the assessments. Also, more robust methodologies for the governance aspects of the assessment could be developed to better take into account differences among sectors.

Finally, it should be noted that maintaining a critical approach to its application, the TRBNA methodology can be used for any transboundary basin and aquifer. However, its applicability to other scales, for example, national or city level has not yet been tested. Intersectoral issues and solutions have been considered across scales while the focus has remained on the transboundary dimension, reflecting the fact that the methodology was developed specifically to facilitate dialogue at this level.

6. Conclusions

The TRBNA methodology is a proposal to put the nexus into practice at transboundary level in a consistent manner. This is the first general approach of its kind that aims to cast its net (and impact) wider than traditional integrated water, energy, land or environmental assessments. It is formal, while at the same time flexible. It provides a useful entry point to motivate why we should look at international cooperation across sectors to make better use of our limited resources.

Compared to IWRM, the TRBNA considers sectors more broadly, with the intention of explicitly including sectoral perspectives and considering a wider range of opportunities for cooperation. Analyzing sectoral goals and priorities beyond the use of the water resource in question differs from the IWRM approach, which considers sectors precisely in function of their water use.

The methodology development was not a purely academic effort and was strongly influenced by the time constraints of the intergovernmental process and by the need to carry out assessments in politically sensitive contexts. It can be concluded that both the methodology and the assessments concretely contributed to pursue the objectives of the UNECE Water Convention, in particular joint assessment in a participatory manner and by exploring benefits of cooperation.

The understanding, interpretation and codification of various nexus concepts for which there are currently no widely accepted definitions was influenced by the scope of the project. A glossary developed ad hoc for this project was crucial to communicate new nexus concepts to stakeholders. Yet this was not validated by a wider academic community, which would be highly valuable not only for future applications of the TRBNA but also to advance with the much needed definition of nexus related concepts.

Refined and improved after its application to different basins, this methodology is applicable to a variety of geographical, socio-economic and political settings. However, future assessments will likely benefit from a critical approach to its application. In this paper, we highlight only a few general, possible improvements because specific ones will depend on the basin and context of application.

Acknowledgments: Many experts and officials contributed to the nexus assessment process but the authors would specifically like to thank Constantinos Taliotis, Dimitris Mentis and Manuel Welsch from KTH Royal Institute of Technology, division of Energy Systems Analysis, for their contribution to the development of the methodology and support throughout the project; Lucie Pluschke from the Food and Agriculture Organization of the United Nations, Holger Rogner from the International Institute for Applied Systems Analysis, Bo Libert from United Nations Economic Commission for Europe and Barbara Janusz-Pawletta from the Kazakh-German University for their input and advice; all colleagues involved in the basin assessments; and the reviewers and editors of this Special Issue.

Author Contributions: Lucia de Strasser, who refined the methodology and consolidated the various basin assessments, is the lead author of this article and integrated the inputs from all the authors. Substantive input was provided by Annukka Lipponen who contributed to shaping the methodology and the nexus assessments as the process and content coordinator. Together with Mark Howells, they have developed the methodology and carried out the assessments. Christian Bréthaut developed the initial approach to the governance methodology, further developed and integrated into the final methodology by Stephen Stec.

Conflicts of Interest: The authors worked for UNECE on the nexus assessment under the Water Convention either as staff (Annukka Lipponen) or as consultants (Christian Bréthaut, Lucia de Strasser, Mark Howells, Stephen Stec).

APPENDIX 1. Indicators

Table A1. Indicators.

Group	Relation to the Methodology Process	Type	Sources
Screening indicators (basin and national level)	Steps 1 and 3	Statistics Geo-spatial (GIS)	National and international statistics; relevant documents such as river basin management plans.
Perspective Indicators	Step 4	Qualitative/Rankin	Opinion based questionnaire.
Assessment-specific indicators	Step 6	Data	Previous studies, experts, authorities, models and estimations.

Notes: Screening Indicators are often available in the same form for each country and less often for each basin: *i.e.*, large basins such as the Syr Darya feature in the FAO database, but small ones such as the Alazani/Ganykh do not.) For this reason, it is not always possible to keep the analysis consistent across different case studies and most of comparable information can be collected only at country level (e.g., energy produced by source, water resources, *etc.*) In Step 1, it is possible to combine basin and country level Screening Indicators to investigate the relation between the basin and its riparian countries—for example, in terms of energy or crops produced in the basin area (as percentages of total produced in each riparian). A non-prescriptive list of sources for international statistics is given in the final publication of the project [23]. In order to obtain Perspective Indicators, each issue listed in the questionnaire is ranked from not relevant to very relevant and each participant states its country of origin and its area of expertise/work (energy, water, food/land, ecosystems). This allows for a comparison of perspectives from different countries and sectors on a same topic. The choice of assessment-specific indicators depends on the choice of interlinkages to be considered (for example, the change in forest area in the basin is an indicator of deforestation).

APPENDIX 2. Structure of the Workshop

1. Introduction of the nexus and relevant explicatory examples (by the analysts).

2. Distribution of the opinion based questionnaire.

3. Introduction to the key sectors, their main characteristics and issues (by selected speakers).

4. Presentation of national sectoral policies, as well as relevant national strategies and targets that may affect the basin (by relevant authorities).

5. Focus on the basin. Discussion on possible future development of the basin (river basin or aquifer management plan, infrastructure plans, sectoral targets, policy priorities, *etc.*).

6. Illustration of possible interlinkages and nexus conditions. Explanation of the working group sessions.

7. First working group session on intersectoral mapping. Stakeholders are divided according to their area of expertise or work (food/land, water, energy, ecosystems). Each group identifies the most important interlinkages (impacts and trade-offs) associated with its component. (For the material used, see Appendix 3).

8. Joint prioritization of the key interlinkages to be considered in the assessment. (For the material used, see Appendix 3).

9. Presentation of official data on climate change and, if available, the predicted impact on the basin.

10. Second working group session on future dimensions. Participants are divided into mixed groups to define a few relevant scenarios and discuss how the key interlinkages will change under those scenarios (see note below).

11. Discussion on synergetic actions for the identified nexus conditions, by means of measures, policies, coordination arrangements and techno-economic solutions. Reflection on the transboundary dimension. Discussion on the benefits and limitations. Identification of who/which actors could advance the actions.

12. Discussion on indicators and sources available.

13. Presentation (by the analysts) of some key findings and preliminary results from the workshop and desk study, in the form of nexus graphs and storylines that will be analysed further and included in the basin assessment.

14. Presentation of next steps in the assessment.

Note: In the early stages of applying the methodology, the joint discussion on trends was not well structured, so that future challenges remained mostly unexplored. This session was then specifically dedicated to discussing future trends in groups. The short time reserved for this exercise (3 hours) however, did not allow for a satisfying outcome.

APPENDIX 3. Material to Facilitate Intersectoral Dialogue

Intersectoral diagrams have been designed specifically to introduce the nexus dialogue and their features deserve some explanation. The idea behind their design is to support the dialogue by moving from the consideration of interlinkages from a sectoral perspective (which is where, for example, IWRM and integrated energy planning (IEP) stop) to their examination in a roundtable where all perspectives are equally represented: a nexus approach. Participants draw arrows between sectors to indicate dependencies and impacts (unidirectional) and trade-offs (bidirectional).

To represent the nexus, we used a triangular scheme where "energy", "water" and "food/land" are located at the apexes and "ecosystem services" is at the center. This was developed for supporting the nexus dialogue in Step 5. For the sectoral groups (Step 4), we rearranged the same scheme by putting at the center the sector that each group is considering. The others are located at its left and right to stimulate the consideration of interlinkages as inputs (needs) and outputs (impacts). (During this working group session we used "ecosystem services" rather than "ecosystems" to explicitly invite participants to think about the interactions of human activities with the environment rather than only their impact on it.).

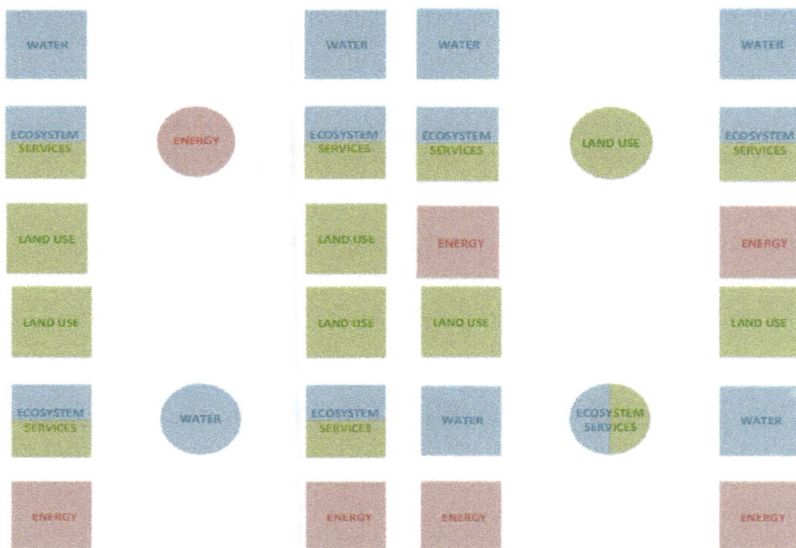

Figure A1. Sector-centered graphs used to facilitate discussion at the workshop [23].

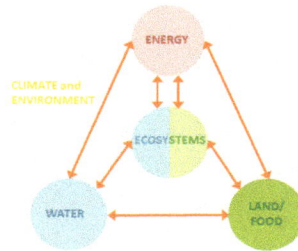

Figure A2. Intersectoral (nexus) graphs used to facilitate discussion at the workshop. Note: Initially—in the first workshop—"ecosystems" constituted a fourth corner in the nexus diagram (Figure A2). This caused confusion because "ecosystems" often logically overlaps with land and water resources (for example, when discussing an impact on water quality participants were unsure if to consider water or ecosystems). The diagram was reviewed accordingly, and "ecosystems" was colored half in blue and half in green to show this overlap [23].

APPENDIX 4. Highlights from the Socio-Economic Analysis (Illustrative of Step 1).

Alazani/Ganykh

(a) State of energy, food, water and environmental security in the basin

In the Georgian side of the basin, population lacks access to affordable energy in rural areas and some heavily rely on fuel wood for household heating and cooking. Water infrastructure, especially irrigation systems, is in poor conditions and lacks maintenance. Frequent flash floods have a devastating effect on local economy, affecting both countries.

(b) Relations within the region, the basin and its riparian countries

The region of Kakheti (practically coinciding with the Georgian part of the basin) is famous for wine production as well as for hosting important biodiversity sites. Water-scarce Azerbaijan recently built a pipeline to transfer groundwater from the basin to its capital, Baku, to supply it with drinking water. Azerbaijan is an important exporter of fossil fuels, and Georgia is a key corridor for natural gas transfer from the Caspian region to Europe. Cooperation between riparian countries is increasing in terms of regional trade agreements and coordinated environmental protection of shared river basins is being built.

(c) Main strategic goals, development policies and challenges

At basin level: interest in advancing cooperation at the Alazani/Ganykh basin level as part of wider international dialogue on the Kura/Ara(k)s basin. At country level: Georgia aims at developing the largely untapped potential of hydropower. Azerbaijan aims at diversifying economy and energy production to reduce dependency from fossil fuels.

Sava

(a) State of energy, food, water and environmental security in the basin

The basin is prone to flood episodes, which can be devastating and—among others—affect energy production and mining sector. Only small areas are irrigated and in times of droughts, large amounts of crops can be lost.

(b) Relations within the region, the basin and its riparian countries

The Sava is an important transportation route. The river and its tributaries are vital for the energy security of riparians as most of their electricity production depends on its water (for hydro and thermal power). Employment in the agro-industry in the basin area is economically relevant for all countries.

(c) Main strategic goals, development policies and challenges

At basin level: Developing and strengthening the mandate of the basin commission (International Sava River Basin Commission), for example, by finalizing a flood control scheme at basin level. At regional level: Strong relation with the European Union (to which some countries are member States). Regional commitments (e.g. EU directives) and development strategies fostering economic cooperation. Targets for increasing the share of renewables (including hydropower), improving energy efficiency and reducing emissions.

Syr Darya

(a) State of energy, food, water and environmental security in the basin

Energy insecurity is an issue in the upstream countries; especially in winter (e.g., winter 2008–2009) households can face severe power shortcuts. This can combine with food insecurity (e.g., high food prices, inadequate transport routes) creating a compound crisis. Seasonally but also due to the high levels of water use, water scarcity affects downstream agricultural production and communities. An aging and inefficient irrigation system is the primary cause of environmental degradation, heavily contributing to seasonal water stress and salinization of agricultural land.

(b) Relations within the region, the basin and its riparian countries

Hydropower production in the basin is the main energy source for Kyrgyzstan. All riparian countries, to a different extent, rely heavily on crop production (and agriculture more broadly)in the basin. The basin area is strategic because of its position in Central Asia, in particular for energy routes (oil and gas pipelines, power transmission lines).

(c) Main strategic goals, development policies and challenges

At basin level: impasse in cooperation over water resources in the Syr Darya basin and in the broader Aral Sea basin. At regional level (Central Asia): Further develop energy production for export and energy transit to neighboring regions (i.e., China, South-Asia, Russia). Challenges exist in all countries to improve livelihoods and resource security in the basin. At country level: Kazakhstan aims at promoting sustainable growth, in particular investing in water and energy saving technologies. Kyrgyzstan and Tajikistan plan to keep on developing hydropower, potentially abundant in both countries. Uzbekistan wants to secure availability of water for irrigated agriculture and modernize related infrastructure.

APPENDIX 5. Map ok Key Organizations and Actors (Illustrative of Step 2)

Table A2. Map ok Key Organizations and Actors.

Basin	Alazani/Ganykh	Sava	Syr Darya
Identification of key actors	Building on earlier intersectoral projects' stakeholder mapping.	Based on a stakeholder analysis for the Sava River Basin Management Plan, seeking to expand the involvement of notably energy and agriculture, which due to the mandate of the International Sava River Basin Commission were less engaged in the basin's management.	The Ministries of Foreign Affairs coordinated the nominations for their country, influencing somewhat the representation of sectors.
Regional and sub-regional level	Intergovernmental Commission for Economic Cooperation	European Union (riparian States have a different status), Energy Community, International Commission for the Protection of the Danube River, Danube (Navigation) Commission.	Commonwealth of Independent States, Eurasian Economic Community; International Fund for Saving the Aral Sea, Interstate Coordination Water Commission (ICWC), Interstate Commission for Sustainable Development, Central Asian Power Council, Central Asian Power System, Coordination dispatching Centre "Energy".
(Transboundary) Basin level	-	International Sava River Basin Commission.	Basin Water Organization Syr Darya (under ICWC).
Central Government	Ministries of energy, agriculture, environment and natural resources, economy (and sustainable development/industry); development and infrastructure; emergency situations, health.	Ministries of trade, economy, energy, agriculture, environment, infrastructure/construction, transport.	Ministries of foreign affairs, economy and trade, energy, agriculture, investment and development, emergency situations, industry, healthcare (and social development or protection)
Entity level	-	Entity level ministries in Federation of Bosnia and Herzegovina and Republika Srpska and District Brcko: energy and industry; agriculture, water management and forestry; physical planning; environment.	-
Government agencies, state committees	National Energy and Water Supply Regulatory Commission; State Committee on Property Issues.	Energy agencies, environmental agencies, national water councils.	State agencies or committees of environmental protection, land management, forestry, geology and mineral resources; Committee for Water Resources (Kazakhstan); Water and Energy Coordination Council (Tajikistan); committee/agency for communal services; state agency, authority or center of hydrometeorology.

Table A2. *Cont.*

Basin	Alazani/Ganykh	Sava	Syr Darya
Companies and utilities (state and private)	Companies on water supply; land reclamation and water resources, renewable energies.	Energy producers	Public utilities for water supply and sanitation; joint stock companies for energy production, transmission or distribution.
Sub-national/provincial level	Regions and districts	Regional or provincial government—absent in Slovenia); (see above regarding entity level in Bosnia and Herzegovina).	Basin inspections, basin water economy administrations, basin councils, basin organizations, basin irrigation authorities.
Local level	Local self-governance institutions, user associations.	Local governments, water supply and sewage companies.	Subsidiary companies of public utilities for water supply and sanitation; local branches of electricity distribution (and transmission); local administrations (city, region and district), water user associations.

APPENDIX 6. Analysis of the Energy Sector in the Sava River Basin (Illustrative of Step 3)

This example illustrates key findings from the analysis of the energy sector (as part of Step 3, which included the other key sectors as well) in the case of the Sava River Basin (SRB) [25]. Such findings informed the dialogue during the participatory workshop, in particular the working group focusing on the energy sector (Step 4).

(a) Key findings from the sector and resource flow analysis:

○ The electricity generation in the Sava countries depends heavily on water from the basin (see Figure A3), as the basin hosts high shares of the total thermal and hydropower capacity installed in the region (i.e., all countries) with some countries more dependent on the Sava than others: e.g., the ratio hydropower capacity in the basin/total national hydropower capacity is 5% in Croatia and 45% in Montenegro.

○ Energy is used for powering the water sector, which includes water pumping, irrigation and treatment.

○ Storage reservoirs help both to balance power demand and supply fluctuations (*i.e.*, providing energy supply to compensate shortfalls from other energy sources) and, together with other water buffer zones such as flood-planes, wetlands and forests, to enhance flood control.

○ Extreme flood events can cause damage to coal mines, affecting security of fuel supply. Recent floods have affected cooling systems and a coal mine.

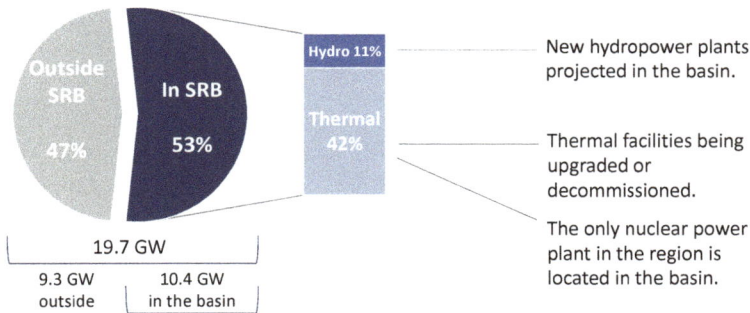

Figure A3. Installed capacity in the region in 2012.

(b) Key findings from the governance analysis:

Targets for renewables and climate mitigation push the countries to develop more hydropower while there are environmental concerns regarding the construction of new dams in environmentally sensitive areas.

Potential conflicts can occur both between upstream and downstream countries based on uses (e.g., hydropower, agriculture) and between local and national authorities within a country (development of energy infrastructure *versus* environmental protection or tourism).

There is a need to improve coordination with the energy sector to achieve representation of the relevant sectors in the work of the International Sava River Basin Commission (ISRBC), and vice versa.

Some of the riparian States do not have well-developed systems for Environmental Impact Assessment (EIA) and Strategic Environmental Assessment (SEA). EIAs and SEAs are effective tools to assess the impact of energy projects on ecosystems and to synchronize competing objectives, as well as to ensure proper public participation.

The geographical and political focus for energy security and related governance is generally at the national level, with only secondary regard for ecosystem boundaries or river basins. Resolution of conflicting water uses related to energy has to take into account such differences in scale and in institutional frameworks as compared to other water uses.

Regional energy frameworks are increasingly influencing energy trade and pricing: The Energy Community Treaty (entered into force in 2006) provides for the creation of an integrated energy market (including electricity and gas) among the European Union (EU) member States and other contracting parties. All the Sava River Basin countries belong to the Energy Community either as EU member States or as parties to the treaty.

The European Union's 2030 Framework for Climate and Energy Policies includes, e.g., reducing greenhouse gas emissions by at least 40% from 1990 levels, increasing the share of renewable energy to at least 27%, increasing energy efficiency by the same amount, and, most importantly perhaps for the SRB countries, proposing a new governance framework based on national plans for competitive, secure and sustainable energy including a set of indicators. Developments in this area are rapidly unfolding and could have serious implications for nexus issues in the SRB.

APPENDIX 7. Intersectoral Issues in the Syr Darya River Basin (Illustrative of Step 4)

As the outcome of sectoral work, each of the four groups (energy, water, food/land, ecosystems) came up with a list of important interlinkages from its perspective. This example includes a compilation of interlinkages between the component "agriculture" of the nexus and the other three components of the nexus, in the case of the Syr Darya River Basin [26]. This list, initially compiled after the workshop as a result of Step 4, has been revised and elaborated on taking into account later analysis. It should be noted that in the case of Syr Darya "agriculture" was prioritized over other land use types.

Table A3. Intersectoral Issues in the Syr Darya River Basin.

Interlinkage	Issue
Water-Agriculture	High water requirements for irrigation (thirsty crops and high losses in irrigation schemes);
	Discharges from agriculture cause diffuse pollution of water, limiting other uses and affecting ecosystems downstream;
	High levels of land degradation and salinization, caused mainly by poor drainage and causing loss of fertile soil.
Energy-Agriculture	River flow regulation optimized for energy generation affecting water availability for agriculture;
	Problem of affordability of energy and food upstream, sometimes combined causing situations of energy and food emergency for rural population;
Ecosystems-Agriculture	Prioritization of productive sectors (namely energy and agriculture) over ecosystems, leaving insufficient water for environmental needs;
	High impact of water scarcity on fish catches and aquaculture, the latter being an important livelihood for local settlements in the middle and low course of the river.

APPENDIX 8. Nexus Dialogue in the Alazani/Ganykh River Basin (Illustrative of Step 5)

This example illustrates a possible outcome of applying Step 5 of the methodology. It comes from the assessment of the Alazani/Ganykh River Basin. During the participatory workshop, a clear nexus "storyline" emerged from the intersectoral dialogue of Step 5, including elements that arose separately in the sectoral group discussions of Step 4 [24].

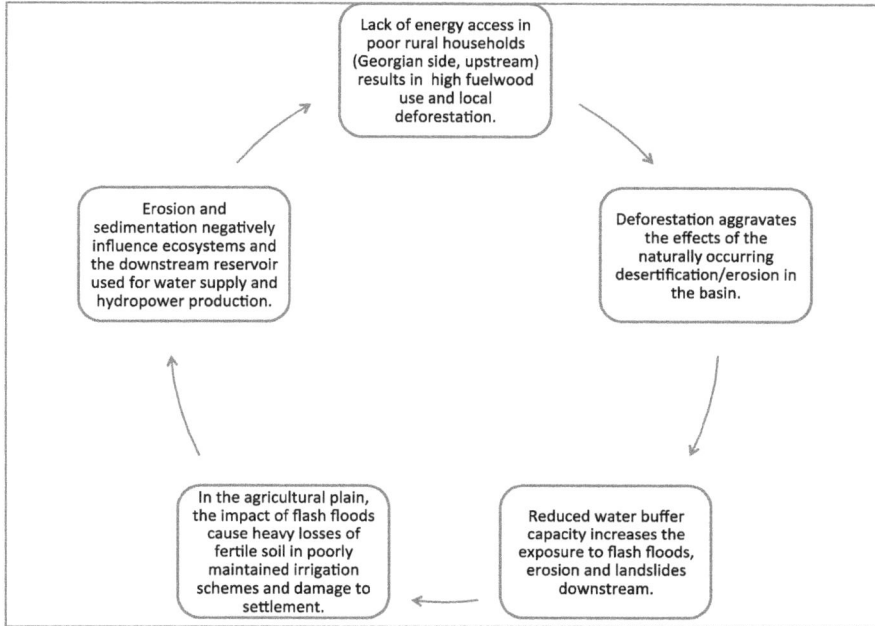

Figure A4. Nexus storyline for the Alazani/Ganykh River Basin.

APPENDIX 9. Solution and Benefits from the Sava River Basin (Illustrative of Step 6)

This example illustrates one specific solution identified in the assessment of the Sava River Basin [25] and the associated table of benefits.

It should be recalled that solutions were not were presented to the countries as "recommendations" but as "potential beneficial actions", the applicability of which was not investigated in detail in the study described in this paper. The emphasis was placed instead on the benefits that the identified nexus solutions would bring across countries. Energy-water modeling results allowed to illustrate the actual benefit of water cooperation at basin level from the perspective of the energy sector.

Selected solution: Coordinate energy and water planning by better involving the energy sector in the Sava consultation process.

Energy planning and climate policies usually take a multi-decadal perspective. However, water planning (driven by the implementation schedule of the EU Water Framework Directive) mostly follows a six-year cycle. Currently, (1) long-term energy planning does not necessarily take into account water constraints, potentially putting long-term investments and policy targets at risk and (2) water management planning marginally reflects important developments on the hydropower sector. While preparation of river basin management plans supports valuable engagement with a broad range of stakeholders at the transboundary level, improving involvement of the energy sector not currently engaged in work of ISRBC at both national and transboundary levels would be beneficial. Furthermore, a consultation process on national and sectoral development strategies, including energy, through the ISRBC, taking into account basin-level impacts, would improve coordination.

A multi-country water-energy model was built to study the electricity system of all riparian countries, in particular hydropower development, allowing to quantify the effects of its deployment on, for example, greenhouse gas emissions and energy imports. One of the findings of this analysis was that hydropower in the Sava Basin is critical for meeting regional targets for renewable energy sources. Coordination is therefore key to balancing between different objectives and constraints, including energy security and de-carbonization of the energy systems of riparian countries as well as maintaining a good status of shared waters. Moreover, cooperation at the basin level would be beneficial to optimize the use of infrastructure, ensuring also robustness in the face of low flows or floods. (For a better description of the model use, see [25]).

○   Example of solution proposed

Coordinate energy and water planning by better involving the energy sector in the Sava consultation process.

○   Benefits

1. At basin/transboundary level: (1) Reduced frictions between sectoral developments of different countries; (2) Improved efficiency of water resource use in the basin; (3) Decreased risk of water-related disasters if the variability of flow is well taken into account.
2. At national level: (1) Higher efficacy of efforts targeting energy security and low-carbon growth (using hydropower); (2) Reduced exposure of energy sector's operations to water scarcity or water related disasters (floods and droughts).

References

1. Hoff, H. *Understanding the Nexus*; Stockholm Environment Institute: Bonn, Germany, 2011.
2. World Economic Forum (WEF), Water Initiative. *Water Security: The Water-Food-Energy-Climate Nexus*; Island Press: Washington, DC, USA, 2011.
3. International Energy Agency (IEA). *Water for Energy*; Excerpt from the World Energy Outlook: Paris, France, 2012.
4. International Water Management Institute (IWMI). *Water for Food, Water for Life: A Comprehensive Assessment of Water Management in Agriculture*; Earthscan and Colombo: International Water Management Institute: London, UK, 2007.
5. Loring, P.; Gerlach, S.C.; Huntington, H. The New Environmental Security: Linking Food, Water, and Energy for Integrative and Diagnostic Social-ecological Research. *J. Agric. Food Syst. Community Dev.* **2013**, 1–7. [CrossRef]
6. Falkenmark, M. The Greatest Water Problem: The Inability to Link Environmental Security, Water Security and Food Security. *Int. J. Water Resour. Dev.* **2001**, *17*, 539–554. [CrossRef]
7. Howells, M.; Rogner, H.-H. Water-energy nexus: Assessing integrated systems. *Nat. Clim. Chang.* **2014**, *4*, 246–247. [CrossRef]
8. United Nations Department of Economic and Social Affairs (UNDESA). *Guidance in Preparing a National Sustainable Development Strategy: Managing Sustainable Development in the New Millennium*; United Nations Department of Economic and Social Affairs (UNDESA): New York, NY, USA, 2002.
9. United Nations Economic and Social Commission for Western Asia (ESCWA). *Conceptual Frameworks for Understanding the Water, Energy and Food Security Nexus*; United Nations Economic and Social Commission for Western Asia (ESCWA): Beirut, Lebanon, 2015.
10. Benson, D.; Gain, A.K.; Rouillard, J.J. Water governance in a comparative perspective: From IWRM to a "nexus" approach. *Water Altern.* **2015**, *8*, 756–773.
11. Bazilian, M.; Rogner, H.; Howells, M.; Hermann, S.; Arent, D.; Gielen, D.; Steduto, P.; Mueller, A.; Komor, P.; Tol, R.S.J.; *et al.* Considering the energy, water and food nexus: Towards an integrated modelling approach. *Energy Policy* **2011**, *39*, 7896–7906.
12. Global Water Partnership (GWP). The challenge. What is IWRM? Available online: http://www.gwp.org/The-Challenge/What-is-IWRM/ (accessed on 21 January 2016).
13. Allouche, J.; Middleton, C.; Gyawali, D. Technical Veil, Hidden Politics: Interrogating the Power Linkages behind the Nexus. *Water Altern.* **2015**, *8*, 610–626.
14. Welsch, M.; Hermann, S.; Howells, M.; Rogner, H.H.; Young, C.; Ramma, I.; Bazilian, M.; Fischer, G.; Alfstad, T.; Gielen, D.; *et al.* Adding value with CLEWS—Modelling the energy system and its interdependencies for Mauritius. *Appl. Energy* **2014**, *113*, 1434–1445. [CrossRef]

15. Phillips, D.; Daoudy, M.; McCaffrey, S.; Ojendal, J.; Turton, A. *Transboundary Water Cooperation as a Tool for Conflict Prevention and Broader Benefit Sharing*; Global Development Studies No. 4; Edita Stockholm: Stockholm, Sweden, 2006.

16. Meadows, D.H.; Meadows, D.L.; Randers, J.; Behrens, W.W.I. *The Limits to Growth: A Report for the Club of Rome's Project on the Predicament of Mankind*; Universe Books: New York, NY, USA, 1972.

17. Foran, T. Node and regime: Interdisciplinary analysis of water-energy-food nexus in the Mekong region. *Water Altern.* **2015**, *8*, 655–674.

18. UN Water. *Transboundary Waters: Sharing Benefits, Sharing Responsibilities*; UN Water: Zaragoza, Spain, 2008.

19. Dale, L.L.; Karali, N.; Millstein, D.; Carnall, M.; Vicuña, S.; Borchers, N.; Bustos, E.; O'Hagan, J.; Purkey, D.; Heaps, C.; et al. An integrated assessment of water-energy and climate change in Sacramento, California: How strong is the nexus? *Clim. Chang.* **2015**, *132*, 223–235. [CrossRef]

20. Bartos, M.D.; Chester, M.V. The Conservation Nexus: Valuing Interdependent Water and Energy Savings in Arizona. *Environ. Sci. Technol.* **2014**, *48*, 2139–2149. [CrossRef] [PubMed]

21. Mekong River Basin Commission (MRC). Cooperation for Water, Energy, and Food Security in Transboundary Basins under Changing Climate. Mekong River Basin Commission (MRC): Vientiane, Laos, 2014.

22. United Nations Economic Commission for Europe (UNECE). *Meeting of the Parties to the Convention on the Protection and Use of Transboundary Watercourses and International Lakes. Report of the Meeting of the Parties on its sixth session. Addendum: Programme of work for 2013–2015*; United Nations Economic Commission for Europe (UNECE): Geneva, Switzerland, 2012.

23. United Nations Economic Commission for Europe (UNECE). *Reconciling Resource Uses in Transboundary Basins: Assessment of the Water-Food-Energy-Ecosystems Nexus*; United Nations Economic Commission for Europe (UNECE): Geneva, Switzerland, 2015.

24. KTH Royal Institute of Technology; UNECE Water Convention Secretariat. *Alazani/Ganykh River Basin Water-Food-Energy-Ecosystems Nexus Assessment. Second Draft Report for Comments by the Concerned Authorities*, Available online: http://www.unece.org/fileadmin/DAM/env/documents/2015/WAT/04Apr_28-29_Geneva/Nexus_assessment_Alazani-Ganikh_2nd_draft_clean_rev_Nov2014.pdf (accessed on 21 January 2016).

25. International Sava River Basin Commission. Assessment of the Water-Food-Energy-Ecosystems Nexus in the Sava River Basin. (forthcoming).

26. UNECE Secretariat, KTH Royal Institute of Technology. *Draft Assessment of the Water-Food-Energy-Ecosystems Nexus In the Syr Darya*; Document presented at the Tenth Meeting of the Working Group on IWRM; United Nations Economic Commission for Europe (UNECE): Geneva, Switzerland, 2015.

27. Bréthaut, C.; Pflieger, G. The shifting territorialities of the Rhone River's transboundary governance: A historical analysis of the evolution of the functions, uses and spatiality of river basin governance. *Reg. Environ. Chang.* **2013**, *15*, 549–558. [CrossRef]

28. Sadoff, C.W.; Grey, D. Beyond the river: The benefits of cooperation on international rivers. *Water Policy* **2002**, *4*, 389–403. [CrossRef]

29. Lipponen, A.; Howells, M. Promoting Policy Responses on the Water and Energy Nexus Across Borders. In *Water Monographies 2: Water and Energy*; World Council of Civil Engineers (WCCE): Madrid, Spain, 2014; pp. 44–55.

30. Qaddumi, H. *Practical Approaches to Transboundary Water Benefit Sharing*; Working Paper 292; Overseas Development Institute: London, UK, 2008.

31. Bach, H.; Bird, J.; Clausen, T.J.; Jensen, K.M.; Lange, R.B.; Taylor, R.; Viriyasakultorn, V.; Wolf, A. *Transboundary River Basin Management: Addressing Water, Energy and Food Security*; Mekong River Commission: Vientiane, Lao PDR, 2012.

32. United Nations Economic Commission for Europe (UNECE). *The Economic Commission for Europe Water Convention and the United Nations Watercourses Convention. An Analysis of their Harmonized Contribution to International Water Law*; United Nations Economic Commission for Europe (UNECE): Geneva, Switzerland, 2015.

33. Howells, M.; Hermann, S.; Welsch, M.; Bazilian, M.; Segerström, R.; Alfstad, T.; Gielen, D.; Rogner, H.; Fischer, G.; van Velthuizen, H.; et al. Integrated analysis of climate change, land-use, energy and water strategies. *Nat. Clim. Chang.* **2013**, *3*, 621–626. [CrossRef]

34. Food and Agriculture Organization of the United Nations (FAO). *The Water-Energy-Food Nexus. A New Approach in Support of Food Security and Sustainable Agriculture*; Food and Agriculture Organization of the United Nations (FAO): Rome, Italy, 2014.

35. Bréthaut, C. A Draft Methodology for Assessing Governance Aspects of the Water-Foodenergy-Ecosystems Nexus. Available online: https://www.unece.org/fileadmin/DAM/env/documents/2014/WAT/09Sept_8-9_Geneva/UNECE_governance_assessment_methodology_forTaskForce_forWeb.pdf (accessed on 21 January 2016).

36. Gerber, J.-D.; Knoepfel, P.; Nahrath, S.; Varone, F. Institutional Resource Regimes: Towards sustainability through the combination of property-rights theory and policy analysis. *Ecol. Econ.* **2009**, *68*, 798–809. [CrossRef]

37. Knoepfel, P.; Nahrath, S.; Varone, F. Institutional Regimes for Natural Resources: An Innovative Theoretical Framework for Sustainability. In *Environmental Policy Analyses*; Environmental Science and Engineering; Springer Berlin Heidelberg: Berlin, Germany, 2007; pp. 455–506.

38. Libert, B. The UNECE Water Convention and the development of transboundary cooperation in the Chu-Talas, Kura, Drin and Dniester River basins. *Water Int.* **2015**, *40*, 168–182. [CrossRef]

39. International Sava River Basin Commission (ISRBC). *Strategy on the Implementation of the Framework Agreement on the Sava River Basin*; International Sava River Basin Commission (ISRBC): Zagreb, Croatia, 2011.

40. United Nations Economic Commission for Europe. *Second Assessment of Transboundary Rivers, Lakes and Groundwaters*; Convention on the Protection and Use of Transboundary Watercourses and International Lakes, United Nations: Geneva, Switzerland, 2011.

41. United Nations Economic Commission for Europe (UNECE). *Working Group on Integrated Water Resource Management—Draft Policy Guidance Note on Identifying, Assessing and Communicating the Benefits of Transboundary Water Cooperation: "Counting Our Gains"*; United Nations Economic Commission for Europe (UNECE): Geneva, Switzerland, 2015.

42. United Nations Economic Commission for Europe (UNECE). *Meeting of the Parties to the Convention on the Protection and Use of Transboundary Watercourses and International Lakes. Draft Programme of Work for 2016–2018*; United Nations Economic Commission for Europe (UNECE): Geneva, Switzerland, 2015.

water

MDPI

Article

Modeling the Hydropower–Food Nexus in Large River Basins: A Mekong Case Study

Jamie Pittock [1,*], David Dumaresq [1] and Andrea M. Bassi [2]

[1] Fenner School of Environment and Society, The Australian National University, 48 Linnaeus Way, Acton 2600, Australia; david.dumaresq@anu.edu.au
[2] KnowlEdge Srl, via San Giovanni Battista 2, Olgiate Olona 21057, Italy; andrea.bassi@ke-srl.com
* Correspondence: Jamie.pittock@anu.edu.au; Tel.: +61-2-6125-5563

Academic Editors: Marko Keskinen and Olli Varis
Received: 5 February 2016; Accepted: 20 September 2016; Published: 28 September 2016

Abstract: An increasing global population and growing wealth are raising demand for energy and food, impacting on the environment and people living in river basins. Sectoral decision-making may not optimize socio-economic benefits because of perverse impacts in other sectors for people and ecosystems. The hydropower–food supply nexus in the Mekong River basins is assessed here in an influence model. This shows how altering one variable has consequent effects throughout the basin system. Options for strategic interventions to maximize benefits while minimizing negative impacts are identified that would enable national and sub-national policy makers to take more informed decisions across the hydropower, water and food supply sectors. This approach should be further tested to see if it may aid policy making in other large river systems around the world.

Keywords: energy; food; water; rivers; nexus

1. Introduction

In this research, we propose the first steps towards a model of the hydropower–food supply nexus in a large river basin. Eventually this may be developed and used to identify key points of leverage where management interventions may change the outcomes. As the global population heads towards nine billion people the demand for energy and food is growing [1,2]. Producing energy and food consumes, converts or has other negative impacts on resources, such as the energy, land and water available for non-human managed ecosystems. Meeting human needs for energy and food while minimizing resource consumption has become a key objective of government policies. These approaches are described as "sustainable development" and "green growth".

In the past decade considerable debate among academics and policy makers has focused on the nexus between an array of key sectors, including energy, food and water. The logic in focusing on only a few sectors in a complex system is that analysis and management interventions may be more tractable for an energy-food nexus, for example, rather than the full range of sectors considered in sustainable development [3].

In this context we examine the hydropower–food supply nexus in large river basins, because the rapid expansion of hydro-electricity production in river basins like the Amazon, Mekong and Yangtze is having perverse impacts on biodiversity and food supply [4–6]. Using influence diagrams, our objective is to identify the first steps towards a model of the variables and their relationships that influence the hydropower–food supply nexus. We argue that such modeling should be further developed to enhance decision-making in river basins subject to hydropower development and intensified food production around the world.

Here, the initial conceptual model is elaborated for the lower Mekong River Basin (LMB) in Southeast Asia because it is a region undergoing rapid development and because the processes of the

Mekong River Commission have enabled the publication of much data that contributes to this analysis. In 2010, a strategic environmental assessment prepared for the Commission was published outlining the benefits, costs and risks of the planned construction of 88 new hydropower dams in the LMB by 2030. While the proposed developments would increase hydroelectric power generation nine-fold, it would diminish wild fish catch by 24%–40% [7] (but with a 10% reservoir fishery gain). Wild fish are a significant source of protein and micro-nutrients for the sixty million people living the LMB [8,9], so the diminution of this fish supply will require the development of alternative sources of protein through trade or local production. LMB production of alternative crop, fish or livestock supplies has opportunity costs, including those associated with the consumption of more water and with land use change, and consequent environmental and socio-economic knock-on effects [10,11].

To make better cross-sectoral decisions in such complex systems, decision makers need to appreciate the relationships between different variables where they are considering change. Here we apply causal loop (or influence) diagrams (CLDs) to map system dynamics in the expectation that this may enable points of intervention to be identified to optimize beneficial changes while minimizing perverse impacts.

2. Methodology

The approach utilized is based on the System Thinking (ST) methodology as its foundation, with the resulting conceptual models serving primarily as knowledge integrators across disciplines and sectors. ST allows us to map systems, visualizing interdependencies across variables (e.g., economic, social and environmental), with the aim of reaching a shared understanding of the underlying functioning mechanisms of the sector analyzed. ST models can be converted into mathematical models for forecasting, using a methodology called System Dynamics (SD) [12–15]. A key characteristic of utilizing ST and SD is that it allows us to integrate the three spheres of sustainable development in its analytical process. The software used to create the CLDs presented in this paper is called Vensim (www.vensim.com) and it is developed by Ventana Simulations.

2.1. Causal Loop Diagrams

A causal loop diagram (CLD) is a map of the system analyzed, or, better, a way to explore and represent the interconnections between the key variables in the analyzed sector or system [16]. In other words, a CLD is an integrated map because it represents different system dimensions of the dynamic interplay. It explores the circular relations (or feedbacks) between the key elements—the main variables—that constitute a given system [17].

By highlighting the drivers and impacts of the issue to be addressed and by mapping the causal relationships between the key variables, CLDs support a systemic decision-making process aimed at designing solutions that last. The creation of a CLD has several purposes. First, it combines the team's ideas, knowledge and opinions. Second, it highlights the boundaries of the analysis. Third, it allows all the stakeholders to achieve basic-to-advanced knowledge of the analyzed issues' systemic properties [17].

Causal loop diagrams include variables and arrows (called causal links), with the latter linking the variables together with a sign (either + or −) on each link, indicating a positive or negative causal relation (see Table 1):

- The causal link from variable A to variable B is positive if a change in A produces a change in B in the same direction.
- The causal link from variable A to variable B is negative if a change in A produces a change in B in the opposite direction.

Circular causal relations between variables form causal, or feedback, loops. The role of feedback loops in the decision making process crucial. It is often the very system we have created that generates the problem, due to external interference, or to a faulty design, which shows its limitations as the system grows in size and complexity. In other words, the causes of a problem are often found within the feedback structures of the system being studied. The indicators are not sufficient to identify these causes and explain the events that led to the creation of the problem.

Table 1. Causal relations and polarity: + = a positive causal relation; − = a negative causal relation.

Variable A	Variable B	Sign
↑	↑	+
↓	↓	+
↑	↓	−
↓	↑	−

Notes: "↑" means increase; "↓" means decrease.

There are two types of feedback loops: positive and negative. A feedback loop is positive when an intervention in the system triggers other changes that amplify the effect of that intervention, thus reinforcing it [18]. A loop is balancing when it tends towards a goal or equilibrium (and hence reducing the rate of change), balancing the forces in the system [18].

For an extended introduction in the application of CLDs see Proust [19], Dyball and Newell [20], and Probst and Bassi [17]. They have been used to model the relationship between climate, energy and water management (see for example [21]). Here they will be used at a very general level to provide static models of basic relationships between a range of ecological and human factors.

2.2. Limitations of CLDs

However, CLDs have two interrelated limitations. Firstly, all CLDs are simplifications of the situation under consideration. As such, all CLDs are only ever partial representations of the actual. This point leads to the second limitation. There are multiple possible CLD representations for any particular situation under consideration. Thus, any particular CLD developed to represent a particular problem situation may be seen by some as incomplete, or as failing to focus on what some particular actor, agent or stakeholder may deem as important. Different actors within the problem situation may hold very different values and worldviews and thus see very different interactions and feedback mechanisms within the system as being important.

The effectiveness of a CLD will be directly related to the system under consideration and its boundaries being clearly identified. Multi-stakeholder perspectives and cross-sectoral knowledge should be incorporated as far as possible to appropriately identify the causes of the problem and design effective interventions. The partial nature of all CLDs along with any errors in creating diagrams may lead to representations that stakeholders do not accept, the generation of policy interventions that are not recognized, and policy implementation that is ineffective or even exacerbates the problem.

The estimation of the strength of causal relations, even if these are appropriately identified, cannot be guaranteed, as CLDs are a qualitative tool. The use of CLDs will be considerably strengthened when used together with a similar integrated and dynamic causal descriptive mathematical simulation model, where such data and agreement on all variables and interrelationships is available.

2.3. Application to River Basins

In this paper we now step through a series of increasingly more complex influence diagrams. We start with the most basic relationships and then add the details needed to capture complexity and identify options for intervention. This modeling of the energy–food–water nexus that occurs in the Lower Mekong Basin resulting from the construction of hydropower dams is based on a set of simple balancing and reinforcing loops set out in Figures 1 and 2.

In Figure 1, the influence model indicates that more of a resource enables greater consumption (+) of that resource while more consumption of that resource in general leads to a diminishing availability

(−) of that resource. This creates a simple balancing loop in which resource use is constrained by the availability of the resource: it is a negative feedback loop.

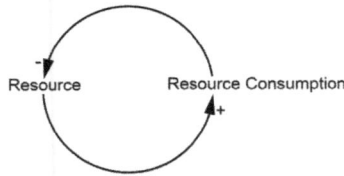

Figure 1. Balancing, negative feedback loop.

In Figure 2, resource use through appropriate management leads to an increase in the availability of that resource, creating a simple reinforcing loop in which the resource is maintained or increased through resource management: a positive feedback loop. When dealing with management of a natural resource such a positive feedback loop will be constrained in time and space as system limits are reached and ultimately fail.

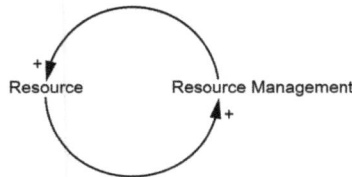

Figure 2. Reinforcing, positive feedback loop.

Using Figures 1 and 2, we develop simple resource use models for the use of the water flow in the Lower Mekong Basin, which supplies food in both the main stem and tributaries of the river and also provides the energy for the development of hydro-electricity generation.

Figure 3 shows the start of the construction of a combined simple model for both resource uses of food supply and hydropower dam construction for energy generation in the LMB. This combined model will be more fully developed in Results below.

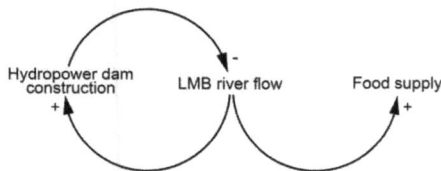

Figure 3. Model of resource uses of food supply and hydropower dam construction for energy generation using lower Mekong River Basin (LMB) river flow.

3. Results

Using the influence model components from Figures 1–3 the basic factors and relationships of the hydropower–food supply nexus in the lower Mekong Basin were identified from the existing literature and public data that is available for the region. Increasingly complex and complete models are set out with variables and causal links identified in increasing detail. We start with a basic combined model at Figure 4, bringing together the relationship between the two resource uses of the same LMB river flow resource: hydropower dams and food supply.

The Strategic Environmental Assessment commissioned by the Mekong River Commission [7] and others indicate that construction of water infrastructure projects in the LMB have major negative effects on the wild fish catch from the river system. While it is easy to see that the construction of hydropower dams change river flow, such a connection with food supply is less obvious: changes in food supply alone do not necessarily change river flow for the LMB or any other river basin. Just how is the management of the resource of the LMB river flow linked to the supply of food and how does that supply of food affect river flow? Further stages in this loop are needed to capture the influences at work. Orr et al. [10], among others, have shown that the development of hydropower projects changes overall food supplies in the lower Mekong Basin, including increasing demand for land-based agricultural production. Changes in land-based food production entail major changes in water infrastructure and water management, for example, changes from once a year flood recession rice crop to two or even three annual crops using managed irrigation [22]. Increasing food supply reduces the water resource, driving more investment in water resource management associated with food production. This forms a basic causal loop, or influence diagram (Figure 4) from which the hydropower–food supply nexus in the lower Mekong Basin can be constructed. The effects of this loop on the system and the nature of its feedback, positive or negative, are further explored below. Information from the literature was drawn on to deduce and graph the relationship between variables.

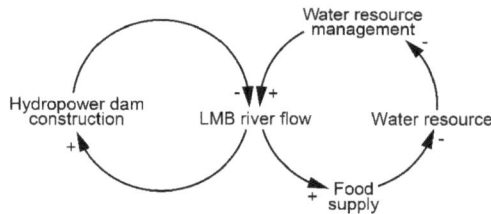

Figure 4. Basic influence model for the hydropower–food supply nexus in the lower Mekong Basin.

3.1. Food Supply Influence Model

In Figure 5, the food supply influence model is stepped out, comprising one feedback loop as follows. The increase in hydroelectricity generation [7], and the upgrading and expansion of irrigated agriculture [22], is leading to water infrastructure construction. Water infrastructure impacts natural river flows in a number of ways, including through diversion of water to crops, loss of water through evaporation from reservoirs, storage and release of water that increases dry season flows and diminishes wet season flows, changing flows of sediments and nutrients, as well as forming barriers to migration and breeding of fish [23]. For example, following construction of the Pak Mun dam in Thailand in the late 1990s, the fish catch directly upstream of the dam declined by 60%–80% [24].

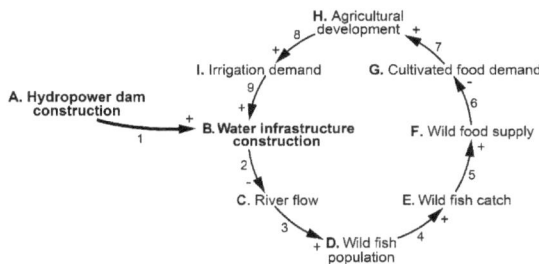

Figure 5. Food supply influence model in the lower Mekong Basin. Details of the variables A to I in Figure 5 and the causal links 1 to 9 are defined in Table 2 below.

Table 2. Definitions of variables and causal links in Figure 5 and later figures.

#	Variable	Definition
A	Hydropower dam construction	There were 46 hydropower dams in the Mekong River basin in 2015 [7]
B	Water infrastructure construction	There were 10,800 irrigation projects in the LMB as at 2009 [25]
C	River flow	The Mekong River has an annual water discharge to the sea of 475 km^3 on average [7]
D	Wild fish population	The Mekong River is one of the most biodiverse river systems in the world with 781 fish species described, including a number of globally endangered species [7]
E	Wild fish catch	The freshwater wild fish catch in the four LMB states was estimated in 2010 to be two million tonnes of fish per year [26]
F	Wild food supply	River flows support a freshwater capture fishery in the four LMB states estimated in 2010 to be two million tonnes of fish per year [26]. In each country, 47%–80% of animal protein for local residents comes from freshwater fisheries, and 90% of this is from capture fisheries [8]. There are other wild foods obtained from freshwater ecosystems as well
G	Cultivated food demand	FAOStat Food Balance Tables (2009) provide a country-by-country analysis of apparent consumption of all foodstuffs with aggregate figures for calorie and protein intake on a daily basis. The total average protein intake for the LMB is 66.7 grams per capita per day (g/c/d). Of this 21.4 g/c/d is from animal protein. Of this animal protein intake the proportions coming from fish are 56% for Cambodia, 38% for Laos, 31% for Thailand and 24% for Vietnam [27]
H	Agricultural development	In one projection, to replace the diminished wild fish catch resulting from 88 hydropower dams in the LMB in 2030, there would need to be an additional 5.483 Mha of crops, or 2.857 Mha of pasture for livestock, or 4.177 Mha for a combination of crops and livestock [11]
I	Irrigation demand	The current area equipped for irrigated agriculture of four million hectares may increase to as much as 10.2 Mha by 2030 under a high development scenario, with a growing portion of the land used for second and third crops during the year [25]. In 2000, irrigated agriculture's consumptive water use was around 72.8 km^3 or 15% of annual average Mekong River discharge, half of which was in the Mekong delta [28]

#	Causal Links	Definitions
1	Dam construction increases water infrastructure	By directly providing infrastructure and by increasing dry season flows, hydropower dam construction is projected to contribute to an expansion of water infrastructure for agriculture, domestic and industrial use [7,28], including for 4000 planned new irrigation projects by 2030 in the LMB [25]
2	Water infrastructure influences river flow	It is projected that river flow will change seasonally and be diminished as water diversion for irrigated agriculture in the Mekong Basin grows to 104.5 km^3 in 2025, or 22% of discharge, and 25%–30% of discharge by 2050 [28]
3	Size of the fish population is influenced by river flows	Seasonally specific river flows support fish populations by enabling migration along the rivers for breeding and access to floodplain habitats in the wet season. The 88 planned hydropower projects by 2030 will increase in active water storage capacity 700% to 69.8 km^3, changing the timing of river flows [7]. This construction of planned dams is projected to reduce wild fish populations and catch by 23.4% to 37.8% by 2030 [10]. River flows are also disrupted when water infrastructure forms barriers to migration. A diminution of the wet season area flooded reduces wild fish populations [26]
4	The wild fish population underpins the fish catch	The size of the wild fish population influences the scale of fish catch [26]
5	Wild fish catch makes a significant contribution to wild food supply	In each LMB country, 47%–80% of animal protein for local residents comes from freshwater fisheries, and 90% of this is from capture fisheries [8]
6	The scale of the wild supply of food influences demand for cultivated foods from aquaculture, crops and livestock	Food demand in the LMB region is rising due to growing populations as well as poverty reduction and growing wealth [27,29]. At the same time, hydropower dam development is project to diminish the wild freshwater fish catch by 23.4% to 37.8% by 2030 and thus supply of important protein and other nutrients [10,11]

Table 2. *Cont.*

#	Causal Links	Definition
7	Demand for food drives agricultural development	If demand is not met this will drive further agricultural development. The options for supplementing food supply include importing food or increasing production of crops, livestock and aquaculture [10,11]. Considerable expansion of food production is projected in the region [25]
8	Irrigation development is an important component of agricultural development	The current area equipped for irrigation of nearly four million hectares in 2009 may increase to as much as 10.2 million hectares by 2030 under a high development scenario, with a growing area used for second and third crops during the year [25]
9	Increased demand for irrigation drives new water infrastructure construction	Around 4000 new irrigation projects are planned by 2030 in the LMB that will require levees, canals, water storages and pumping stations [25]. It is projected that water diversion for irrigated agriculture in the Mekong Basin will grow to 104.5 km^3 in 2025, or 22% of discharge, and 25%–30% of discharge by 2050 [28]

As a consequence, wild fish populations in the Mekong River system are diminishing, and a net reduction of in the wild fish catch for food of 23.4% to 37.8% by 2030 is projected from the construction of 88 hydro dams in the Lower Mekong Basin due to the barrier effect alone [7,10]. This is anticipated to diminish the supply of protein per country as indicated in Table 3, and consequently, adding to increasing demand for protein [11].

Table 3. Projected loss in fish protein due to hydropower development to 2030 as a proportion (%) of all national protein supply. The range reflects uncertainty as to the portion of the Mekong wild fish harvest caught in each country (after [11]).

Min/Max	Cambodia (%)	Laos (%)	Thailand (%)	Vietnam (%)
Minimum	6.5	2.6	1.6	0.7
Maximum	23.1	8.2	3.3	2.1

The diminution of the wild fish catch, along with population growth and government policies favoring exports are creating greater demand for food [10,22], leading to increased demand for irrigation [30–32]. As a consequence, each of the LMB national governments have policies for agricultural development that emphasize greater production, including through expansion of irrigated agriculture, facilitated through water infrastructure construction [22,33–36]. In turn this is leading to an expansion of water infrastructure, including water storages, canals and flood protection dykes, which exacerbate loss of wild fish [37,38].

This is a reinforcing feedback loop with the external influence of hydropower dam construction increasing water infrastructure construction leading to changes in the river flow. This then reduces the stock of wild fish with a subsequent decrease in the wild fish catch, decreasing the food supply. This adds to increasing demand for food, requiring further agricultural development, giving rise to an ever increasing demand for further water infrastructure construction, leading to further loss of wild fish stock and loss of food supply.

3.2. Hydropower Supply Influence Model

The hydropower supply influence model is shown in Figure 6. Increased demand for electricity is driving still further hydropower construction in the region [39]. Southeast Asia's energy demand is projected to increase by 80% or more between 2013 and 2035 [11,39]. Programs for grid interconnections among the countries of Southeast Asia will enable countries like Laos to further export electricity to neighboring countries where there is a high demand, including China, Thailand and Vietnam. The Strategic Environmental Assessment for the LMB calculates that there is 53,000 MW potentially feasible hydropower generation available based on flows in the Mekong River basin while the eleven main stem dams would produce 14,697 MW of this total [7].

New hydropower dams are increasing power generation, which is supporting industrial development. Access to electricity ranges from near-universal in Thailand down to 66% of the population in Cambodia [32,39]. Cambodia has experienced over a decade of strong economic growth focused in the garment manufacture, tourism and construction industries. The high cost of energy has constrained agricultural processing, leading to a call for access to cheaper electricity from neighboring countries [40]. Laos has also had a prolonged period of economic growth driven by investments in the hydropower, mining, agriculture, transport and tourism supported by new roads and rural electrification: it sees itself becoming the hydropower "battery" for the region [41]. There have been three decades of strong economic growth in Vietnam's agricultural sector and industry, and the economy is changing from a focus on natural resource utilization to services. However, poor infrastructure and power cuts constrain the economy at this time [42]. Although Vietnam has recently expressed concern over the impacts of hydropower development in Laos on the Mekong delta, its state-owned corporations are among the investors in proposed hydropower projects in Laos [43]. Thailand has experienced four decades of economic growth with the economy transitioning from an agricultural focus to one based on manufacturing and services, including tourism. Extensive imports of energy, including hydropower from Laos, fuel the Thai economy [44]. In turn, this industrial development further increases demand for electricity.

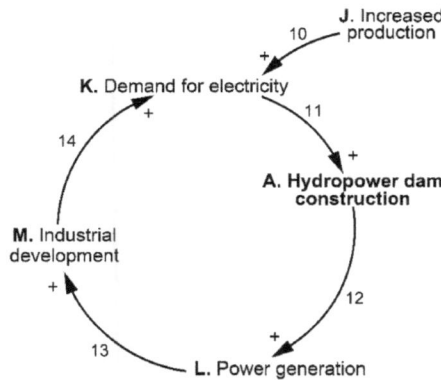

Figure 6. Hydropower influence model for the lower Mekong Basin. Details of the variables J to N and the causal links 10 to 14 for Figure 6 are defined in Tables 2 and 4.

Table 4. Definitions of variables and causal links in Figure 6 and later figures, following on from data in Table 3.

#	Variable	Definition
J	Increased production	In one projection, the area of crop production in the lower Mekong River Basin would need to increase by between 6% and 59% per country to replace the diminished wild fish catch resulting from 88 hydropower dams in the LMB in 2030 [11]
K	Demand for electricity	In 2015 annual demand for electricity was around 400,000 GWHr in the LMB [7]
L	Power generation	There were 5574 MW of hydropower generation operating or under construction in the LMB in 2010, and 90% of electricity generation in the region is from fossil fuels [7]

Table 4. Cont.

#	Variable	Definition
M	Industrial development	"In common with other countries in the wider East Asia region, over the last two decades the four countries of the LMB have experienced rapid economic development. This has largely been driven by industrial growth, and in particular growth in manufacturing production for export" (p. 54) [7]

#	Causal links	Definition
10	Increased agricultural production increases demand for electricity	There are relatively few options for gravity-fed expansion of irrigated agriculture in the region, with plans for new projects involving large scale lift irrigation or smaller scale pumping from local canals and water bodies [38,45]. Pumping water requires significant amounts of energy, in addition to the energy requirements for producing farm chemicals, transport and processing crops [46]
11	Increased demand for electricity drives further hydropower construction	The Mekong Basin has 53,000 MW of potentially feasible hydropower generation available based on flows in the Mekong River basin [7]. There are 42 new hydropower dams likely to be constructed in the Mekong Basin by 2030 [7]
12	Increasing hydropower construction increases generation	Generating capacity of 29,760 MW is possible from identified, planned LMB hydropower projects [7]
13	More power generation drives industrial development	More power generation drives industrial development. In addition to expanding current growth in manufacturing production for export, a number of new, energy intensive industries depend in large part on hydropower development. Most notable is the proposed mining and smelting of an estimated 300 million tons of exploitable bauxite to produce alumina in Laos. Production of 0.5 M tons of alumina per year would require around 150 MW for smelting and 600–800 MW for downstream processing (and a lot of water) [47]. Production of other mineral products is also likely
14	Industrial development drives further demand for electricity	All LMB countries show high average annual demand growth rates between 2010 and 2025 of between 5.5% in Thailand and 11.6% in Cambodia based on official forecasts [7]. Economic growth in the region is expected to increase 240% between 2005 and 2030 [7]

3.3. Hydropower–Food Supply Nexus Influence Model for the LMB

In Figure 7, the food supply and hydropower feedback loops are connected with two major links, as follows. In Loop 1, construction of water infrastructure is diminishing wild fish stocks. In Loop 2, hydro dam construction increases demand for electricity. Details of the variables A to M and the causal links 1 to 14 are found in the lists provided for Figures 5 and 6.

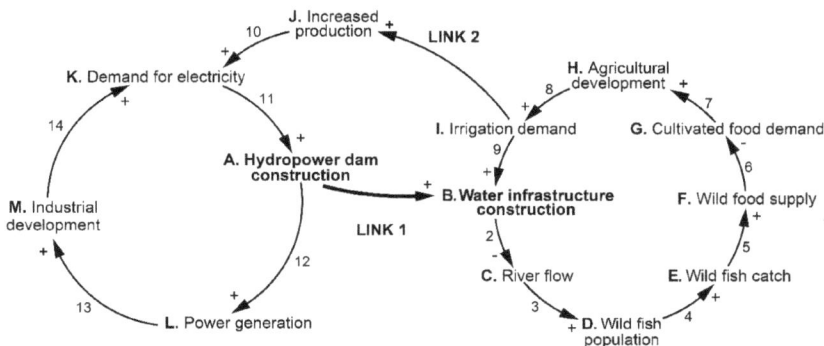

Figure 7. Influence model for the hydropower–food supply nexus in the lower Mekong Basin.

Connecting these two reinforcing loops to create a generic river basin model, in Link 1, hydro dam construction is projected to lead to a net reduction in wild fish stock, diminishing the supply of protein (Table 1), and consequently, adding to increasing demand for protein [11]. Agricultural development in part is being driven by demand for protein rich foods in countries with emerging economies, including Thailand and Vietnam [48–50]. In Link 2, agricultural development requires considerable energy for pumping water, for manufacture of farm chemicals and for transport, among other aspects [51].

Arguably, a third link would show that increased irrigation diminishes power generation by diverting water upstream of hydropower dams. For simplicity this link is not included. While the precise degree to which hydropower generation would be diminished by irrigation diversions is unknown, the major areas of irrigation in the Basin are currently downstream of planned dams. Further, for one of the major proposed lift irrigation projects the immediate downstream loss of hydropower potential is 5.2% in the wet season and double that in the dry season [52].

Consequently, agricultural intensification is increasing demand for more energy and thus power generation. Balancing loops are also present in this system, such as related to the dynamics of fish population (catch affects the stock) and ecosystem health. On the other hand, the two reinforcing feedback loops mentioned above emerge as the dominant ones.

In this way, the water infrastructure–food production and the hydropower–industrial development loops are connected and self-reinforcing. To test this generic model, it was applied to the hydropower–food supply nexus in the Lower Mekong Basin to provide the starting point to construct a model of the energy food water nexus in that region. Elaborating the model, Figure 8 gives our initial version of an elaborated influence model for the hydropower–food supply nexus in the lower Mekong Basin. This elaborated model also includes other major, external influences that affect major variables of the generic models, for example, fossil fuel inputs into energy as an individual factor in agricultural development [53]. The darker shaded overlap between the energy and food loops in Figure 8 highlights that water, agricultural resources use, biodiversity and hydropower dams are the nexus in this system.

It also starts to identify consequences for a range of natural resources availability as ecosystem services are increasingly used for energy and agricultural production. These include impacts on forest cover, biodiversity and atmospheric greenhouse gas composition.

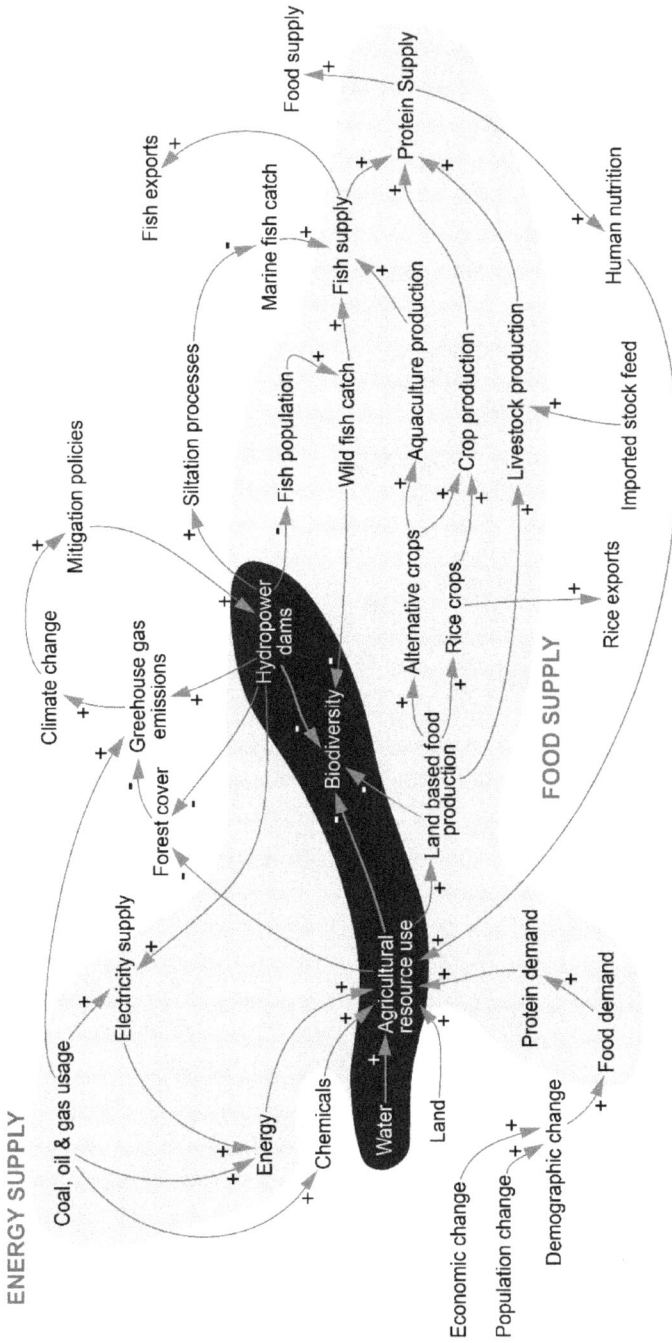

Figure 8. Elaborate influence model for the hydropower–food supply nexus in the lower Mekong Basin. The shaded food supply loop on the right is drawn from Figure 5, whereas the energy loop on the left is based on Figure 6. The darker shaded overlap between these two loops highlights that water, agricultural resources use, biodiversity and hydropower dams are the nexus in this system.

4. Discussion

This influence model for the hydropower–food supply nexus in the lower Mekong Basin illustrates two important attributes of this system, namely that it is a complex system that has reinforcing loops that tend to increase resource consumption at the expense of ecosystem health, and also that achieving desired outcomes requires strategic interventions.

4.1. A Complex System

Our model (Figure 8) highlights that the hydropower–food supply nexus in the lower Mekong Basin is part of a complex system where an intervention in one aspect, such as hydro dam construction, will have cascading consequences for a wide range of other sectors in the socio-ecological system. This raises a number of challenges for governance institutions. The key implication is that sectoral decision making to maximize outputs of one variable can have negative consequences on other variables in other sectors. A key example in this case is that hydro dam development leads to a loss in wild fish stock among a range of negative impacts on other variables.

More conventional assessment approaches misrepresent the complexity of the system being impacted, for example, by investment in hydropower dams. As an example, the 2015 technical review report on "Prior Consultation for the Proposed Don Sahong Hydropower Project" in Laos focuses almost exclusively on potential local impacts without considering the broader range of values influenced in the Basin system [54]. Further Matthews and Motta [55] detail the gap between Chinese policies to manage the water, energy and food nexus versus the lack of application by Chinese state-owned enterprises constructing dams in the Mekong region. We contend that the influence modeling detailed here could inform better nexus decisions in both these cases were the relevant decision makers motived to improve outcomes across sectors.

An important finding from this influence model is that the current development trajectory in the lower Mekong Basin is self-reinforcing. To elaborate: hydro dam development in diminishing wild fish stocks adds to the demand for alternative protein-rich food supplies from intensive agriculture, which in turn requires industrial inputs such as farm chemicals and water pumping. This further increases demand for electricity, hydropower dam construction and further undermines environmental integrity. In other words, without high-level policy intervention across sectors, such as by central institutions in a national government, the current development path is not sustainable and will continue to diminish fisheries, increase land use conversion to agriculture, consume more water and drive biodiversity loss.

Decisions in single sectors, especially those that are more economically and politically powerful such as in energy generation and agriculture, are unlikely to maximize benefits and minimize impacts for sustainable development. This suggests that institutions such as the Mekong River Commission, whose mandate does not cover all of the key variables in this system, cannot catalyze sustainable development without the active support of senior leaders in its member national governments [56]. Further, the river system links the lower Mekong Basin nations, such that (for example) hydropower dam development in Laos and Vietnam impacts negatively on variables downstream in Cambodia (such as fish supply) regardless of the Cambodian Government's policies and practices.

There may be two approaches to managing these transboundary water use conflicts within the Mekong River basin. As Belinskij [57] identifies, the 1995 Mekong Agreement has a number of gaps (such as regulating use of tributaries) and emphasizes the sovereignty of member states more than international water law. Were the Mekong Agreement to be revised, influence modeling could be used to inform the scope of a new mandate consistent with international water law that would strengthen institutions for managing the nexus. Further, the lower Mekong Basin national governments could make better cross-sectoral decisions at national and lower Basin scales using influence modelling. There are governance processes that with enhancements could be harnessed to this end, for instance, the national five-year planning process in Vietnam [42], and the Basin Development Planning process of the Mekong River Commission [58]. The influencing modeling could be applied, for example, to underpin and increase the robustness of alternative scenarios for policy development, such as those proposed in the Mekong region by Kesninen et al. [59].

Many of the causal links illustrated in Figure 8 are not new. A key question is why will another analytical or planning tool be used by decision makers, given that so many nexus decisions taken in the Mekong have perverse impacts? What our influence model provides is a framework for policy makers to explore how to maximize benefits and minimize perverse outcomes across sectors, in order for them to make more informed decisions. Over time, we argue that particularly impacted countries and key governmental agencies will be motived to use influence models to solve acute problems. One example is the recent decisions by the government in Vietnam to cease rice intensification programs in the Mekong delta in favor of restoring traditional floating rice and diversifying food production because of a number of negative impacts of triple crop rice production.

4.2. Strategic Interventions

This model (Figure 8) then prompts the question of whether government and other decision makers can intervene to generate alternative outcomes, for instance, to enhance food security or maintain forest cover, without incurring unacceptable negative impacts. A number of the variables in this model are very difficult to change. For example, alone a national government in the Mekong region will have almost no influence over climate change, as the atmosphere is a global commons, and demographic change has long lag times that mean it would take decades to effect a different outcome were this desired. Yet this model suggests that there are a number of strategic opportunities for government and other decision makers to intervene to significantly change the outcomes. Here we suggest a three such interventions to illustrate this potential: in particular, alternative electricity supplies, agricultural land use intensification and better fisheries management.

Supplying electricity differently could significantly change outcomes. If non-hydro renewable power generation were used to forestall new hydropower generation capacity, there would be the potential to conserve wild fish populations, biodiversity and forest cover, while reducing greenhouse gas emissions [52]. For instance, a number of studies suggest that significant wind and biofuel generation is feasible in the region, as well as demand management through energy trade and also energy efficiency in the industrial sector [60–62]. These interventions would require policies that favor investment in non-hydro renewable power generation. Alternatively, if there were a desire to continue hydropower development but reduce the scale of its perverse impacts, a number of methods exist to select projects with lower social and environmental impacts as opposed to the current ad hoc commissioning processes [63,64]. These methods would, for instance, enable projects to be selected that have lower impacts on fish populations or inundate smaller areas of land. More centralized planning of hydropower concessions would be required.

Theoretically, managing existing agricultural land more intensively and sustainably has the potential to produce more food with no further loss of forest cover and biodiversity, while reducing greenhouse gas emissions [65]. Sustainable intensification of agriculture is the focus of relevant global programs, including of the UN Food and Agricultural Organization [66] and the Cooperative Group on International Agriculture Research [67]. Nevertheless there are risks, including of increased agricultural productivity driving demand for agricultural expansion in the landscape [68,69]. Governments would need to better regulate land use and invest in agricultural support for this intervention to succeed.

Better wild fisheries management has the potential to increase the sustainable catch and thus the protein supply, reducing demand for this food from aquaculture, crops and livestock. Measures such as fish conservation areas, enforcement of fisheries regulations and removal of non-hydropower barriers to migration may all increase fish stocks and thus the sustainable yield [70–72]. This intervention requires stronger, day-to-day regulation at a local scale if it is to succeed.

The three interventions described above could be considered largely complementary and could be led by sectoral government agencies, such as those for energy, agriculture, forests and fisheries. In this sense these interventions are not overly complex governance challenges. We suggest that the primary challenges for societies and governments to implement this sort of approach lie in leadership and transparent implementation. Political leadership is required for these types of strategic intervention to receive priority and funding for implementation from central government planning and financial agencies. As these types of interventions involve favoring some kinds of vested interests over others, such as solar panel suppliers over hydropower developers, then transparency of decision making and enforcement will be needed at national and sub-national scales.

4.3. Analytical Value

Our analysis of the hydropower–food supply nexus in the lower Mekong Basin through progressively more elaborate influence models has mapped a complex system at a level of detail that may be interrogated to identify strategic interventions that may maximize desirable outcomes while minimizing perverse impacts. We argue that this approach can be used to identify the most important, generic variables that may apply to the hydropower–food supply nexus to inform policy making for other large river systems globally. Around the world, planning is advanced to develop hydropower on the last remaining free flowing rivers, for example, with the Amazon, Amur, Brahmaputra, Congo and Irrawaddy rivers [73]. This model needs to be tested to assess its applicability to such large rivers around the world. At either end of a development spectrum, two places where there may be sufficient data to inform such an analysis are the heavily developed Yangtze River in China [74] and the Tapajos River tributary of the Amazon that is little impacted to date (Bagossi, personal communication). This type of analysis would enable national and sub-national policy makers to take more informed development and management decisions across a range of sectors.

4.4. Novelty

This research adds new findings and methodological lessons for managing the water, energy and food nexus in the Mekong River basin and transboundary river basins more generally. The methodological novelty lies in the systematic scope of these influence models, which show how any one intervention has knock-on effects on others. More conventional assessment approaches, such as the Don Sahong hydropower project environmental impact assessment, misrepresent the complexity of the system and are thus likely to result in negative impacts. Application of these models is not only an analytical tool, but can also be a focus for dialogue among actors across sectors, and inform governance.

A key finding from this analysis of the Mekong River basin is that the current development trajectory for more industrial energy and food production is self-reinforcing. This means that for proponents of changed management in a particular sector, that they will need to engage other sectors and at greater scales if their proposed change is to last. In the case of the Mekong, the analysis presented here highlights policy decisions that need to be taken by national leaders if greater positive synergies with interventions across sectors are to be realized.

5. Conclusions

This analysis of the hydropower–food supply nexus in the lower Mekong Basin through the application of influence models has shown how changing one variable can be seen to have knock on effects throughout a complex system. This modeling also enables identification of options for strategic interventions to maximize benefits while minimizing negative impacts. This kind of analysis requires political leadership and would enable national and sub-national policy makers to take more informed cross-sectoral decisions. We propose that this approach be tested to see if it may identify the generic variables that may apply to the hydropower–food supply nexus to inform policy making for other large river systems globally. The different levels of industrial use of the waters of large transboundary rivers offers the prospect that systematically applying this model may enable lessons on avoiding perverse impacts and seizing positive synergies to be shared to enhance river management.

In the context of a "nexus approach" to transboundary river basins [75], the use of influence models aids systematic analysis by identifying links across sectors. Developing and using influence models enables discourse among actors on trade-offs and synergies across water, energy, food and other sectors. Finally, by making costs and benefits more explicit among different actors, this modeling can enhance governance with better cross-sectoral collaboration and policy coherence.

Acknowledgments: The authors' research was supported by the Luc Hoffmann Institute for conservation research through its Navigating the Nexus program. Clive Hilliker, ANU, prepared the figures in this paper.

Author Contributions: Jamie Pittock proposed the research, was lead writer, and assessed the variables. Jamie Pittock and David Dumaresq drafted the initial influence diagrams and text. Andrea M. Bassi contributed expertise on influence modelling methods, and revised the diagrams and text.

Conflicts of Interest: The authors declare no conflict of interest.

References

1. Molden, D. (Ed.) *Water for Food, Water for Life a Comprehensive Assessment of Water Management in Agriculture*; Earthscan & International Water Management Institute: London, UK; Colombo, Sri Lanka, 2007.
2. International Energy Agency (IEA). *World Energy Outlook 2012*; IEA: Paris, France, 2012.
3. Hussey, K.; Pittock, J.; Dovers, S. Justifying, extending and applying "nexus" thinking in the quest for sustainable development. In *Climate, Energy and Water*; Pittock, J., Hussey, K., Dovers, S., Eds.; Cambridge University Press: Cambridge, UK, 2015; pp. 1–5.
4. Richter, B.D.; Postel, S.; Revenga, C.; Scudder, T.; Lehner, B.; Churchill, A.; Chow, M. Lost in development's shadow: The downstream human consequences of dams. *Water Altern.* **2010**, *3*, 14–42.
5. Vorosmarty, C.J.; McIntyre, P.B.; Gessner, M.O.; Dudgeon, D.; Prusevich, A.; Green, P.; Glidden, S.; Bunn, S.E.; Sullivan, C.A.; Liermann, C.R.; et al. Global threats to human water security and river biodiversity. *Nature* **2010**, *467*, 555–561. [CrossRef] [PubMed]
6. World Commission on Dams (WCD). *Dams and Development: A New Framework for Decision-Making. The Report of the World Commission on Dams*; Earthscan: London, UK, 2000.
7. International Center for Environmental Management (ICEM). *MRC Strategic Environmental Assessment (SEA) of Hydropower on the Mekong Mainstream: Final Report*; ICEM: Hanoi, Vietnam, 2010.
8. Hortle, K.G. *Consumption and the Yield of Fish and Other Aquatic Animals from the Lower Mekong Basin*; Mekong River Commission: Vientiane, Laos, 2007.
9. Ziv, G.; Baran, E.; Nam, S.; Rodríguez-Iturbe, I.; Levin, S.A. Trading-off fish biodiversity, food security, and hydropower in the Mekong river basin. *Proc. Natl. Acad. Sci. USA* **2012**, *109*, 5609–5614. [CrossRef] [PubMed]
10. Orr, S.; Pittock, J.; Chapagain, A.; Dumaresq, D. Dams on the Mekong River: Lost fish protein and the implications for land and water resources. *Glob. Environ. Chang.* **2012**, *22*, 925–932. [CrossRef]
11. Pittock, J.; Dumaresq, D.; Orr, S. The Mekong River: Trading off hydropower, fish and food. *Reg. Environ. Chang.* **2016**, submitted.
12. Meadows, D.H. The unavoidable a priori. In *Elements of the System Dynamics Method*; Randers, J., Ed.; Productivity Press: Portland, OR, USA, 1980; pp. 23–57.
13. Randers, J. *Elements of System Dynamics Method*; Productivity Press: Portland, OR, USA, 1980.
14. Richardson, G.P.; Pugh, A.I., III. *Introduction to System Dynamics Modeling with Dynamo*; Productivity Press: Portland, OR, USA, 1981.
15. Forrester, J.W.; Lux, N.; Stuntz, L. *Road Maps: A guide to Learning System Dynamics*; System Dynamics Group, Sloan School of Management, MIT: Cambridge, MA, USA, 2002.
16. Sterman, J.D. *Business Dynamics: Systems Thinking and Modeling for a Complex World*; Irwin/McGraw-Hill: Boston, MA, USA, 2000.
17. Probst, G.; Bassi, A. *Tackling Complexity. A Systems Approach for Decision Makers*; Greenleaf Publishing: London, UK, 2014.
18. Forrester, J.W. *Industrial Dynamics*; Productivity Press: Portland, OR, USA, 1961.
19. Proust, K.; Newell, B. *Catchment and Community: Towards a Management-Focused Dynamical Study of the Act Water System*; The Australian National University: Canberra, Australia, 2006.

20. Dyball, R.; Newell, B. *Understanding Human Ecology: A Systems Approach to Sustainability*; Earthscan/Routledge: London, UK, 2014.

21. Proust, K.; Dovers, S.; Foran, B.; Newell, B.; Steffen, W.; Troy, P. *Climate, Energy and Water: Accounting for the Links*; Land and Water Australia: Canberra, Australia, 2007; p. 68.

22. Johnston, R.; Try, T.; de Silva, S. *Agricultural Water Management Planning in Cambodia*; International Water Management Institute and Australian Centre for International Agricultural Research: Colombo, Sri Lanka, 2013.

23. Postel, S.; Richter, B. *Rivers for Life: Managing Water for People and Nature*; Island Press: Washington, DC, USA, 2003; p. 253.

24. Amornsakchai, S.; Annez, P.; Vongvisessomjai, S.; Choowaew, S.; Kunurat, P.; Nippanon, J.; Schouten, R.; Sripapatrprasite, P.; Vaddhanaphuti, C.; Vidthayanon, C.; et al. *Pak Mun Dam, Mekong River Basin, Thailand. A WCD Case Study Prepared as an Input to the World Commission on Dams*; World Commission on Dams: Cape Town, South Africa, 2000.

25. Basin Development Plan Programme (BDPP). *Regional Irrigation Sector Review for Joint Basin Planning Process*; Basin Development Plan Programme, Mekong River Commission: Vientiane, Laos, 2009.

26. International Center for Environmental Management (ICEM). *MRC SEA for Hydropower on the Mekong Mainstream. Fisheries Baseline Assessment Working Paper*; ICEM: Hanoi, Vietnam, 2010.

27. Food and Agricultural Organization of the United Nations. *FAOstat*; FAO: Rome, Italy, 2009.

28. Pech, S. Water sector analysis. In *The Water-Food-Energy Nexus in the Mekong Region*; Smajgl, A., Ward, J., Eds.; Springer: New York, NY, USA, 2013; pp. 19–60.

29. Godfray, H.C.J.; Beddington, J.R.; Crute, I.R.; Haddad, L.; Lawrence, D.; Muir, J.F.; Pretty, J.; Robinson, S.; Thomas, S.M.; Toulmin, C. Food security: The challenge of feeding 9 billion people. *Science* **2010**, *327*, 812–818. [CrossRef] [PubMed]

30. Floch, P.; Molle, F. Irrigated agriculture and rural change in Northeast Thailand: Reflections on present developments. In *Governing the Mekong: Engaging in the Politics of Knowledge*; Daniel, R., Lebel, L., Manorom, K., Eds.; Strategic Information and Research Development Centre: Petaling Jaya, Malaysia, 2013; pp. 185–198.

31. Phengphaengsy, F.; Noble, A. Reconsidering irrigation management transfer in Laos. In *Governing the Mekong: Engaging in the Politics of Knowledge*; Daniel, R., Lebel, L., Manorom, K., Eds.; Strategic Information and Research Development Centre: Petaling Jaya, Malaysia, 2013; pp. 137–162.

32. Thuon, T. Localizing development and irrigation management in Cambodia. In *Governing the Mekong: Engaging in the Politics of Knowledge*; Daniel, R., Lebel, L., Manorom, K., Eds.; Strategic Information and Research Development Centre: Petaling Jaya, Malaysia, 2013; pp. 163–184.

33. Brooks, J. *Agricultural Policy Choices in Developing Countries: A Synthesis*; OECD: Paris, France, 2010.

34. Ministry for Agriculture and Forestry. *Strategy for Agricultural Development 2011 to 2020*; Lao PDR Ministry for Agriculture and Forestry: Vientiane, Laos, 2010.

35. Tran, C.T. *Overview of Agriculture Policy in Vietnam*; Food Fertilizer and Technology Center for the Asian and Pacific Region: Taipei, Taiwan, 2014.

36. Singhapreecha, C. *Economy and Agriculture in Thailand*; Food Fertilizer and Technology Center for the Asian and Pacific Region: Taipei, Taiwan, 2014.

37. Trung, N.H.; Tuan, L.A.; Trieu, T.T. Multi-level governance and adaptation to floods in the Mekong delta. In *Governing the Mekong: Engaging in the Politics of Knowledge*; Daniel, R., Lebel, L., Manorom, K., Eds.; Strategic Information and Research Development Centre: Petaling Jaya, Malaysia, 2013; pp. 111–126.

38. De Silva, S.; Johnston, R.; Senaratna Sellamuttu, S. *Agriculture, Irrigation and Poverty Reduction in Cambodia: Policy Narratives and Ground Realities Compared*; CGIAR Research Program on Aquatic Agricultural Systems: Penang, Malaysia, 2014.

39. International Energy Agency (IEA). *Southeast Asia Energy Outlook. World Energy Outlook Special Report*; IEA: Paris, France, 2013.

40. World Bank. *Clear Skies. Cambodia Economic Update*; World Bank Group: Washington, DC, USA, 2014.

41. World Bank. *Country Partnership Strategy Progress Report for the Lao People's Democratic Republic*; Report No. 90281-la; World Bank Group: Washington, DC, USA, 2014.

42. World Bank. *Country Partnership Strategy for the Socialist Republic of Vietnam for the Period fy12-fy16*; Report No. 65200-vn; World Bank Group: Washington, DC, USA, 2011.

43. The Mekong River Commission (MRC). *Mekong Basin Hydropower Database—Master*; The Mekong River Commission Secretariat: Vientiene, Laos, 2009.
44. Central Intelligence Agency (CIA). *Thailand*; CIA: Langley, VA, USA, 2015.
45. Fullbrook, D. Food security in the wider Mekong region. In *The Water-Food-Energy Nexus in the Mekong Region*; Smajgl, A., Ward, J., Eds.; Springer: New York, NY, USA, 2013; pp. 61–104.
46. Mushtaq, S.; Maraseni, T.N.; Maroulis, J.; Hafeez, M. Energy and water tradeoffs in enhancing food security: A selective international assessment. *Energy Policy* **2009**, *37*, 3635–3644. [CrossRef]
47. Lazarus, K. Mining in the Mekong region. In *The Water-Food-Energy Nexus in the Mekong Region*; Smajgl, A., Ward, J., Eds.; Springer: New York, NY, USA, 2013; pp. 191–208.
48. Walker, P.; Rhubart-Berg, P.; McKenzie, S.; Kelling, K.; Lawrence, R.S. Public health implications of meat production and consumption. *Public Health Nutr.* **2005**, *8*, 348–356. [CrossRef] [PubMed]
49. Bender, A. *Meat and Meat Products in Human Nutrition in Developing Countries*; FAO: Rome, Italy, 1992.
50. High Level Panel of Experts on Food Security and Nutrition (HLPE). *Note on Critical and Eerging Issues for Food Security and Nutrition. Prepared for the Committee on World Food Security*; FAO: Rome, Italy, 2014.
51. Faidley, L. Energy and agriculture. In *Energy in Farm Production*; Fluck, R., Ed.; Elsevier: Amsterdam, The Netherlands, 1992; pp. 1–12.
52. Foran, T. Impacts of natural resource-led development on the Mekong energy system. In *The Water-Food-Energy Nexus in the Mekong Region*; Smajgl, A., Ward, J., Eds.; Springer: New York, NY, USA, 2013; pp. 105–142.
53. Conforti, P.; Giampietro, M. Fossil energy use in agriculture: An international comparison. *Agric. Ecosyst. Environ.* **1997**, *65*, 231–243. [CrossRef]
54. The Mekong River Commission (MRC). *Technical Review Report on Prior Consultation for the Proposed Don Sahong Hydropower Project*; The Mekong River Commission Secretariat: Vientiane, Laos, 2015.
55. Matthews, N.; Motta, S. Chinese state-owned enterprise investment in Mekong hydropower: Political and economic drivers and their implications across the water, energy, food nexus. *Water* **2015**, *7*, 6269–6284. [CrossRef]
56. Dore, J.; Lazarus, K.; Molle, F.; Foran, T.; Kakonen, M. De-marginalizing the Mekong River Commission. In *Contested Waterscapes in the Mekong Region: Hydropower, Livelihoods and Governance*; Molle, F., Foran, T., Kakonen, M., Eds.; Earthscan: Abingdon, UK, 2012; pp. 357–382.
57. Belinskij, A. Water-energy-food nexus within the framework of international water law. *Water* **2015**, *7*, 5396–5415. [CrossRef]
58. The Mekong River Commission (MRC). *Mekong Basin Planning. The Story Behind the Basin Development Plan. The BDP Story*; MRC: Vientienne, Laos, 2013.
59. Keskinen, M.; Someth, P.; Salmivaara, A.; Kummu, M. Water-energy-food nexus in a transboundary river basin: The case of Tonle Sap Lake, Mekong River basin. *Water* **2015**, *7*, 5416–5436. [CrossRef]
60. Watcharejyothin, M.; Shrestha, R.M. Regional energy resource development and energy security under CO_2 emission constraint in the greater Mekong sub-region countries (GMS). *Energy Policy* **2009**, *37*, 4428–4441. [CrossRef]
61. Foran, T. Action research to improve Thailand's electricity planning processes. In *Governing the Mekong: Engaging in the Politics of Knowledge*; Daniel, R., Lebel, L., Manorom, K., Eds.; Strategic Information and Research Development Centre: Petaling Jaya, Malaysia, 2013; pp. 49–70.
62. Karki, S.K.; Mann, M.D.; Salehfar, H. Energy and environment in the ASEAN: Challenges and opportunities. *Energy Policy* **2005**, *33*, 499–509. [CrossRef]
63. Opperman, J.J.; Hartmann, J.; Harrison, D. Hydropower within the climate, energy and water nexus. In *Climate, Energy and Water*; Pittock, J., Hussey, K., Dovers, S., Eds.; Cambridge University Press: Cambridge, UK, 2015; pp. 79–107.
64. The Mekong River Commission (MRC). *Basin-Wide Rapid Sustainability Assessment Tool*; MRC: Vientiene, Laos, 2010.
65. Godfray, H.C.J.; Garnett, T. Food security and sustainable intensification. *Phil. Trans. R. Soc. B* **2014**, *369*, 20120273. [CrossRef] [PubMed]
66. UN Food and Agricultural Organisation. *Strategic Framework 2010–2019*; UN Food and Agricultural Organisation: Rome, Italy, 2009.

67. Water Land and Ecosystems (WLE). *Sustainable Intensification of Agriculture: Oxymoron or Real Deal*; CGIAR Research Program on Water, Land and Ecosystems: Colombo, Sri Lanka, 2015.

68. Tscharntke, T.; Klein, A.M.; Kruess, A.; Steffan-Dewenter, I.; Thies, C. Landscape perspectives on agricultural intensification and biodiversity—Ecosystem service management. *Ecol. Lett.* **2005**, *8*, 857–874. [CrossRef]

69. Perfecto, I.; Vandermeer, J. The agroecological matrix as alternative to the land-sparing/agriculture intensification model. *Proc. Natl. Acad. Sci. USA* **2010**, *107*, 5786–5791. [CrossRef] [PubMed]

70. Dugan, P.J.; Barlow, C.; Agostinho, A.A.; Baran, E.; Cada, G.F.; Chen, D.; Cowx, I.G.; Ferguson, J.W.; Jutagate, T.; Mallen-Cooper, M. Fish migration, dams, and loss of ecosystem services in the Mekong basin. *Ambio* **2010**, *39*, 344–348. [CrossRef] [PubMed]

71. Biodiversity and Fisheries Management Opportunities in the Mekong River Basin. Available online: http://69.90.183.227/doc/nbsap/fisheries/Coates.pdf (accessed on 22 September 2016).

72. Cowx, I.; Almeida, O.; Bene, C.; Brummett, R.; Bush, S.; Darwall, W.; Pittock, J.; Van Brakel, M. Value of river fisheries. In Proceedings of the Second International Symposium on the Management of Large Rivers for Fisheries, Volume I; Welcomme, R., Petr, T., Eds.; FAO Regional Office for Asia & the Pacific: Bangkok, Thailand, 2004; pp. 1–20.

73. World Wide Fund (WWF). *Free-Flowing Rivers—Economic Luxury or Ecological Necessity*; WWF: Gland, Switzerland, 2006.

74. Anonymous. *Yangtze Conservation and Development Report*; Chinese Academy of Sciences: Beijing, China, 2007.

75. Keskinen, M.; Guillaume, J.; Kattelus, M.; Porkka, M.; Räsänen, T.; Varis, O. The water-energy-food nexus and the transboundary context: Insights from large Asian rivers. *Water* **2016**, *8*, 193. [CrossRef]

water

MDPI

Article

The Water-Energy-Food Nexus and the Transboundary Context: Insights from Large Asian Rivers

Marko Keskinen *, Joseph H. A. Guillaume, Mirja Kattelus, Miina Porkka, Timo A. Räsänen and Olli Varis

Water & Development Research Group, Aalto University, PO Box 15200, 00076 Aalto, Finland; joseph.guillaume@aalto.fi (J.H.A.G.); mirja.kattelus@aalto.fi (M.K.); miina.porkka@aalto.fi (M.P.); timo.rasanen@aalto.fi (T.A.R.); olli.varis@aalto.fi (O.V.)
* Correspondence: marko.keskinen@aalto.fi; Tel.: +358-50-3824626

Academic Editor: Davide Viaggi
Received: 31 January 2016; Accepted: 26 April 2016; Published: 10 May 2016

Abstract: The water-energy-food nexus is a topical subject for research and practice, reflecting the importance of these sectors for humankind and the complexity and magnitude of the challenges they are facing. While the nexus as a concept is not yet mature or fully tested in practice, it has already encouraged a range of approaches in a variety of contexts. This article provides a set of definitions recognizing three perspectives that see the nexus as an analytical tool, governance framework and as an emerging discourse. It discusses the implications that an international transboundary context brings to the nexus and *vice versa*. Based on a comparative analysis of three Asian regions—Central Asia, South Asia and the Mekong Region—and their related transboundary river basins, we propose that the transboundary context has three major implications: diversity of scales and perspectives, importance of state actors and importance of politics. Similarly, introducing the nexus as an approach in a transboundary context has a potential to provide new resources and approaches, alter existing actor dynamics and portray a richer picture of relationships. Overall, the significance of water-energy-food linkages and their direct impacts on water allocation mean that the nexus has the potential to complement existing approaches also in the transboundary river basins.

Keywords: transboundary water-energy-food nexus; transboundary rivers; water resources management; water security; energy security; food security; Central Asia; South Asia; Southeast Asia; Mekong

1. Introduction

Nexus-nexus everywhere—the water-energy-food nexus ("the nexus") has been recognized, promoted and also criticized by a variety of actors (e.g., [1–12]) during the past 5 years. As a result, the number of nexus studies, articles and even special issues is rapidly increasing. And for good reason, given how critical the three 'nexus sectors' of water, energy and food are for the very existence of humankind.

Yet, no commonly agreed definition or conceptual framework for the nexus has emerged and therefore different organisations and authors—intentionally or not—interpret its essence quite differently. Even the very name of the approach is not consistent, with the three nexus sectors written in differing order and the term "security" both included and excluded from the term. The actual number of nexus sectors also differs, focusing sometimes only on two sectors (e.g., [13,14]) or extended to additional sectors such as climate change [15,16], ecosystems [17] and livelihoods [18]. The terms change too, for example, food is sometimes replaced by land [19]. The actual context where nexus is applied varies greatly as well, ranging from cities [2,20,21] to transboundary river basins crossing international borders [17,22–25]. Despite all these publications, the actual added value provided by the nexus approach in different contexts remains partly unclear and also contested.

This article—published as part of the Water journal's "Water-Energy-Food Nexus in Large Asian River Basins" special issue—aims to provide some clarity and structure for the current nexus discussion in the specific context of transboundary river basins. This also means that we focus on the nexus from a water management point of view. We start by synthesizing definitions for the nexus concept based on interpretation of its historical emergence and reflections on its popularity, building on analysis of relevant nexus literature. This provides a foundation for an illustrative comparative nexus analysis for three regions centred around selected transboundary Asian river basins. Based on this analysis as well as the relevant findings from the other articles published in this Special Issue [17,21,26–33], we then discuss what the nexus means in the specific context of transboundary river basins, noting the implications that transboundary context bring to nexus approaches, and *vice versa*.

2. Water-Energy-Food Nexus: Defining the "Nexus Approach"

2.1. The Emergence of Water-Energy-Food Nexus

Water, energy and food are the key prerequisites for our existence, there is no life without them. They are also closely linked with water acting commonly as an enabler and in the areas of water scarcity, limiter for food and energy production [34–36]. Globally, food production comprises the great majority of consumptive water use [37,38], while the most important renewable energy production types in particular, including hydropower, bioenergy and waste, are dependent on water [39]. Hence, the idea of looking at these three aspects and their connections together is in no way new (e.g., [40–42]). In fact, it can be argued that scientists, public officials and other practitioners have been well aware of such interactions for many decades; it has just not happened under the specific term "nexus".

What then, has led to the rapid increase in nexus-related literature since 2011 (a search in Google Scholar for "water energy food nexus" generates just 7 results for year 2011 but 254 results for year 2015)? We argue that the reason is two-fold. The first is increased awareness about the economic risks included in the nexus. The second is the drive to promote the nexus as a new framework for global policy debate about the linkages between resource use and development and ultimately, about the means to facilitate sustainable development.

For the former, the key publication is the World Economic Forum's Global Risks 2011 report [4], which concluded that water, food and energy security are chronic impediments to economic growth and social stability, emphasised their interrelatedness and noted the importance of addressing the nexus trade-offs. For the latter, the discussion emerged first and foremost thanks to the Bonn2011 Conference on water, energy and food security nexus [1,2]. While the conference failed in its aim to include the nexus into the outcome document of 2012 United Nations Conference on Sustainable Development, it was successful in entering the nexus into global policy discussion more broadly. The conference linked the nexus strongly to hunger, poverty reduction and a human rights and recognised two major aspects for the nexus: one related to poverty reduction and the other to sustainable development and growth [1]. The methodological emphasis is similar to that of World Economic Forum, with a focus on trade-offs and synergies.

Other relevant publications and processes have pushed the nexus agenda further as well. In addition to the publications referred to in Introduction for example, the 2011/2012 European Report on Development [43] looked at linkages between water, energy and land, the annual World Water Week in Stockholm has discussed the nexus in many of its sessions in recent years and the UNECE has established a water-food-energy-ecosystems nexus task force that looks at the nexus in selected transboundary river basins [17]. Yet, even with all these publications and processes, there is no general agreement on what the nexus is and what a "nexus approach" actually means and requires. Actual empirical evidence on the benefits and caveats of applying a nexus approach is also relatively thin, although rapidly increasing [17,19,21,30,44].

2.2. Three Perspectives and Definitions for the Nexus

We argue that there are three interlinked perspectives and related definitions for the water-energy-food nexus, stemming from differing interpretations and expectations. First of all, many (see e.g., [2,3,45]) consider the nexus predominantly as an analytical approach or tool, intended to look at the three nexus sectors and their interconnections, often mainly quantitatively based on existing data and related models. This could include consideration of other sectors that have a major influence on one or more of water, energy and food.

The nexus also emerges as a governance framework, intended to facilitate cross-sectoral collaboration and enhance policy coherence across planning and management of the nexus sectors (see e.g., [1,43,46]). The third perspective can be seen to provide a foundation for the others, where the nexus emerges as a boundary concept [47,48] that is used to form a new discourse related to the use and management of water, energy and food [8,49–51]. The nexus is used as a means of establishing a new way of framing problems [52] related to water-energy-food linkages, complementing dominant, often more water-centred and technical (e.g., [22,53]) discourses, for example surrounding Integrated Water Resources Management (IWRM) [54].

Based on the discussion above, we can now provide our definitions on what a "nexus approach" is according to each perspective:

- **A nexus-based analysis** is a systematic process that explicitly includes consideration of water, energy, food and other linked sectors in either quantitative or qualitative terms with a view to better understanding their relationships and hence providing more integrated information for planning and decision making in these sectors.
- **A nexus approach to governance** is one that explicitly focuses on linkages between water, energy, food and linked sectors as well as their related actors in order to enhance cross-sectoral collaboration and policy coherence and ultimately promote sustainability, win-win solutions and resource use efficiency.
- **The nexus is an emerging discourse** that emphasizes the trade-offs and synergies across water-energy-food connections and encourages actors to cross their sectoral and disciplinary boundaries (*i.e.*, acting as a boundary concept).

Each of these broad definitions can and has been followed independently and can be expressed in more precise terms in many different ways, which explains the broad range of nexus approaches and definitions that have been observed to date. Without denying the legitimacy of other interpretations, we argue that an ideal nexus approach would integrate these three definitions and hence draws on disciplines across four levels: value, normative, pragmatic and empirical [55]. The value level corresponds to disciplines interested in values and ethics; the normative level to laws, politics and planning; the pragmatic to management and technological disciplines; and empirical to understanding of the biophysical and social world (Figure 1). The different perspectives of the nexus together span these levels, but in the ideal nexus approach the individual perspectives are transcended and integrated to form a broader view. For the complex problems related to water, energy and food security, it is as important to understand cross-level linkages between disciplines and between nexus perspectives as it is to understand cross-sectoral linkages themselves.

An ideal *nexus approach therefore consists of a systematic process for both analysis and policy-making that focuses on the linkages between water, energy, food and other linked sectors to promote sustainability, synergies and resource use efficiency.* Quantitative analysis helps to understand interconnections, particularly focusing on critical dependencies and means of reducing trade-offs. Analysis of the governance context identifies key actors, policies and legislative frameworks relevant to these connections. It should pay particular attention to politics as the lack of cross-sectoral collaboration is usually not merely due to lack of appropriate mechanisms and approaches, but relates first and foremost to the fight over influence and power between different sectors and actors [8,51].

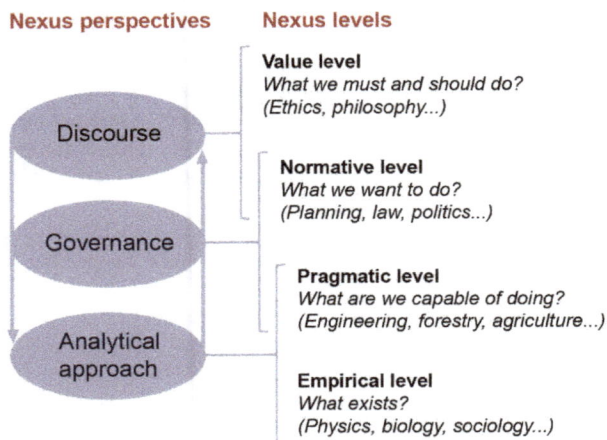

Figure 1. Nexus perspectives and their relation to disciplines of different levels (levels adapted from [55]). In an ideal nexus approach the individual perspectives are integrated to form a broader view, necessary to tackle the complex problems related to the water, energy and food nexus. In addition to cross-sectoral linkages, cross-level linkages between disciplines and between nexus perspectives should be understood.

Analyses should also remain open to changes in scope, recognizing that the nexus is also a dynamic boundary concept that encourages interaction on these key issues. For example, it should, where relevant, include related issues such as non-food, non-energy commodity agriculture that does not strictly fit within water-energy-food linkages. Together, these analyses inform efforts of the actors involved to manage interactions between sectors, facilitating policy coherence and cross-sectoral collaboration and potentially creating new ways to look at water-energy-food linkages.

Taken together, the different perspectives and levels of the nexus emphasise the need to have an integrated view on the nexus. Yet, very few nexus analyses have to our knowledge considered all these different perspectives, with one of the exceptions being the UNECE's Task Force on the Water-Food-Energy-Ecosystems Nexus. As described by another article in this special issue [17], the work of the task force builds on a coherent analytical approach that also addresses the governance context and aims ultimately to revise the way the UNECE views and assesses its transboundary waters. While recognising that it is difficult to achieve this level of integration under any single process or project, we see this as an ideal worth striving for—while leaving room for "better" visions and definitions of the nexus as they emerge.

In relation to transboundary contexts specifically, the linkages in question may occur across boundaries and on large scales. Thus, information must typically be integrated from disparate sources, collaboration must occur between politically independent entities and perspectives on the nexus are likely to differ markedly. The comparative analysis in the next section and the subsequent discussion aim to further highlight these issues.

3. Comparative Nexus Analysis in Three Asian Regions with Large Transboundary River Basins

This section presents a comparative nexus analysis for three regions, namely Central Asia, South Asia and the Mekong Region in Southeast Asia and their related transboundary river basins (Appendix A). The discussion section draws on this comparative analysis to identify key implications of the transboundary river basin context for the nexus and vice versa. Consistent with our approach to look at the nexus from a transboundary water management perspective, transboundary river basins are used as focal points. However, broader regions are used for this comparative analysis, recognising that energy and food interconnections are not limited by river basin boundaries (Appendix B).

The analysis builds on the concept of a "nexus triangle" (Figure 2), which summarizes key nexus interconnections and the relevance of each nexus sector in each region. As the name suggests, the purpose of the comparative analysis is to provide a general comparison of how the nexus manifests itself in different transboundary regions with large river basins. It builds on our previous research in these regions (e.g., [27,30,56–66]) as well as key literature referred to in the text. Comparative analysis is not intended to provide an in-depth analysis of a nexus situation in any particular region or river basin.

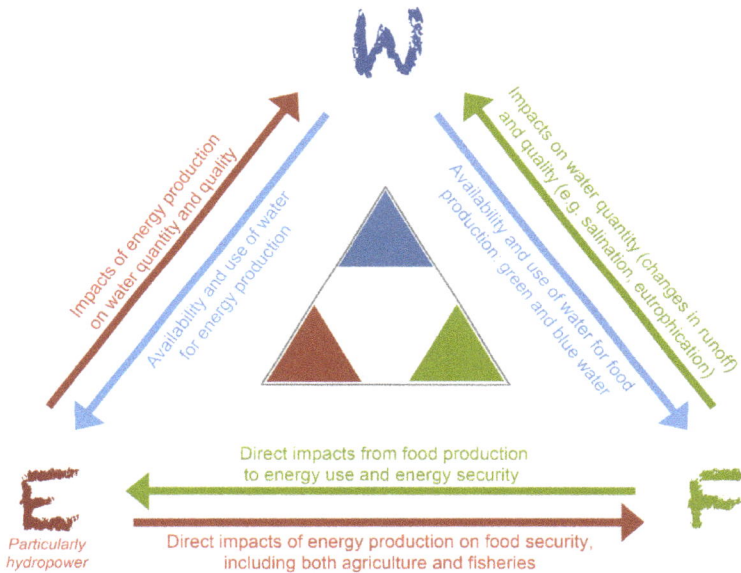

Figure 2. The logic of our "nexus triangle", indicating how we have defined the key interconnections and impacts between the three nexus sectors of water (W), energy (E) and food (F). In the region-specific nexus triangles, we used three different thicknesses for the arrow, corresponding with our subjective judgment of the significance of the connection/impact comparatively within and between regions: the thicker the arrow, the more significant the connection or impact to be. The "status triangle" in the middle similarly indicates the current levels of water security (blue), energy security (red) and food security (green) comparatively for the three regions using three different sizes for the area: the larger the coloured area, the higher level that type of security in the region is seen to be.

3.1. Central Asia

Central Asia here refers to Kazakhstan, Kyrgyzstan, Tajikistan, Turkmenistan and Uzbekistan, who notably share the Aral Sea basin. Key aspects of the history of the nexus in Central Asia are the need for irrigation in arid areas, winter heating in mountainous areas and environmental impacts from water abstraction and hydropower. Oil and gas production have an important role in Kazakhstan, Turkmenistan and Uzbekistan for national and regional energy security.

During the Soviet era, the water, energy and food sectors were developed in a closely connected manner. Downstream riparian states focused on developing agricultural production (particularly cotton) and provided fossil fuel for heating, while upstream states stored water to facilitate agricultural production downstream as well as to produce hydropower energy [26,67]. The Amu Darya and Syr Darya Rivers were seen as resources to be utilised to allow development, in what has been described as a "hydraulic mission" [68]. Unsustainably high water use, particularly for irrigation, has led to high water stress and resulting environmental consequences, most visibly on the Aral Sea [27,69–72]. Today, key aspects and challenges of the nexus in Central Asia are linked to the historic infrastructural interdependences via water storages, dams and power grids, and high profile competition over the water resources [26–28,67] (Figure 3).

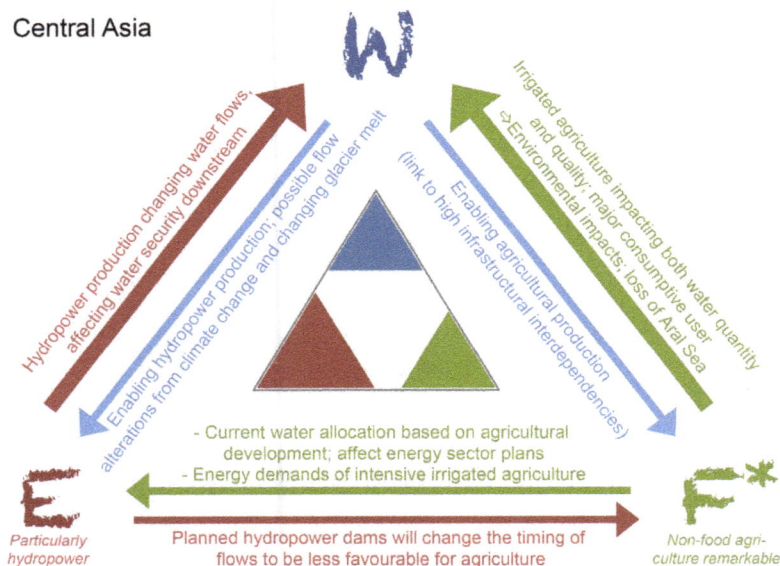

Figure 3. The nexus triangle for Central Asia, with key connections and impacts as well as current levels of security described with text. The F* indicates that food sector also includes a very important non-food crop *i.e.*, cotton. See Figure 2 and text for explanation.

After becoming independent, the Central Asian countries were left in a state of high level interdependency with the individual states increasing their focus on national interests [26]. The view of the nexus shifted to emphasise agricultural development and food security. The expansion of irrigation has resulted in demands by all to increase the historic water quotas [67,71], despite already unsustainably high water use. Water availability for agriculture is also affected by infrastructural interdependencies. For example, Uzbekistan and Turkmenistan have a very tense relationship over water use, as the water storage infrastructure providing water to the region's largest water user Uzbekistan is located in Turkmenistan [67].

Meanwhile, upstream countries are highly dependent on hydropower production (Appendix A) and therefore defend their right to use the hydropower potential of the rivers, much opposed by the downstream countries that are dependent on reliable quantity and quality of water [70]. Instead of emptying the reservoirs in the summer for the irrigation of downstream cotton fields, as they used to do, the upstream riparians now have an interest in storing the water to use for hydroelectric power in the winter [71]. Efforts have been made to maintain or renew Soviet-era bilateral barter arrangements as well as establish regional bodies for cooperation and coordination, such as the Interstate Commission on Water Coordination, International Fund to Save the Aral Sea and Basin Water Management Associations. Difficulties have however arisen over disputed prices, breach of agreements, lack of confidence in regional bodies and mutual distrust [73]. Strategic water management is further undermined by inaccurate and politicised national water use data which skews the discussion on possible cooperative strategies on water management in one way or another [67].

Going forward, successful management of the nexus in Central Asia and its transboundary rivers depends on fostering trust and cooperation, revitalisation and capacity building within public sector organisations [28,73], shared understanding of limits (e.g., environmental needs and water availability) and norms (e.g., contractual obligations) [27], and agreeing on compromises that maximise total benefits across the water-energy-food nexus rather than national benefits that often focus on one of the nexus sectors alone [29] (Figure 3).

3.2. South Asia

South Asia and its major transboundary rivers, the Indus, Ganges and Brahmaputra, cover seven riparian countries: Afghanistan, Bangladesh, Bhutan, China, India, Nepal and Pakistan. Across this vast and very densely populated region (Appendix A), water plays a key role for both food and energy security (Figure 4). Agriculture, particularly in the Ganges and the Indus, is very dependent on intensive irrigation [74,75]. Water for irrigating crops is largely pumped from groundwater, requiring a lot of energy, such that energy use is intensifying: there was a sixfold increase of electricity consumption per 1000 hectares of cultivated land from the year 1980/81 to 1999/2000 [76]. The groundwater extraction lowers the groundwater tables, thus exacerbating current water scarcity and impacting downstream water availability [77].

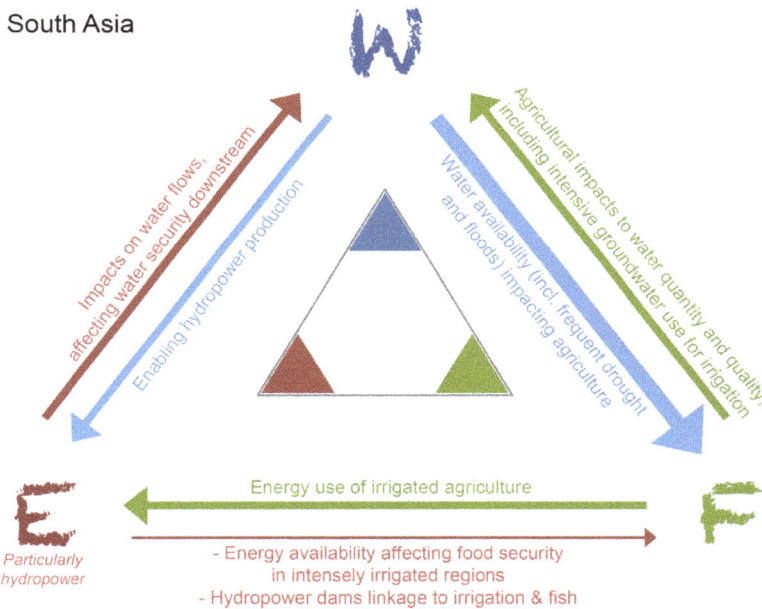

Figure 4. The nexus triangle for South Asia, with key connections and impacts as well as current levels of security described with text. See Figure 2 and text for explanation.

Both the Indus and the Ganges are under extreme pressure. The Indus basin faces severe water scarcity during eight months of the year, while the Ganges River is experiencing decreasing flows during the dry season [75,78,79] (Appendix A). The downstream riparians are highly vulnerable to upstream water use, while having no control of upstream water flows [61,80]. As a large part of the region is energy and water deficient, food production through intensive and energy-consuming irrigation has a significant impact on overall water and energy security.

These complex interdependencies combined with increasing pressures to intensify food and energy production have led to ambitious plans for ensuring water, energy and food security. These include massive plans to tap the abundant hydropower potential of the rivers [80–82]. The proposed dams will have potential transboundary environmental and social impacts, which will create pressures for the other nexus sectors. In terms of energy, the region is also highly dependent on bioenergy such as fuelwood, crop residues and animal dung as a residential energy source [83]. Bioenergy has an important role for the nexus as well, since it is potentially a very water- and land-intensive energy source and can increase the need for energy-intensive chemical fertilizers as animal wastes are diverted to fuel [84]. The shortage of energy has also led to increasing dependence on imported oil in Pakistan, India and Bangladesh, with oil imports doubling in India during 2003–2013 [83]. The region is also extremely dynamic when it comes to urbanization, population growth and industrialization, creating expanding demands and additional challenges for water, energy and food security and their interlinkages [76].

In terms of governance, the region's water, food and energy nexus constitutes a highly complex entity. In India and Pakistan, even the coordination of water and agricultural activities at the provincial level has proven to be quite challenging [85,86]. A number of transboundary water treaties have been signed on sharing the main rivers of the region. The most important among them include the 1960 Indus Water Treaty between India and Pakistan, as well as the 1996 Mahakali Treaty between Nepal and India and the 1996 Ganges Water Treaty between India and Bangladesh [87]. While they provide a fairly good basis for transboundary governance of the Indus and Ganges basins, in practice the water related policy-making is overly nationalistic and integration of policies of nations and sectors has proven to be difficult [88]. The history of Pakistan, India, Nepal and Bangladesh is riddled with distrust and tensions [89], and China's absence from the transboundary treaties makes governance particularly difficult in the Brahmaputra basin.

3.3. Mekong Region

The Mekong Region consists of the six countries sharing the Mekong River: China, Myanmar, Lao PDR, Thailand, Cambodia and Vietnam. The region is urbanising and developing rapidly, notably with massive plans for hydropower in several rivers [90,91], which has significant implications for environment and natural resources. On the other hand the current levels of water stress in the region's major transboundary rivers—Mekong, Irrawaddy and Salween—are remarkably less than for the rivers in other two study region (Appendix A). Yet, the transboundary Mekong River presents a topical case for the water-energy-food nexus due to its rapid hydropower development and the major importance of water and related resources—including fish and rice, the region's staple food—for regional food security (Figure 5) [32,66,92–95]. The high availability of water has enabled the region to become one of the world's largest rice producers, and the Mekong River itself is considered to be one of the world's most productive inland fisheries [96]. The water-related activities are an important part of the regional economy and provide livelihoods and food security for millions of people [97].

Rapid regional development has led to increasing demand for energy, which has resulted in extensive hydropower development [98,99]. In the Mekong Basin alone, there are currently 57 large hydroelectric dams (height > 15 m) and plans for over 100 more [100,101]. This would remarkably increase the hydropower production capacity in the region: while the existing dams have a capacity of 18,960 MW, the planned dams have a total capacity of 54,830 MW. A large share of the ongoing hydropower development occurs in upstream China and Lao PDR. The construction of hydroelectric dams in China is driven by increasing energy demand, particularly in the eastern parts of China and by the need to reduce emissions from energy production [102,103]. In Lao PDR, the domestic need for electricity is relatively low and the hydroelectric dams are built to generate revenue by exporting electricity to neighbouring countries, such as Thailand [104]. The hydropower development in the region is also fuelled by investments between the countries and outside the region [90,105].

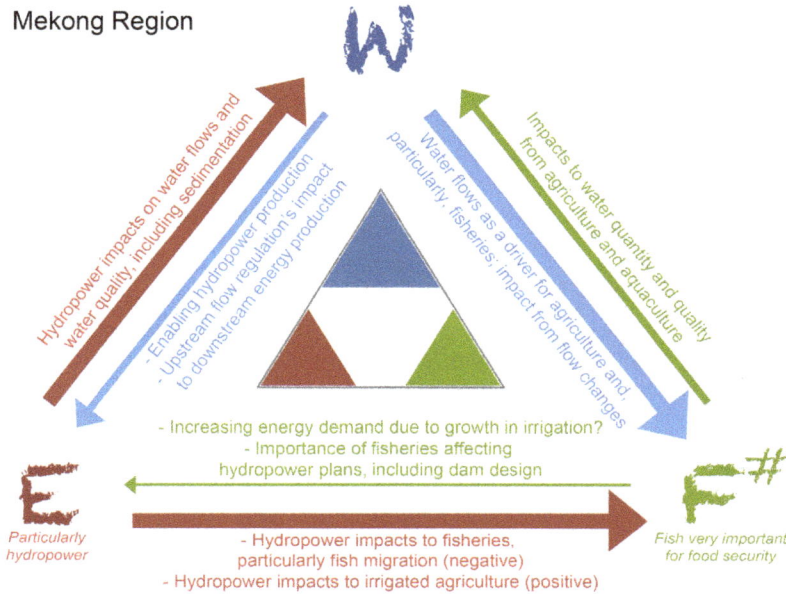

Figure 5. The nexus triangle for the Mekong Region, with key connections and impacts as well as current levels of security described with text. The F# indicates that fisheries' role for food security is remarkable. See Figure 2 and text for explanation.

The hydropower development in the region will improve regional energy security but it will have direct implications on food security. While the dams are expected to improve irrigation opportunities for agriculture [106], they will have major negative impacts on the aquatic ecosystems [96,107,108] and thus on livelihoods and food security [97]. In the Mekong River, the dams will affect the flow regimes [109], sediment [110] and nutrient transport [111] and fish migration [112], which are all key factors of the immense fisheries of the river. The annual wild fish harvest in the lower Mekong Basin is estimated to be 2.2 million tonnes and the value of the fisheries at retail markets is estimated to be 4.3–7.8 billion US$ [108,113]. The interplay between water, energy and food security has strong cross-scale characteristics: while hydropower development is largely taking place upstream and facilitated by regional energy grids, its negative implications, particularly on ecosystem productivity, are accumulating at local scales downstream [30].

In addition to hydropower, bioenergy and imported oil, gas and coal are important sources of energy for Southeast Asia [114]. Bioenergy continues to be the main source of energy for residential energy consumption, while imported oil required for energy production and transport (including those of agricultural products) has been steadily increasing [83,107]. The production of biofuels can have major sustainability implications related to land use, food security, rural livelihoods as well as water and land security [115,116].

The Mekong has existing regional cooperation mechanisms related to water, energy and food, most notably the economic- and energy-focused Greater Mekong Subregion (GMS) Programme and the river basin-focused Mekong River Commission (MRC). The MRC includes the four Lower Mekong riparians as its member countries and structures its activities according to the IWRM principles. On the other hand, the MRC has noted that "the practice of true involvement, and ownership, of the food and energy sectors in IWRM is often lagging behind, which is what the water, energy and food security nexus approach aims to solve" [22]. As a result, the MRC has also recognised the relevance of the nexus approach for its work [22,117]—and for a good reason, given that it has so far been largely sidelined from the massive hydropower development plans in the region [32,118,119].

Finally, it must be noted that there are naturally many "flows" other than the river itself connecting the countries in our study regions and the volumes and directions of transboundary flows of water, energy and food differ remarkably from each other. Appendix B provides an example of such flows for the Mekong Region. As can be seen, some of these flows also have important connections beyond the region itself. For example, given that one of the key questions in the region is how to replace the fish protein lost due to hydropower development [120], the significance of global food trade in ensuring regional food security is likely to increase.

4. Discussion: Implications of Transboundary Water-Energy-Food Nexus

In this article, our view on the nexus is at regional, *i.e.*, transnational, level and on transboundary river basins in particular. The preceding comparative analysis and the findings from other articles in this Special Issue demonstrate the challenges and opportunities for the transboundary water-energy-food nexus. In this section, we discuss these from two opposite viewpoints; first from the viewpoint of what the transboundary context brings to the nexus and then from the viewpoint of what nexus approaches bring to transboundary contexts (Figure 6). Given that nexus approaches are used in different contexts, it is useful to highlight distinctive features in a transboundary context. Given that the nexus is one of many ways to tackle transboundary issues, its particular contributions need to be highlighted. There is a clear link between these two viewpoints, as the contributions of the nexus can only be achieved if the distinctive features of the transboundary context are adequately addressed.

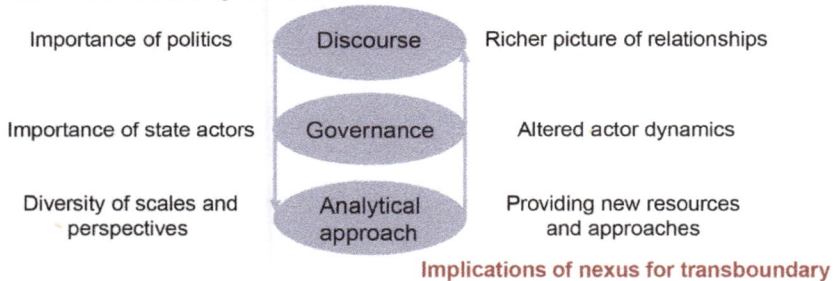

Implications of transboundary for nexus

Importance of politics	Discourse	Richer picture of relationships
Importance of state actors	Governance	Altered actor dynamics
Diversity of scales and perspectives	Analytical approach	Providing new resources and approaches

Implications of nexus for transboundary

Figure 6. The key findings for transboundary water-energy-food nexus along the three perspectives, looking at the implications of the transboundary context for the nexus (left) and *vice versa* (right).

4.1. Transboundary Context's Implications for the Nexus

We see that the transboundary context impacts the implementation of the water-energy-food nexus in three main ways, which are particularly relevant to the use of the nexus respectively for analysis, for governance and for discourse formulation. Firstly, the transboundary context requires an increased emphasis on multiple geographical scales and on diversity of different perspectives. Secondly, international transboundary contexts have distinctive key players in the nexus, notably nation-states. Thirdly, these two factors emphasise the importance of looking at the political aspects related to the nexus, which may help to broaden the nexus as a concept in that direction.

First of all, looking at the nexus in transboundary river basins and their regions necessitates the consideration of multiple geographical—and related institutional and political—scales, from local to national and to regional scales [121,122]. This kind of multi-scale context brings several challenges, ranging from the difficulty to obtain relevant information to increased complexity of nexus linkages that cross scales. In the Mekong, for example, our comparative analysis indicated that while the most relevant scale to look at food security is regional, even global, the most appropriate scale for energy security is regional and for water security, the basin scale (Mekong) and even local scale (Tonle Sap). More broadly, the characteristics of water, energy and food sectors vary significantly, including the importance of private and public ownership and trade in each of them. Analysis at the transboundary context may therefore face different problems depending on the sector. The multi-scale character of transboundary settings can thus also create scale mismatches [123], where the scale for managing the nexus is different to the (multi-scale) processes that are being managed.

The transboundary context also requires consideration of a *greater diversity of perspectives*, as multiple institutional and political scales usually mean less social cohesion and more heterogeneity on views towards the nexus when compared to a river basin within a single country. The transboundary context unavoidably includes several countries, all of them with their own view and interests towards water, energy and food. This, in turn, is likely to lead to increasing resource politics and even resource and water securitisation. Such a situation is visible in Central Asia, where the change from the centralized Soviet era to independent states inserted an additional institutional and political layer to the transboundary context, leading to very complex and politicized collaboration. In such a setting, the countries view the nexus and its sectors differently: in upstream Tajikistan, water enhances its energy security, while Uzbekistan and Turkmenistan see water as vital for their agriculture. From the nexus point of view, it is also noteworthy that downstream agriculture in the Amu Darya and the Syr Darya basins revolves around a non-food crop, cotton. This reminds us that the nexus may need to be broadened beyond a narrow focus on just water, energy and food.

Turning to the second issue, the transboundary context influences which actors are most prominent and exert power over the nexus. Given that international transboundary river basins by definition cross national boundaries, the *transboundary context easily implies state-centrism, the key actors on negotiating and deciding on the nexus being nation states and their representatives*. While such state-centrism has its problems [124,125], it also means that there are often already existing organisations and institutional arrangements facilitating transboundary cooperation. In our comparative analysis, these include legal agreements such as the Mekong Agreement [31,126] as well as the bilateral treaties in the South Asian river basins [87]. While the existing organisations are commonly water-focused (e.g., river basin organisations), they often extend their activities towards (water-relevant) parts of food and energy sectors (e.g., [17,97,117]).

This brings forward an interesting question: in the context of transboundary waters, should the nexus be implemented through already existing organisations (mainly River Basin Organisations *i.e.*, RBOs), or is it better to establish a novel cooperation mechanism for implementing the nexus? The answer to this question depends also on how we view the nexus; as a water-centred approach to support water resources management or as a process linking the three sectors together on broader and more equal terms. Given that the majority of current nexus analyses are water-centred and the nexus literature emphasises the importance of building on existing institutional settings [2], different river basin organisations have been actively engaged in nexus discussions (e.g., [17,22]). While the nexus can help to refocus the RBOs' activities and extend their understanding and even political leverage [25], we nevertheless feel that the nexus as a multisectoral and multilevel process should preferably not have a single owner. Rather, it would be important to implement the nexus through a multi-stakeholder process [127–129], where key actors from food and energy sectors also take initiative and ownership of the process. It could be, for example, initiated by executive branches of government or by a third party (for example, see [17,46]). As a flow-on effect, such a process could encourage sharing of analyses and data between sectors as well as countries. This may be an opportunity for building trust and transparency but could also be constrained by existing political as well as technical difficulties, as discussed above (see also [130–134]). These changes are consistent with the dominant nexus discourse, in which each sector is expected to take into account their effect on the others. The implication is that the sectors are expected to share their responsibilities (and hence power), at least to some extent.

Thirdly and related to the above, the transboundary contexts are inherently political. Yet, key publications (e.g., [2,4] establishing the nexus as a concept are relatively silent about politics, although several authors have emphasized its importance (e.g., [8,30,51,135]. In our opinion, without consideration of the often highly politicised nature of resource use, the analytical approach to the nexus is at risk of being naïve. As shown in our comparative analysis, *applying nexus in transboundary contexts thus puts an increased emphasis on the importance of politics; this in turn, can help to revise the existing understanding of nexus as a concept to better acknowledge politics.* More broadly, the transboundary context can encourage us to re-consider the "discourse-setting" realities of the nexus. While the nexus may indeed act as a handy boundary concept to challenge the existing discourses on water-energy-food connections, it is by no means neutral. Instead, it also forms a tool that is used to shape such discourses. This is particularly important in areas like South Asia that are experiencing multiple pressures with widespread impacts on all parts of the nexus, such as urbanisation, population growth and industrialisation. Given the current water-centrism of the nexus literature, we should also allow food and energy sectors to forge their own view of the nexus and the way it shapes the discourse related to water-energy-food connections. This may lead to varying conceptualisations of the nexus (in terms of all three perspectives), depending on the context.

4.2. Nexus' Implications for Transboundary Issues

Based on our comparative analysis, we argue that the nexus has three main implications for water, food and energy issues in transboundary river basins and their regions. Similarly to the preceding section, these implications can be linked to using the nexus approaches from the perspective of analysis, governance, and discourse-setting. Firstly, the nexus may provide new resources and approaches for transboundary cooperation. Secondly, the nexus alters actor dynamics by engaging key actors from energy and food sectors. Thirdly, the nexus can help to reset transboundary cooperation between the riparian countries by extending thinking about cooperation from water resources management to broader aspects of water, energy and food security.

The nexus can bring *new resources and approaches for transboundary cooperation.* In addition to setting the focus on selected sectors only, the nexus also emphasises the importance to look at the interconnections and related trade-offs between those sectors. In South Asia, problems of groundwater over-extraction are closely linked with energy pricing policy and sustainability of livelihoods. The comparative analysis shows that nexus approaches are already being used in practice. There is a need and opportunity for tools to mature to provide additional support and help make better cross-sector decisions. This need is doubly pressing given that nexus sectors and their relationships are changing, as with the Mekong shift from subsistence to market economy-based food supply. It becomes more difficult to predict new trade-offs, conflicts and synergies that may emerge. Emerging literature on nexus approaches and methods (e.g., [6,21,136,137]) provides a fruitful ground to reflect and possibly, revise the existing methods and tools used to support transboundary cooperation. The majority of the current discussion on the nexus is however rather water-centred and -driven and we thus see that there is great potential for cross-fertilisation between the three sectors, if water, energy and food researchers and practitioners are brought together (as has already been happening in terms of water-food and water-energy nexus). In addition, the current enthusiasm about the nexus can mobilise additional funding and expertise to look at the interconnections between water, energy and food in transboundary contexts, both through specific funding calls (e.g., [138–141]) and in framing funding proposals.

Active engagement of energy and food sectors can also change the existing actor dynamics in transboundary cooperation by including more actors in its governance. The riparian countries' views towards water-energy-food linkages are by no means monolithic but different ministries may harbor quite differing interests. In the Mekong River Commission (MRC) for example, each of the four member countries is currently represented in the MRC Council only by the water and/or environment ministry [142]. An alternative high-level nexus working group could include representatives also from energy or fisheries-related ministries. Such a setting could result in novel discussions and even new kinds of cross-state alliances where for example, the representatives responsible for energy or fisheries would find common ground. As was also found in a workshop setting [30], explicitly naming the nexus sectors can help to bring the actors together. In this way, the nexus can also complement and even challenge existing power structures related to transboundary governance. However, achieving these outcomes requires existing and evolving political realities to be acknowledged and the food and energy sectors must also buy into the nexus approach, as argued above. Otherwise, some key actors may prefer to act unilaterally, as is the case with Laos and its energy plans in the Mekong.

Cooperation about management of transboundary river basins typically revolves around water and related sectors. While such a focus is obvious, it may also hinder the countries from reaching an agreement about the development of the river by focusing on difficult upstream-downstream relations, as illustrated by all comparative case studies. By extending the discussion into energy and food security and their regional linkages, *the nexus paints a richer picture of the relations between countries and, ideally, a more integrated understanding of the range of linkages between the countries.* It emphasises the need for assessing the total impacts on water, energy and food (including those to economy and the environment) and raises awareness that alternative approaches to water, energy and food issues may exist. In this way, the nexus may open political deadlocks and encourage regional cooperation. In focusing on water, energy and food, it provides a more concrete context to examine benefit-sharing options [29,133,143].

Broadening scope can however carry its own difficulties, introducing new dependencies [27] and requiring new approaches, as argued above. Otherwise, focusing on the nexus may actually make the current situation more difficult rather than easier. For example, in Central Asia, nexus-wide infrastructure and economic arrangements developed in the Soviet era were dependent on centralized economic planning and barter arrangements [68], for which the now-independent states are still struggling to find a durable alternative [27]. In the Mekong Region, the recognition of impacts of dams spanning the nexus has not yet been accompanied by potential solutions that span the nexus, which has paralysed collective action and allowed unilateral action to proceed [59,118,119]. These examples confirm that the possible changes in discourse prompted by the nexus need to be adequately supported by related adjustments in governance and analytical approaches.

These issues of scope extend beyond the nexus: the concepts of sustainable development and IWRM also try to provide a richer picture of relationships, with varying success. Sustainable development encourages cross-sectoral collaboration while emphasising triple bottom line outcomes on environmental, economic and social aspects [144]. IWRM—which has been recognised by several UN processes such as the Sustainable Development Goals (SDGs)—builds on the concept of sustainable development and emphasises how other sectors relate to water management [54]. The nexus provides an intermediate view that explicitly focuses on water, energy and food sectors and their actors and can even be seen to have arisen as a more focused, tractable response to the broad and ambiguous nature of sustainable development [145]. As shown in the comparative analysis, it can be useful to frame problems as arising specifically from the interaction between actors from different sectors rather than as issues of sustainability or integrated water management more generally. It has been argued that the nexus as a concept therefore provides a complementary view, contributing to the diverse ways of thinking about sustainability and water, energy and food issues (e.g., [16,22,30,49]). In the specific context of transboundary river basins, the nexus can also increase political leverage to address complex water-energy-food connections, which IWRM processes and RBOs tend to lack.

5. Conclusions

In this article, we examined the nexus approach to water, energy and food issues, focusing on its application in the context of transboundary river basins. The literature reviewed for the article shows the diversity of nexus-related publications, but also indicates that there is still little practical evidence on nexus and that there is not even commonly agreed definition for the nexus.

We introduced three definitions for the nexus, corresponding to three complementary perspectives, which emphasize the nexus' role as an *analytical approach, governance framework* and as an *emerging discourse.* While the nexus discussion often focuses on the linkages between three nexus sectors, we see that due to complexity of water-energy-food connections, it is actually as important to understand cross-level linkages between disciplines and between nexus perspectives as it is to understand cross-sectoral linkages themselves. Hence, an ideal nexus approach would integrate all three of these perspectives. Indeed, given that water-energy-food issues have been well recognised already for decades, integration of the perspectives is important for the nexus to provide added value for current planning and management practices.

We then focused on the implications that transboundary context brings to the nexus as an approach and *vice versa.* Based on our comparative analysis of three Asian regions and related transboundary river basins, we concluded that the transboundary context has three major implications: *the diversity of scales and perspectives, the importance of state actors, and the importance of politics.* On the other hand, introducing the nexus approach in a transboundary context has the potential to *provide new resources and approaches, alter existing actor dynamics, and portray a richer picture of relationships.* These two sets of implications are linked, and also span the three perspectives on the nexus, demonstrating the usefulness of an integrated view (Figure 6).

According to our assessment, the use of transboundary water-energy-food nexus is worth pursuing but there are still issues that need to be addressed in its implementation. The nexus is not a mature concept and experiences on its benefits and caveats—in transboundary and other contexts, and in relation to other approaches—need to be strengthened. One fundamental issue is that the possibility to alter existing actor dynamics means that the ownership of the nexus should not rest on any single, sectoral organisation (such as a River Basin Organisation) but it should be implemented through a multisectoral, multistakeholder process. Such a process would also encourage interest in the nexus approach by food and energy sectors. Given that the nexus is still largely water sector driven, such buy-in is crucial for the success of the nexus approach.

At the same time, we should remember the facts on the ground: the pressures of food and energy production on water and the importance of the river-basin scale for understanding water flow, allocation and use. The issues on which the nexus focuses are fundamental and any effort to better understand and manage them should be encouraged—particularly in transboundary situations, where various approaches related to water, energy and food have been applied but which still regularly remain complicated and 'deadlocked'. Admittedly, the fact that the nexus is not as established a concept as, for example, IWRM means that it may still fade from fashion. In the meantime, however, the nexus could make some researchers and practitioners think about relationships between food, energy and water in new ways. With its focus on three critical sectors and their interlinkages, the nexus has potential as a complementary approach to support water resources management, transboundary cooperation and, ultimately, sustainable development.

Acknowledgments: The work for this article and the related Special Issue was supported by the Academy of Finland funded Nexus Asia project (#269901). T.A.R. received funding also from Maa- ja vesitekniikan tuki ry. The ideas and support provided by Hanna-Mari Juvonen, Aura Salmivaara and Shokhrukh-Mirzo Jalilov helped us to set the focus for the article, while Matti Kummu helped with maps and data. The analyses provided by other authors of this Special Issue helped our comparative analysis and discussion: thank you to all of you. Thank you also for the support of our colleagues at Aalto University's Water & Development Research Group as well as for the well-formulated comments by our three reviewers.

Author Contributions: M.Ke. and O.V. developed the original idea for the article, with all authors contributing to its further development. M.Ke. and J.G. had the main responsibility for structuring and writing the actual article, with M.Ka., M.P. and T.A.R. contributing particularly to comparative analysis and O.V. to discussion and conclusions. M.Ke., J.G. and T.A.R. wrote the majority of the Section 2, M.P. and M.Ke. the majority of Appendix A, and M.Ke. the majority of Appendix B. All authors contributed to finalising the text of the article.

Conflicts of Interest: The authors declare no conflict of interest.

Appendix A. Selected Water-Energy-Food Data on Three Study Regions

The tables below show selected water, energy and food-related data for the three study regions, namely Central Asia (Table A1), South Asia (Table A2) and the Mekong Region in Southeast Asia (Table A3). The data aims to provide a quantitative indication of the issues that are qualitatively described in the nexus triangles. The data is grouped according the countries and/or key transboundary river basins. The definitions and sources for the data are listed in Table A4.

In the tables, hydroelectric production indicates electricity production from hydroelectric sources [% of total], while water shortage is calculated as renewable water resources per capita per year (moderate if 1000–1700 m^3/cap/yr, high if <1000 m^3/cap/yr) and water stress as water consumption/renewable water resources (moderate if 20%–40%, high if >40%). The letters in the parentheses under each transboundary river basin refer to the riparian countries.

Appendix B. Multiple "Transboundarities": Example from the Mekong

The comparative analysis notes that countries in our study regions are connected in multiple ways and are therefore transboundary in multiple ways, when viewed from a nexus perspective. The maps included in Figure A1 show examples of these multiple 'transboundarities' in the Mekong Region.

The Mekong riparian countries are connected by the flow of the river itself, its tributaries and its shared basin from upstream to downstream (a), but this relationship is not only unidirectional, as a result of important fish migration patterns from downstream to upstream (b). From an energy and food perspective, the connections are different again, as shown by the connecting power transmission lines (c) and available data on the regional food trade exports (d). For map d), kindly note that the data on Laos, Myanmar and Vietnam is missing and the size of the arrows and circles is indicative only: for actual values, see the numbers.

Table A1. Selected water-, energy- and food-related data for the countries and transboundary river basins in Central Asia.

Central Asia

		Population	Net Energy Imports	Hydro-Electric Production	Access to Electricity	Energy Used in Agriculture and Forestry	Food Supply	Food Self-Sufficiency	Area Equipped for Irrigation		Water Shortage	Water Stress	Share of Water Withdrawals			
													Irrigation	Electricity	Manufactu	Domestic
		[millions]	[% of energy use]	[% of total]	[% of population]	[% of energy use]	[kcal/cap/d]	[% of kcal requirements produced domestically]	[% of total land area]	(ha/capita)	[m³/year/ca]	[%]				
Countries	Afghanistan A	28.4	-	9%	43%	-	-	-	5%	0.106						
	Kazakhstan Kz	15.9	−120%	-	100%	2.5%	3225	196%	1%	0.170						
	Kyrgyz Rep. Kg	5.3	58%	93%	100%	5.1%	2771	79%	6%	0.151						
	Tajikistan Tj	7.6	26%	99%	100%	17.9%	2134	36%	6%	0.099						
	Turkmenistan Tk	5.0	−166%	0%	100%	2.2%	2802	102%	5%	0.345						
	Uzbekistan U	27.8	−18%	19%	100%	5.5%	2502	83%	9%	0.153						
Basins	Amu Darya (A, Tj, Tk, U)	30.2							7%	0.140	1636	47%	92%	3%	1%	4%
	Syr Darya (Kz, Kg, Tj, U)	26.6							5%	0.163	1209	52%	77%	12%	3%	7%

Table A2. Selected water-, energy- and food-related data for the countries and transboundary river basins in South Asia. GBM = Ganges-Brahmaputra-Meghna river basin.

South Asia

		Population	Net Energy Imports	Hydro-Electric Production	Access to Electricity	Energy Used in Agriculture and Forestry	Food supply	Food Self-Sufficiency	Area Equipped for Irrigation		Water Shortage	Water Stress	Share of Water Withdrawals			
													Irrigation	Electricity	Manufactu	Domestic
		[millions]	[% of energy use]	[% of total]	[% of population]	[% of energy use]	[kcal/cap/d]	[% of kcal requirements produced domestically]	[% of total land area]	(ha/capita)	[m³/year/ca]	[%]				
Countries	Bangladesh Bg	151.1	18%	2%	60%	5.1%	2,413	85%	37%	0.032						
	Bhutan Bh	0.7	-	-	76%	-	-	-	1%	0.017						
	China Ch	1360	13%	15%	100%	2.1%	2,915	85%	6%	0.047						
	India I	1206	31%	12%	79%	3.9%	2,252	94%	20%	0.053						
	Nepal N	26.8	16%	100%	76%	1.2%	2,333	79%	7%	0.039						
	Pakistan P	173.1	23%	30%	94%	1.3%	2,315	101%	20%	0.099						
Basins	GBM (Bg, Bh, Ch, I, N)	705.0							20%	0.048	1,604	16%	90%	1%	5%	3%
	Indus (Ch, I, P)	243.4							21%	0.100	836	58%	97%	1%	1%	2%

Table A3. Selected water, energy and food-related data for the countries and transboundary river basins in the Mekong Region of Southeast Asia.

		Population	Net energy imports	Hydro-Electric Production	Access to Electricity	Energy Used in Agriculture And Forestry	Food Supply	Food Self-Suffiency	Area Equipped for Irrigation		Water Shortage	Water Stress	Share of Water Withdrawals			
													Irrigation	Electricity	Manufactu	Domestic
		[millions]	[% of energy use]	[% of total]	[% of population]	[% of energy use]	[kcal/cap/d]	[% of kcal requirements produced domestically]	[% of total land area]	(ha/capita)	[m³/year/ca]	[%]				
Countries																
Cambodia	Cm	14.4	28%	4%	31%	2.5%	2280	109%	3%	0.040						
China	Ch	1360	13%	15%	100%	2.1%	2915	85%	6%	0.047						
Lao PDR	L	6.4	-	-	70%		2258	117%	2%	0.055						
Myanmar	M	51.9	−47%	70%	52%	1.0%	2374	111%	3%	0.044						
Thailand	T	66.4	40%	5%	100%	4.5%	2587	218%	12%	0.097						
Vietnam	V	89.0	−7%	30%	99%	1.1%	2754	79%	14%	0.051						
Basins																
Irrawaddy (M + Ch, I)		33.0							4%	0.057	17,352	1%	94%	2%	1%	3%
Mekong (Cm, Ch, L, M, T, V)		70.6							6%	0.066	6057	4%	79%	10%	4%	6%
Salween (Ch, M, T)		7.9							1%	0.035	13,518	1%	93%	2%	0%	5%

Table A4. The descriptions and sources for the data listed in Tables A1–3.

Data	Description	Data Source
Country population	UN estimates of total population by country, year 2010	[146]
Basin population	Gridded population data for year 2010, based on HYDE 3.1 and IIASA Population projections	[147]
Energy imports and access to electricity	Net energy imports (% of energy use); Access to electricity (% of population) for year 2012	[148]
Hydroelectricity	Electricity production from hydroelectric sources, by country	[148]
Water withdrawals and consumption	Gridded data on sectoral water withdrawals and consumption	[149]
Energy used in agriculture and forestry	FAO data on energy use related to different sectors.	[150]
Domestic food production & food supply	Dietary energy production by country, year 2005; based on FAOSTAT data.	[151]
ADER	Average dietary energy requirement in kcal/cap by country, years 2004-2006	[152]
Area equipped for irrigation	Gridded data on area equipped for irrigation (AEI); product AEI_HYDE_FINAL_CP.	[153]
Renewable freshwater resources	Calculations based on average annual surface runoff computed from monthly modelled runoff for years 1950–2000	[154]

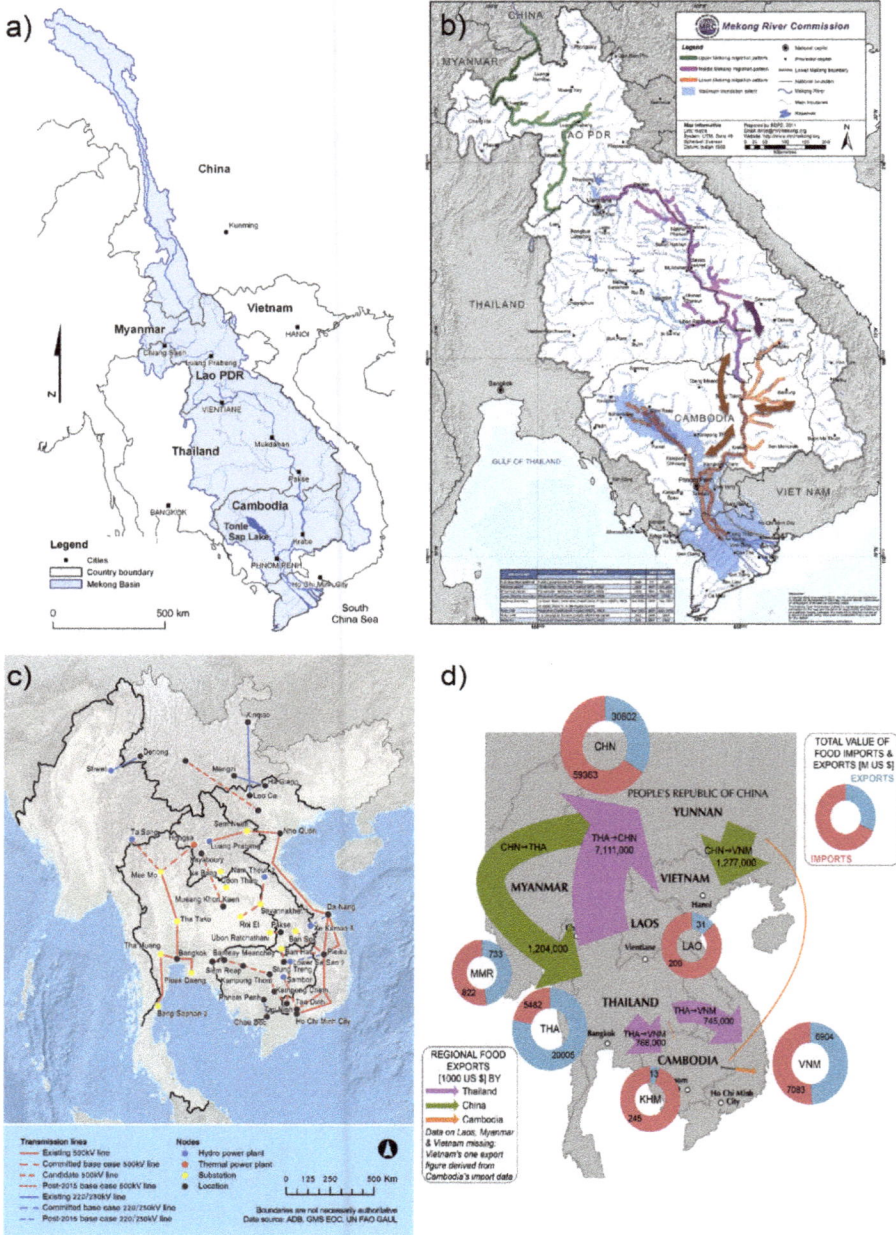

Figure A1. Maps showing example of different transboundary flows in the Mekong Region. (**a**) river basin (map by Matti Kummu, used with permission); (**b**) fish migration patterns (modified from [155]); (**c**) power transmission lines and key power plants [156]; (**d**) indicative information on regional food export flows (arrows) and the total size of food imports and experts per country (blue-red circles). Map d by the authors, based on data available in FAOSTAT Country Profiles [157].

References

1. What's the NEXUS? Messages and Policy Papers. Available online: http://www.water-energy-food.org/en/whats_the_nexus/messages_policy_recommendations.html (accessed on 20 April 2016).
2. Hoff, H. Understanding the Nexus, Background Paper for the Bonn 2011 Conference. In Proceedings of the Bonn2011 Conference, The Water, Energy and Food Security Nexus: Solutions for the Green Economy, Bonn, Germany, 16–18 November 2011; Stockholm Environment Institute: Stockholm, Sweden, 2011; p. 51.
3. Bazilian, M.; Rogner, H.; Howells, M.; Hermann, S.; Arent, D.; Gielen, D.; Steduto, P.; Mueller, A.; Komor, P.; Tol, R.S.J.; *et al.* Considering the energy, water and food nexus: Towards an integrated modelling approach. *Energy Policy* **2011**, *39*, 7896–7906. [CrossRef]
4. World Economic Forum. *Global Risks 2011—Sixth Edition*; World Economic Forum: Cologny, Switzerland, 2011.
5. Atakan, E.; Bellet, L.; Bird, J.; Cramwinckel, J.; Dupont, A.; Fernandez, M.-I.; Marre, F.; Mohtar, R.; Newton, J.; Nguyen-Khoaman, S.; *et al.* High-Level Panel: Water, Energy & Food Security. In Proceedings of the 6th World Water Forum, Marseille, France, 12–17 March 2012.
6. Granit, J.; Fogde, M.; Hoff, H.; Joyce, J.; Karlberg, L.; Kuylenstierna, J.L.; Rosemarin, A. Unpacking the water-energy-food nexus: Tools for assessment and cooperation along a continuum. In *Cooperation for A Water Wise World—Partnerships for Sustainable Development*; Jägerskog, A., Clausen, T.J., Lexén, K., Holmgren, T., Eds.; Stockholm International Water Institute: Stockholm, Sweden, 2013; pp. 45–50.
7. United Nations, Economic and Social Commission for Asia and the Pacific (UN ESCAP). *Water, Food and Energy Nexus in Asia and the Pacific*; ESCAP: Bangkok, Thailand, 2013.
8. Allouche, J.; Middleton, C.; Gyawali, D. Technical veil, hidden politics: Interrogating the power linkages behind the nexus. *Water Altern.* **2015**, *8*, 610–626.
9. Muller, M. The "nexus" as a step back towards a more coherent water resource management paradigm. *Water Altern.* **2015**, *8*, 675–694.
10. Allan, T.; Keulertz, M.; Woertz, E. The water-food-energy nexus: An introduction to nexus concepts and some conceptual and operational problems. *Int. J. Water Resour. Develop.* **2015**, *31*, 301–311. [CrossRef]
11. Tickner, D. Is the Water Debate Suffering from a Language problem? Available online: http://www.theguardian.com/sustainable-business/water-debate-suffering-language-problem (accessed on 20 April 2016).
12. Leck, H.; Conway, D.; Bradshaw, M.; Rees, J. Tracing the water-energy-food nexus: Description, theory and practice. *Geogr. Compass* **2015**, *9*, 445–460. [CrossRef]
13. Hussey, K.; Pittock, J. The Energy-water nexus: Managing the links between energy and water for a sustainable future. *Ecol. Soc.* **2012**, *17*, 31. [CrossRef]
14. Kouangpalath, P.; Meijer, K. Water-energy nexus in Shared River Basins: How hydropower shapes cooperation and coordination. *Chang. Adapt. Socio-Ecol. Syst.* **2015**, *2*, 85–87. [CrossRef]
15. Waughray, D. *Water Security: The Water-Food-Energy-Climate Nexus*; Island Press: Washington, DC, USA, 2011.
16. Pittock, J.; Hussey, K.; Dovers, S. *Climate, Energy and Water: Managing Trade-Offs, Seizing Opportunities*; Cambridge University Press: New York, NY, USA, 2015.
17. De Strasser, L.; Lipponen, A.; Howells, M.; Stec, S.; Bréthaut, C. A methodology to assess the water energy food ecosystems nexus in transboundary river basins. *Water* **2016**, *8*, 59. [CrossRef]
18. Biggs, E.M.; Bruce, E.; Boruff, B.; Duncan, J.M.A.; Horsley, J.; Pauli, N.; McNeill, K.; Neef, A.; van Ogtrop, F.; Curnow, J.; *et al.* Sustainable development and the water-energy-food nexus: A perspective on livelihoods. *Environ. Sci. Policy* **2015**, *54*, 389–397. [CrossRef]
19. Howells, M.; Hermann, S.; Welsch, M.; Bazilian, M.; Segerstrom, R.; Alfstad, T.; Gielen, D.; Rogner, H.; Fischer, G.; van Velthuizen, H.; *et al.* Integrated analysis of climate change, land-use, energy and water strategies. *Nature Clim. Chang.* **2013**, *3*, 621–626. [CrossRef]
20. Villarroel Walker, R.; Beck, M.B.; Hall, J.W.; Dawson, R.J.; Heidrich, O. The energy-water-food nexus: Strategic analysis of technologies for transforming the urban metabolism. *J. Environ. Manag.* **2014**, *141*, 104–115. [CrossRef] [PubMed]
21. Endo, A.; Burnett, K.; Orencio, P.; Kumazawa, T.; Wada, C.; Ishii, A.; Tsurita, I.; Taniguchi, M. Methods of the water-energy-food nexus. *Water* **2015**, *7*, 5806–5830. [CrossRef]

22. Bach, H.; Bird, J.; Jonch Clausen, T.; Morck Jensen, K.; Baadsgarde Lange, R.; Taylor, R.; Viriyasakultorn, V.; Wolf, A. *Transboundary River Basin Management: Addressing Water, Energy and Food Security*; Mekong River Commission: Vientiane, Laos, 2012.

23. Kibaroglu, A.; Gürsoy, S.I. Water-energy-food nexus in a transboundary context: The Euphrates-Tigris river basin as a case study. *Water Int.* **2015**, *40*, 824–838. [CrossRef]

24. Scott McLachlan, N. Implementing the water-energy-food nexus at various scales: Trans-boundary challenges and solutions. *Chang. Adapt. Socio-Ecol. Syst.* **2015**, *2*, 94–96. [CrossRef]

25. Lawford, R.; Bogardi, J.; Marx, S.; Jain, S.; Wostl, C.P.; Knüppe, K.; Ringler, C.; Lansigan, F.; Meza, F. Basin perspectives on the water-energy-food security nexus. *Curr. Opin. Environ. Sustain.* **2013**, *5*, 607–616. [CrossRef]

26. Soliev, I.; Wegerich, K.; Kazbekov, J. The costs of benefit sharing: Historical and institutional analysis of shared water development in the Ferghana Valley, the Syr Darya Basin. *Water* **2015**, *7*, 2728–2752. [CrossRef]

27. Guillaume, J.; Kummu, M.; Eisner, S.; Varis, O. Transferable principles for managing the nexus: Lessons from historical global water modelling of Central Asia. *Water* **2015**, *7*, 4200–4231. [CrossRef]

28. Wegerich, K.; van Rooijen, D.; Soliev, I.; Mukhamedova, N. Water security in the Syr Darya Basin. *Water* **2015**, *7*, 4657–4684. [CrossRef]

29. Jalilov, S.-M.; Varis, O.; Keskinen, M. Sharing benefits in transboundary rivers: An experimental case study of Central Asian water-energy-agriculture nexus. *Water* **2015**, *7*, 4778–4805. [CrossRef]

30. Keskinen, M.; Someth, P.; Salmivaara, A.; Kummu, M. Water-energy-food nexus in a transboundary river basin: The case of Tonle Sap Lake, Mekong River Basin. *Water* **2015**, *7*, 5416–5436. [CrossRef]

31. Belinskij, A. Water-energy-food nexus within the framework of international water law. *Water* **2015**, *7*, 5396–5415. [CrossRef]

32. Matthews, N.; Motta, S. Chinese state-owned enterprise investment in Mekong hydropower: Political and economic drivers and their implications across the water, energy, food nexus. *Water* **2015**, *7*, 6269–6284. [CrossRef]

33. Pittock, J.; Dumaresq, D.; Bassi, A. Modelling the hydropower-food nexus in large river basin. *Water* **2016**, in press.

34. Kummu, M.; Gerten, D.; Heinke, J.; Konzmann, M.; Varis, O. Climate-driven interannual variability of water scarcity in food production potential: A global analysis. *Hydrol. Earth Syst. Sci.* **2014**, *18*, 447–461. [CrossRef]

35. Porkka, M.; Gerten, D.; Schaphoff, S.; Siebert, S.; Kummu, M. Causes and trends of water scarcity in food production. *Environ. Res. Lett.* **2016**, *11*, 015001. [CrossRef]

36. Olsson, G. Water, energy and food interactions—Challenges and opportunities. *Front. Environ. Sci. Eng.* **2013**, *7*, 787–793. [CrossRef]

37. Oki, T.; Kanae, S. Global Hydrological Cycles and World Water Resources. *Science* **2006**, *313*, 1068–1072. [CrossRef] [PubMed]

38. Yoshihide, W.; van Beek, L.P.H.; Niko, W.; Marc, F.P.B. Human water consumption intensifies hydrological drought worldwide. *Environ. Res. Lett.* **2013**, *8*, 034036.

39. Varis, O. Water demands for bioenergy production. *Int. J. Water Resour. Develop.* **2007**, *23*, 519–535. [CrossRef]

40. Harris, G. Energy, Water and Food-Scenarios on the Future of Sustainable Development. 2002. Available online: http://ericbestonline.com/bestpartners/pdfs/partners/gerald_harris/g_harris_energy_water_and_food_scenarios.pdf (accessed on 20 April 2016).

41. Hellegers, P.; Zilberman, D.; Steduto, P.; McCornick, P. Interactions between water, energy, food and environment: Evolving perspectives and policy issues. *Water Policy* **2008**, *10*, 1–10. [CrossRef]

42. Mushtaq, S.; Maraseni, T.N.; Maroulis, J.; Hafeez, M. Energy and water tradeoffs in enhancing food security: A selective international assessment. *Energy Policy* **2009**, *37*, 3635–3644. [CrossRef]

43. European Union. *Confronting Scarcity: Managing Water, Energy and Land for Inclusive and Sustainable Growth*; European Report on Development 2011–2012; European Commission, International Cooperation and Development (DG DEVCO): Brussels, Belgium, 2012.

44. Smajgl, A.; Ward, J. *The Water-Food-Energy Nexus in the Mekong Region: Assessing Development Strategies Considering Cross-Sectoral and Transboundary Impacts*; Springer Science & Business Media: Berlin, Germany, 2013.

45. Food and Agriculture Organization of the United Nations (FAO). *An Innovative Accounting Framework for the Food-Energy-Water Nexus—Application of the MuSIASEM Approach to Three Case Studies*; FAO: Rome, Italy, 2013.

46. International Union for Conservation of Nature (IUCN) and International Water Association (IWA). *Nexus Dialogue on Water Infrastructure Solutions: Building Partnerships for Innovation in Water, Energy and Food Security*; IUCN & IWA.: Gland, Swizerland; London, UK, 2012.

47. Star, S.L.; Griesemer, J.R. Institutional ecology, "translations" and boundary objects: Amateurs and professionals in Berkeley's Museum of Vertebrate Zoology. *Soc. Stud. Sci.* **1989**, *19*, 387–420. [CrossRef]

48. Mollinga, P. *For a Political Sociology of Water Resources Management*; ZEF Working Paper Series 31; ZEF Center for Development Research, Universität Bonn: Bonn, Germany, 2008.

49. Benson, D.; Gain, A.K.; Rouillard, J.J. Water Governance in a Comparative Perspective: From IWRM to a "Nexus" Approach? *Water Altern.* **2015**, *8*, 756–773.

50. Juvonen, H.-M. Nexus for What? Challenges and Opportunities in Applying the Water-Energy-Food Nexus. Master's Thesis, Aalto University School of Engineering, Espoo, Finland, 2015.

51. Foran, T. Node and Regime: Interdisciplinary Analysis of Water-Energy-Food Nexus in the Mekong Region. *Water Altern.* **2015**, *8*, 655–674.

52. Gasper, D.; Apthorpe, R. Introduction: Discourse Analysis and Policy Discourse. In *Arguing Development Policy: Frames And Discourses*; Apthorpe, R., Gasper, D., Eds.; Frank Cass and Company Limited: London, UK, 1996.

53. Keskinen, M. Bringing back the Common Sense? Integrated Approaches in Water Management: Lessons Learnt from the Mekong. Ph.D. Thesis, Aalto University, Espoo, Helsinki, Finland, 2010.

54. Global Water Partnership (GWP). *Integrated Water Resources Management*; Technical Advisory Committee Background Paper #6; GWP: Stockholm, Sweden, 2000.

55. Max-Neef, M.A. Foundations of transdisciplinarity. *Ecol. Econ.* **2005**, *53*, 5–16. [CrossRef]

56. Varis, O.; Kummu, M.; Salmivaara, A. Ten major rivers in monsoon Asia-Pacific: An assessment of vulnerability. *Appl. Geogr.* **2012**, *32*, 441–454. [CrossRef]

57. Varis, O.; Kummu, M. The Major Central Asian River Basins: An Assessment of Vulnerability. *Int. J. Water Resour. Develop.* **2012**, *28*, 433–452. [CrossRef]

58. Keskinen, M. Water resources development and impact assessment in the Mekong Basin: Which way to go? *Ambio* **2008**, *37*, 193–198. [CrossRef]

59. Keskinen, M.; Kummu, M.; Kakonen, M.; Varis, O. Mekong at the crossroads: Next step for impact assessment of large dams. *Ambio* **2012**, *41*, 319–324. [CrossRef] [PubMed]

60. Salmivaara, A.; Kummu, M.; Keskinen, M.; Varis, O. Using Global Datasets to Create Environmental Profiles for Data-Poor Regions: A Case from the Irrawaddy and Salween River Basins. *Environ. Manag.* **2013**, *51*, 897–911. [CrossRef] [PubMed]

61. Kattelus, M.; Kummu, M.; Keskinen, M.; Salmivaara, A.; Varis, O. China's southbound transboundary river basins: A case of asymmetry. *Water Int.* **2014**, *40*, 113–138. [CrossRef]

62. Rahaman, M.M.; Varis, O. Integrated water management of the Brahmaputra basin: Perspectives and hope for regional development. *Nat. Resour. Forum* **2009**, *33*, 60–75. [CrossRef]

63. Varis, O.; Rahaman, M.M.; Kajander, T. Fully connected Bayesian belief networks: A modeling procedure with a case study of the Ganges river basin. *Integr. Environ. Assess. Manag.* **2012**, *8*, 491–502. [CrossRef] [PubMed]

64. Kattelus, M.; Rahaman, M.M.; Varis, O. Myanmar under reform: Emerging pressures on water, energy and food security. *Nat. Resour. Forum* **2014**, *38*, 85–98. [CrossRef]

65. Keskinen, M.; Chinvanno, S.; Kummu, M.; Nuorteva, P.; Snidvongs, A.; Varis, O.; Västilä, K. Climate change and water resources in the Lower Mekong River Basin: Putting adaptation into the context. *J. Water Clim. Chang.* **2010**, *1*, 103–117. [CrossRef]

66. Mehtonen, K.; Keskinen, M.; Varis, O. The Mekong: IWRM and institutions. In *Managemeng of Transboundary Rivers and Lakes*; Varis, O., Tortajada, C., Biswas, A.K., Eds.; Springer-Verlag Berlin: Berlin, Germany, 2008.

67. Wegerich, K. Hydro-hegemony in the Amu Darya basin. *Water Policy* **2008**, *10*, 71–88. [CrossRef]

68. Abdullaev, I.; Rakhmatullaev, S. Transformation of water management in Central Asia: From State-centric, hydraulic mission to socio-political control. *Environ. Earth Sci.* **2013**, *73*, 849–861. [CrossRef]

69. Stucki, V.; Sojamo, S. Nouns and Numbers of the Water–Energy–Security Nexus in Central Asia. *Int. J. Water Resour. Develop.* **2012**, *28*, 399–418. [CrossRef]

70. Granit, J.; Jägerskog, A.; Lindström, A.; Björklund, G.; Bullock, A.; Löfgren, R.; de Gooijer, G.; Pettigrew, S. Regional Options for Addressing the Water, Energy and Food Nexus in Central Asia and the Aral Sea Basin. *Int. J. Water Resour. Develop.* **2012**, *28*, 419–432. [CrossRef]

71. Allouche, J. The Governance of Central Asian Waters: National Interests versus Regional Cooperation. *Disarm. Forum* **2007**, *4*, 45–56.

72. Varis, O.; Rahaman, M.M. The Aral Sea keeps drying out but is Central Asia short of water? In *Central Asian Waters: Social, Economic, Environmental and Governance Puzzle*; Rahaman, M.M., Varis, O., Eds.; Water & Development Publications—Helsinki University of Technology: Helsinki, Finland, 2008; pp. 3–9.

73. Abdolvand, B.; Mez, L.; Winter, K.; Mirsaeedi-Gloßner, S.; Schütt, B.; Rost, K.T.; Bar, J. The dimension of water in Central Asia: Security concerns and the long road of capacity building. *Environ. Earth Sci.* **2014**, *73*, 897–912. [CrossRef]

74. Cook, S.; Fisher, M.; Tiemann, T.; Vidal, A. Water, food and poverty: Global- and basin-scale analysis. *Water Int.* **2011**, *36*, 1–16. [CrossRef]

75. Sharma, B.; Amarasinghe, U.; Xueliang, C.; de Condappa, D.; Shah, T.; Mukherji, A.; Bharati, L.; Ambili, G.; Qureshi, A.; Pant, D.; *et al.* The Indus and the Ganges: River basins under extreme pressure. *Water Int.* **2010**, *35*, 493–521. [CrossRef]

76. Rasul, G. Food, water, and energy security in South Asia: A nexus perspective from the Hindu Kush Himalayan region. *Environ. Sci. Policy* **2014**, *39*, 35–48. [CrossRef]

77. Rodell, M.; Velicogna, I.; Famiglietti, J.S. Satellite-based estimates of groundwater depletion in India. *Nature* **2009**, *460*, 999–1002. [CrossRef] [PubMed]

78. Shahid, S.; Behrawan, H. Drought risk assessment in the western part of Bangladesh. *Nat. Hazard.* **2008**, *46*, 391–413. [CrossRef]

79. Hoekstra, A.Y.; Mekonnen, M.M.; Chapagain, A.K.; Mathews, R.E.; Richter, B.D. Global Monthly Water Scarcity: Blue Water Footprints *versus* Blue Water Availability. *PLoS ONE* **2012**, *7*, e32688. [CrossRef] [PubMed]

80. Rahaman, M.M. Hydropower ambitions of South Asian nations and China: Ganges and Brahmaputra Rivers basins. *Int. J. Sustain. Soc.* **2012**, *4*, 131–157. [CrossRef]

81. Dharmadikhary, S. *Mountains of Concrete: Dam Building in the Himalayas*; IDEAS: Berkeley, CA, USA, 2008.

82. Grumbine, R.E.; Pandit, M.K. Threats from India's Himalaya Dams. *Science* **2013**, *339*, 36–37. [CrossRef] [PubMed]

83. International Energy Agency (IEA). *IEA Statistics*; IEA: Paris, France, 2016.

84. Lal, R. Soil degradation and environment quality in South Asia. *Int. J. Ecol. Environ. Sci.* **2007**, *33*, 91–103.

85. Yang, Y.C.E.; Brown, C.; Yu, W.; Wescoat, J.; Ringler, C. Water governance and adaptation to climate change in the Indus River Basin. *J. Hydrol.* **2014**, *519*, 2527–2537. [CrossRef]

86. Condon, M.; Kriens, D.; Lohani, A.; Sattar, E. Challenge and response in the Indus Basin. *Water Policy* **2014**, *16*, 58–86. [CrossRef]

87. Rahaman, M.M. Principles of Transboundary Water Resources Management and Ganges Treaties: An Analysis. *Int. J. Water Resour. D* **2009**, *25*, 159–173. [CrossRef]

88. Rahaman, M.M. Integrated Ganges basin management: Conflict and hope for regional development. *Water Policy* **2009**, *11*, 168–190. [CrossRef]

89. Uprety, K.; Salman, S.M.A. Legal aspects of sharing and management of transboundary waters in South Asia: Preventing conflicts and promoting cooperation. *Hydrol. Sci. J.* **2011**, *56*, 641–661. [CrossRef]

90. Urban, F.; Nordensvärd, J.; Khatri, D.; Wang, Y. An analysis of China's investment in the hydropower sector in the Greater Mekong Sub-Region. *Environ. Dev. Sustain.* **2013**, *15*, 301–324. [CrossRef]

91. Grumbine, R.E.; Xu, J. China Shakes the World—and Then What? *Conserv. Biol.* **2009**, *23*, 513–515. [PubMed]

92. Grumbine, R.E.; Xu, J. Mekong hydropower development. *Science* **2011**, *332*, 178–179. [CrossRef] [PubMed]

93. Salmivaara, A.; Kummu, M.; Varis, O.; Keskinen, M. Socio-economic changes in Cambodia's unique Tonle Sap Lake area: A spatial approach. *Appl. Spat. Anal. Policy* **2015**. [CrossRef]

94. Hortle, K.G. *Consumption and the Yield of Fish and Other Aquatic Animals from the Lower Mekong Basin*; MRC Technical Paper #16; Mekong River Commission: Vientiane, Laos, 2007; p. 87.

95. Winemiller, K.O.; McIntyre, P.B.; Castello, L.; Fluet-Chouinard, E.; Giarrizzo, T.; Nam, S.; Baird, I.G.; Darwall, W.; Lujan, N.K.; Harrison, I.; *et al.* Balancing hydropower and biodiversity in the Amazon, Congo, and Mekong. *Science* **2016**, *351*, 128–129. [CrossRef] [PubMed]

96. Baran, E.; Myschowoda, C. Dams and fisheries in the Mekong Basin. *Aquat. Ecosyst. Health Manag.* **2009**, *12*, 227–234. [CrossRef]

97. Mekong River Commission (MRC). *State of the Basin Report 2010*; MRC: Vientiane, Laos, 2010.

98. Hennig, T.; Wang, W.; Feng, Y.; Ou, X.; He, D. Review of Yunnan's hydropower development. Comparing small and large hydropower projects regarding their environmental implications and socio-economic consequences. *Renew. Sustain. Energy Rev.* **2013**, *27*, 585–595. [CrossRef]

99. Mekong River Commission (MRC). *Assessment of Basin-wide Development Scenarios: Main Report*; MRC: Vientiane, Laos, 2011; p. 254.

100. Mekong River Commission Secretariat. *Hydropower Database*; Mekong River Commission Secretariat: Vientiane, Laos, 2015.

101. CGIAR Research Program on Water, Land and Ecosystems. *Mekong Dam Database*; CGIAR Research Program on Water, Land and Ecosystems: Vientiane, Laos, 2015.

102. Chang, X.; Liu, X.; Zhou, W. Hydropower in China at present and its further development. *Energy* **2010**, *35*, 4400–4406. [CrossRef]

103. Chen, W.; Li, H.; Wu, Z. Western China energy development and west to east energy transfer: Application of the Western China Sustainable Energy Development Model. *Energy Policy* **2010**, *38*, 7106–7120. [CrossRef]

104. Matthews, N. Water Grabbing in the Mekong Basin—An Analysis of the Winners and Losers of Thailand's Hydropower Development in Lao PDR. *Water Altern.* **2012**, *5*, 392–411.

105. Merme, V.; Ahlers, R.; Gupta, J. Private equity, public affair: Hydropower financing in the Mekong Basin. *Glob. Environ. Chang.* **2014**, *24*, 20–29. [CrossRef]

106. Räsänen, T.A.; Joffre, O.; Someth, P.; Cong, T.T.; Keskinen, M.; Kummu, M. Model-based assessment of water, food and energy trade-offs in a cascade of multi-purpose reservoirs: Case study from transboundary Sesan River. *J. Water Resour. Plan. Manag.* **2015**, *141*, 05014007. [CrossRef]

107. Arias, M.E.; Cochrane, T.A.; Kummu, M.; Lauri, H.; Koponen, J.; Holtgrieve, G.; Piman, T. Impacts of hydropower and climate change on drivers of ecological productivity of Southeast Asia's most important wetland. *Ecol. Model.* **2014**, *272*, 252–263. [CrossRef]

108. Dugan, P.; Barlow, C.; Agostinho, A. Fish migration, dams, and loss of ecosystem services in the Mekong Basin. *Ambio* **2010**, *39*, 344–348. [CrossRef] [PubMed]

109. Lauri, H.; Moel, H.D.; Ward, P.; Räsänen, T.; Keskinen, M.; Kummu, M. Future changes in Mekong River hydrology: Impact of climate change and reservoir operation on discharge. *Hydrol. Earth Syst. Sci.* **2012**, *16*, 4603–4619. [CrossRef]

110. Kummu, M.; Lu, X.; Wang, J.; Varis, O. Basin-wide sediment trapping efficiency of emerging reservoirs along the Mekong. *Geomorphology* **2010**, *119*, 181–197. [CrossRef]

111. Maavara, T.; Parsons, C.T.; Ridenour, C.; Stojanovic, S.; Dürr, H.H.; Powley, H.R.; van Cappellen, P. Global phosphorus retention by river damming. *Proc. Natl. Acad. Sci. USA* **2015**, *112*, 15603–15608. [CrossRef] [PubMed]

112. Ziv, G.; Baran, E.; Nam, S.; Rodriquez-Iturbe, I.; Levin, S. Trading-off fish biodiversity, food security, and hydropower in the Mekong River Basin. *PNAS* **2012**, *109*, 5609–5614. [CrossRef] [PubMed]

113. Hortle, K.G. Fisheries of the Mekong River Basin. In *The Mekong: Biophysical Environment of a Transboundary River*; Campbell, I.C., Ed.; Elsevier: New York, NY, USA, 2009.

114. Asian Development Bank. *Assessment of the Greater Mekong Subregion Energy Sector Development: Progress, Prospects, and Regional Investment Priorities*; Asian Development Bank: Mandaluyong City, Philippines, 2013.

115. Phalan, B. The social and environmental impacts of biofuels in Asia: An overview. *Appl. Energy* **2009**, *86* (Suppl. 1), S21–S29. [CrossRef]

116. Sasaki, N.; Knorr, W.; Foster, D.R.; Etoh, H.; Ninomiya, H.; Chay, S.; Kim, S.; Sun, S. Woody biomass and bioenergy potentials in Southeast Asia between 1990 and 2020. *Appl. Energy* **2009**, *86* (Suppl. 1), S140–S150. [CrossRef]

117. Bach, H.; Glennie, P.; Taylor, R.; Clausen, T.J.; Holzwarth, F.; Jensen, K.M.; Meija, A.; Schmeier, S. *Cooperation for Water, Energy and Food Security in Transboundary Basins under Changing Climate*; Mekong River Commission: Vientiane, Laos, 2014.

118. Dore, J.; Lazarus, K. Demarginalizing the Mekong River Commission. In *Contested Waterscapes in the Mekong Region: Hydropower, Livelihoods and Governance*; Molle, F., Foran, T., Käkönen, M., Eds.; Earthscan: London, UK, 2009; pp. 357–382.

119. Grumbine, R.E.; Dore, J.; Xu, J. Mekong hydropower: Drivers of change and governance challenges. *Front. Ecol. Environ.* **2012**, *10*, 91–98. [CrossRef]

120. Orr, S.; Pittock, J.; Chapagain, A.; Dumaresq, D. Dams on the Mekong River: Lost fish protien and the implications for land and water resources. *Glob. Environ. Chang.* **2012**, *22*, 925–932. [CrossRef]

121. Kummu, M. Spatio-Temporal Scales of Hydrological Impact Assessment in Large River Basins: The Mekong Case. Ph.D. Thesis, Helsinki University of Technology, Helsinki, Finland, 2008.

122. Dore, J.; Lebel, L. Deliberation and scale in Mekong Region water governance. *Environ. Manag.* **2010**, *46*, 60–80. [CrossRef] [PubMed]

123. Cumming, G.S.; Cumming, D.H.M.; Redman, C.L. Scale Mismatches in Social-Ecological Systems: Causes, Consequences, and Solutions. *Ecol. Soc.* **2006**, *11*, 14.

124. Blatter, J.; Ingram, H. States, Markets and Beyond: Governance of Transboundary Water Resources. *Nat. Resour. J.* **2000**, *40*, 439–473.

125. Dore, J. An agenda for deliberative water governance arenas in the Mekong. *Water Policy* **2014**, *16*, 194–214. [CrossRef]

126. Mekong River Commission (MRC). *Agreement on the Cooperation for the Sustainable Development of the Mekong River Basin*; MRC: Chiang Rai, Thailand, 1995.

127. Warner, J. *Multi-Stakeholder Platforms for Integrated Water Management*; Ashgate: Aldershot, UK, 2007.

128. Dore, J. Multi-stakeholder platforms (MSPs): Unfulfilled potential. In *Democratizing Water Governance in the Mekong Region*; Lebel, L., Dore, J., Daniel, R., Yang, S.K., Eds.; Mekong Press: Chiang Mai, Thailand, 2007; pp. 197–226.

129. Mohtar, R.H.; Daher, B. Water-Energy-Food Nexus Framework for facilitating multi-stakeholderdialogue. *Water Int.* **2016**. [CrossRef]

130. Timmerman, J.G.; Langaas, S. Water information: What is it good for? The use of information in transboundary water management. *Reg. Environ. Chang.* **2005**, *5*, 177–187. [CrossRef]

131. UN-Water. Transboundary Waters: Sharing Benefits, Sharing Responsibilities. 2008. Available online: http://www.unwater.org/downloads/UNW_TRANSBOUNDARY.pdf (accessed on 20 April 2016).

132. Gerlak, A.K.; Lautze, J.; Giodarno, M. *Greater Exchange, Greater Ambiguity: Water Resources Data and Information Exchange in Transboundary Water Treaties*; Global Water Forum (GWF): Canberra, Australia, 2013.

133. Sadoff, C.; Grey, D. A cooperation on international rivers: A continuum for securing and sharing benefits. *Water Int.* **2005**, *30*, 1–7. [CrossRef]

134. Mirumachi, N. *Transboundary Water Politics in the Developing World*; Routledge: London, UK, 2015.

135. Middleton, C.; Allouche, J.; Gyawali, D.; Allen, S. The Rise and Implications of the Water-Energy-Food Nexus in Southeast Asia through and Environmental Justice Lens. *Water Altern.* **2015**, *8*, 627–654.

136. Flammini, A.; Puri, M.; Pluschke, L.; Dubois, O. *Walking the Nexus Talk: Assessing the Water-Energy-Food Nexus in the context of the Sustainable Energy for All Initiative*; Food and Agriculture Organisation of the United Nations (FAO): Rome, Italy, 2014.

137. Byers Edward, A. Tools for tackling the water-energy-food nexus. *Chang. Adapt. Socio-Ecol. Syst.* **2015**, *2*, 109–111.

138. National Science Foundation. Innovations at the Nexus of Food, Energy and Water Systems (INFEWS). Available online: http://www.nsf.gov/pubs/2016/nsf16524/nsf16524.htm (accessed on 31 March 2016).

139. European Commission. Horizon 2020 Topic: Integrated approaches to food security, low-carbon energy, sustainable water management and climate change mitigation. Available online: https://ec.europa.eu/research/participants/portal/desktop/en/opportunities/h2020/topics/240-water-2b-2015.html (accessed on 31 March 2016).

140. The Nexus Network. Funding available. Available online: http://www.thenexusnetwork.org/funding-available/ (accessed on 31 March 2016).

141. Engineering and Physical Sciences Research Council. Sandpit: Water Energy Food Nexus. Available online: http://www.epsrc.ac.uk/funding/calls/sandpitwaterenergyfoodnexus/ (accessed on 31 March 2016).

142. MRC. Organisational structure. Available online: http://www.mrcmekong.org/about-mrc/organisational-structure/ (accessed on 26 January 2016).

143. Sadoff, C.; Grey, D. Beyond the river: The benefits of cooperation on international rivers. *Water Policy* **2002**, *4*, 389–403. [CrossRef]

144. WCED. *Our Common Future*; Report of the World Commission on Environment and Development; Oxford University Press: Oxford, UK, 1987.

145. Hussey, K.; Pittock, J.; Dovers, S. Justifying, extending and applying "nexus" thinking in the quest for sustainable development. In *Climate, Energy and Water: Managing Trade-Offs, Seizing Opportunities*; Pittock, J., Hussey, K., Dovers, S., Eds.; Cambridge University Press: New York, NY, USA, 2015; pp. 1–5.

146. UN. The 2012 Revision of World Population Prospects. Population Division of the Department of Economic and Social Affairs of the United Nations (UN) Secretariat: New York, NY, USA, 2013.

147. Klein Goldewijk, K.; Beusen, A.; Janssen, P. Long-term dynamic modeling of global population and built-up area in a spatially explicit way: HYDE 3.1. *The Holocene* **2010**, *20*, 565–573. [CrossRef]

148. World Development Indicators, The World Bank. Available online: http://data.worldbank.org/indicator (accessed on 21 March 2016).

149. WATCH. EU WATCH: Water and Global Change. Available online: http://www.eu-watch.org (accessed on 18 July 2011).

150. FAO. FAOSTAT: FAO Database for Food and Agriculture. Food and Agriculture Organisation of the United Nations (FAO), Rome. Available online: http://faostat3.fao.org/home/E (accessed on 21 March 2016).

151. Porkka, M.; Kummu, M.; Siebert, S.; Varis, O. From Food Insufficiency towards Trade Dependency: A Historical Analysis of Global Food Availability. *PLoS ONE* **2013**, *8*, e82714. [CrossRef] [PubMed]

152. FAO. Food security indicators, February 09, 2016 revision. Food and agriculture Organization of the United Nations (FAO), Rome. Available online: http://www.fao.org/economic/ess/ess-fs/ess-fadata/en/ (accessed on 21 March 2016).

153. Siebert, S.; Kummu, M.; Porkka, M.; Döll, P.; Ramankutty, N.; Scanlon, B.R. A global data set of the extent of irrigated land from 1900 to 2005. *Hydrol. Earth Syst. Sci.* **2015**, *19*, 1521–1545. [CrossRef]

154. GWSP Digital Water Atlas. *Map 38: Mean annual surface runoff 1950–2000, v. 1.0*; Center for Development Research, University of Bonn: Bonn, Germany, 2008.

155. Mekong River Commission (MRC). *Map on Fish Migration Patterns*; MRC: Vientiane, Laos, 2010.

156. Asian Development Bank (ADB). Overview Map of GMS Crossborder Power Transmission. In *Greater Mekong Subregion Atlas of the Environment*, 2nd ed.; ADB: Manila, Philippines, 2012.

157. Food and Agriculture Organization of the United Nations (FAO). *FAOSTAT Country Profiles*; FAO: Rome, Italt, 2015.

MDPI AG

St. Alban-Anlage 66

4052 Basel, Switzerland

Tel. +41 61 683 77 34

Fax +41 61 302 89 18

http://www.mdpi.com

Water Editorial Office

E-mail: water@mdpi.com

http://www.mdpi.com/journal/water

www.ingramcontent.com/pod-product-compliance
Lightning Source LLC
Chambersburg PA
CBHW051727210326
41597CB00032B/5633